Flow and reactions in permeable rocks

Flow and reactions in permeable rocks

O. M. PHILLIPS, F.R.S.
DECKER PROFESSOR OF SCIENCE AND ENGINEERING
AT THE JOHNS HOPKINS UNIVERSITY

The right of the University of Cambridge to print and sell all manner of books was granted by Henry VIII in 1534. The University has printed and published continuously since 1584.

CAMBRIDGE UNIVERSITY PRESS
CAMBRIDGE
NEW YORK PORT CHESTER MELBOURNE SYDNEY

Published by the Press Syndicate of the University of Cambridge
The Pitt Building, Trumpington Street, Cambridge CB2 1RP
40 West 20th Street, New York, NY 10011, USA
10 Stamford Road, Oakleigh, Melbourne 3166, Australia

First published 1991

Printed in the United States of America

Library of Congress Cataloging-in-Publication Data

Phillips, O. M. (Owen M.), 1930–

Flow and reactions in permeable rocks/O.M. Phillips.
p. cm.
ISBN 0-521-38098-7
1. Fluid dynamics. 2. Rocks – Permeability. I. Title.

TA357.P48 1991 90-44208
552 – dc20 CIP

British Library Cataloguing in Publication Data

Phillips, O. M. (Owen Martin) *1930–*

Flow and reactions in permeable rocks.
1. Hydrology
I. Title
551.48

ISBN 0-521-38098-7

To
Francis Pettijohn
and to the memory of
Hans Eugster

Contents

Preface

The aim of this book is to provide an account of the patterns of flow through permeable rocks and rock structures in the earth, patterns whose imprint is reflected in the distributions of mineralogical changes that have taken place over eons past. The nature of the changes themselves, including postdepositional formation of ore bodies and hydrocarbon deposits, is the proper concern of mineralogy and geochemistry, but the patterns, and in many instances the rates, of formation have been controlled by the flow, whether episodic and relatively rapid or continuing but slow, that occurred over these time intervals. Though the topic of flow through permeable media is rooted in the preceding century, its application to geochemical and mineralogical questions is fairly recent. Contributions to this area of inquiry are scattered in the geological, hydrological, and fluid mechanics literature. Some are important and penetrating; others, alas, needlessly obscure, misleading, or irrelevant. Part of the material presented here is being published for the first time, and much more remains to be discovered. This book is offered in the hope that it will provide a basis for future work or possibly a target for dissent; in either case understanding will be enhanced.

It is intended primarily for geologists, especially the well-trained graduate students who are emerging today, though I am conscious that it may present conceptual and technical difficulties for those whose background is more deeply geochemical than is mine. The subject necessarily involves a level of mathematics that may be uncomfortable to many geologists, but for those with a recent undergraduate background to the level of differential equations and vector calculus and with a spirit of mathematical adventure, I hope it will not be daunting. In the analytical developments, I have tried to eschew mathematical generality for its own sake, seeking instead to be specific to the matter at hand. Sufficient detail is given in the derivations that the logical steps can be followed with perhaps moderate effort or, for those prepared to trust, skipped over. Students of hydrology or the nascent area of geological fluid mechanics may, in contrast, find the discussion of basic flow properties elementary and unnecessarily detailed, but I must beg their patience until a common basis of understanding is developed. In an area such as this, the greatest rewards will come by a combination of insights from disciplines usually studied separately, and what is elementary to some may be complex to others.

The broad subject of fluid mechanics has always involved the art of approximation, and this context provides no exception. Approximations must, however, be firmly based and clearly specified – there is little place for fanciful models that have sometimes intruded into the geological literature. In discussing the results, I have tried to explain in words not only what they mean, but also what geometrical simplifications and dynamical approximations have been involved and under what conditions they apply. To use results in an inappropriate context is to invite confusion, but if the context is clear, the ultimate reward can be the development of an understanding, an intuition, a feeling for the role that physical and chemical processes associated with flow have had, and are having, in remaking the rocks beneath us, what the flow can and cannot do, what patterns it can produce, and where to look for its manifestations.

In writing this book, I have relied heavily on the guidance and advice of many colleagues, friends, and students who listened with patience, criticized thoughtfully, and pointed in new directions. In particular, I must thank Lucas Baumgartner, Edwin Clawsey, John Ferry, Grant Garven, Gu Daifang, Lee Hagee, Lawrence Hardie, Eugene Ilton, Barry Lester, Sakiko Olsen, Haydee Salmun, Dimitri Sverjensky, and Edith Wilson. It was James R. Wood who, quite some time ago, suggested the need for such a book and who gave generously of his time and his deep insights in reviewing an earlier draft. I am greatly indebted to Elizabeth Hendrickson for her skillful preparation of the manuscript and for her endless patience in revision, revision, and revision again. Most of all, I am grateful to Francis Pettijohn and the late Hans Eugster, from whom I have learned so much and to whom, and to whose memory, this book is dedicated with affection and respect.

Baltimore O.M.P.
September 1990

1 Introduction

It has been evident since the days of James Hutton in the late eighteenth century that this solid earth has been in upheaval throughout a history longer than could easily be visualized. Recurrent and overlapping cycles of igneous intrusion and deposition, deformation and folding, with erosion by air and water, redistribution and sedimentation have shaped the surface of the earth and hinted at great tectonic forces within. The grandeur of the canvas painted by geology, the science of the earth, has excited the imagination of those who have the capacity to wonder and has drawn in the talents of biologists, physicists, and chemists such as Darwin, Rayleigh, Jeffreys, Elsasser, Urey, and many more in the preceding century and this. New discoveries in the basic sciences and new techniques of observation have again and again revolutionized our view of the earth – it has been barely twenty years since measurement of magnetic anomalies over midoceanic ridges forced the acceptance as reality of the process of sea-floor spreading leading to continental drift, a notion previously regarded as close to heresy.

But no more. Abundant evidence now indicates that the crust of the earth is spreading from the midocean ridges that gird the globe and is disappearing in the subduction zones of the deep ocean trenches carrying the continental blocks in a grandly and gradually changing pattern. Deep-mantle convection is presumably responsible, with material solid enough to transmit seismic S-waves behaving also as a gradually deforming fluid. The worldwide distribution of volcanism is intimately associated with these tectonic motions. In recent years, the fluid behavior of magmas, igneous intrusions, and melts has drawn the attention of such imaginative and powerful scientists as Herbert Huppert, Dan McKenzie, and Stewart Turner, and this field is sparkling with new ideas.

As applications of physics and chemistry brought new insights to geology, they also posed new questions. The development of geochemistry in the past fifty years has led to the clarification of many aspects of metamorphism, the term used to describe the transformation of minerals from one form to another at depth under conditions of high temperature and pressure. Volatiles may have had to escape so that the reaction would not reverse itself, but there has been plenty of time to accomplish this. Yet a number of mysteries remain. Natural processes are, in the large, dispersive, as the second law of thermodynamics insists. How is it that local concentrations of minerals aggregate deep inside the earth, leading to the accumulation of economically valuable resources? The Mississippi Valley lead–zinc deposits were clearly formed after the lithification of the host rock, but where did they come from, how did they get there, why did they deposit where they did and not somewhere else? The dolomitization of limestone requires the importation of large quantities of magnesium; in these and other geochemical processes there is an evident need to account for the transport of minerals, presumably dissolved, through existing assemblages in quantities much larger than can be associated with compaction of a fluid-saturated matrix and at rates much faster than can be produced by diffusion. No reader of Francis Pettijohn's *Sedimentary Rocks* can remain unaware of the puzzle, but only fairly recently has the idea gained acceptance that these chemical transports must be associated with fluid flow through rocks that a hydrologist may consider hardly permeable, over time intervals of millions or hundreds of millions of years. When pressure or buoyancy forces are present, the latter usually associated with temperature gradients, interstitial fluid may flow through the matrix if it is reasonably permeable or along cracks and fractures if it is less so. What are the large-scale and small-scale patterns of flow? In which direction does the fluid move, how fast, where from, and where to? What governs the distributions of reaction, of dissolution, and of deposition that, over eons, led to the distributions of mineralization that we see today?

The search for answers to these questions involves a combination of structural geology, which defines the geometry and characteristics of the matrix; geochemistry, which specifies the nature of the reactions that can occur; and fluid mechanics, which governs the flow and transport of dissolved chemicals from one place to another. The fundamental rules of fluid percolation through rocks or other permeable media have been known since Darcy's time, but the variety of flow patterns in geological structures and their textural and geochemical consequences are far from having been fully explored. The ranges of pertinent parameters are often extreme – the permeability of an evaporite bed may be ten orders of magnitude smaller than that of a gravel bank. Even in one material, limestone, field measurements may indicate a permeability five orders of magnitude larger than that inferred from laboratory samples because of mesoscale fracturing or dolomitization. Chemical reactions may proceed almost instantaneously or take thousands of years to approach equilibrium. In many instances, the appropriate parameters are simply not known even to an order of magnitude, and this provides an opportunity and a trap – an opportunity in that careful analyses of the consequences of flow can reduce or place limits on the uncertainty, and a trap in that one can often "pick a number" within a wide range to give any desired numerical agreement.

Even the basic three-dimensional geometrical structure of a basin is seldom known precisely. Much of the information in structural geology has been derived from two-dimensional exposures and surficial mapping. Drilling is expensive and cores give only widely separated vertical traverses. The techniques of seismic reflection now make it possible to obtain much greater detail, though even in well-explored regions, the disciplined imagination of the structural geologist is still a necessity. The difficulties are compounded as we reach farther into the past. Not only have the geological structures evolved with the relentless sequence of tectonic upheaval, erosion, and deposition, with the consequent evolution of the patterns of flow, temperature, and pressure, but also the physical and chemical nature of the rocks themselves has changed with metamorphosis, dissolution, and cementation. The correct interpretation of these changes requires all the skill of the geochemist and the petrologist, and an appreciation of the influence of interstitial flow is but one of the tools that must be employed. It is an important one, though; flow provides fluxes of chemical species, slow but enduring throughout geological time, that allows reactions to continue and petrological changes to accumulate.

It is not the purpose of this book to construct a detailed model of the flow-associated changes that have occurred in any particular basin or other geological structure, but rather to develop, from the basic physical and chemical balances, a set of derived or secondary rules that can guide geological interpretation in a variety of contexts. We seek associations among the general characteristics of geological structures, the characteristic patterns of flow that they allow, and the patterns and rates of physical and chemical changes that occur as a consequence and are reflected in the distributions of mineralization. A number of examples that exhibit these connections are given, but they are far too few. Ideally, a development such as this would be guided by geochemical flow experiments in permeable media, but this would require extrapolation from achievable laboratory flow rates, time scales, temperatures, and pressures to those that occur deep in the earth over millions of years. A number of important experiments by Elder and others, however, have illustrated aspects of convective flow patterns in some simple geometries. The art of numerical simulation of basin-wide flow, temperature, and reaction patterns has developed rapidly in recent years, notable contributions having been made by Garven and his collaborators, but these inevitably require a specific configuration and choice of the numerical values of the parameters involved; it is not always clear how specific features of the solution depend on these choices. The results presented here can, nonetheless, provide insights by relating these features to the local or overall flow characteristics. The two approaches are, indeed, complementary. A conceptual framework elucidating the relations among flow characteristics, driving forces, structure, and reaction patterns enables us not only to understand the results of numerical modeling more clearly, but to check them. (Numerical calculations *can* converge to a grid-dependent limit, and artifacts of a solution *can* be numerical rather than geological.) Numerical modeling provides a quantitative description and synthesis of a basin-wide flow in far greater detail than would be feasible analytically. The combination of the two techniques is a much more powerful research tool than either alone.

In the next chapter are specified the general geometrical characteristics and basic physical and chemical balances that underlie the developments in the remainder of the book. These are the rules by which the game is to be played, but within these rules, many kinds of play are possible. Chapters 3 and 4 develop a variety of these and illustrate the closeness of the relationships among structure (the distribution and characteristic geometry of permeability variations), flow within the structures, and distributions of chemical reactions of several types associated with this flow. When the interstitial fluid viscosity is uniform and buoyancy effects are negligible, the flow solutions are unique, but if temperature or salinity variations are significant, a given initial distribution may be dynamically unstable and evolve into some other. Some conditions for stability or instability are illustrated in Chapter 5 by typical geological examples. The two great driving forces for large-scale circulations through rocks are pressure differences, hydraulically induced, and variations in basement temperature or heat flux; their flow, thermal, and mineralogical signatures are described in the last two chapters, again by means of several examples in which direct observations or numerical experiments are sufficiently complete to make comparison possible. One again laments that there are not yet enough examples such as these; it is to be hoped that in future years more will become available to provide support (or otherwise) for existing results and point in new directions. Only thus can this area of geology advance.

2 The governing principles

2.1 THE PHYSICAL NATURE OF PERMEABLE ROCKS

The geological materials with which we are concerned usually lie at one extreme of the range of "porous media" encountered in nature and technology. The porosity, the volume fraction of connected voids that allow fluid movement, may approach unity in slag-wool insulation material; it may be as large as 0.4 or 0.5 in a well-sorted sandbank (Pryor, 1973). The skeletal remains of corals that abound in tropical reefs contain myriad interstices on scales of up to a centimeter or so and may have a similarly large porosity; Figure 2.1 shows a sample from Bermuda at approximately half-scale. This kind of structure, however, is at the high-porosity extreme of those generally encountered. Compaction by overlying sediments, the infilling of interstices by finer grains, and the precipitation of cements from solution can reduce the porosity by an order of magnitude or more and reduce the permeability, as we shall see, by four orders of magnitude or more.

Still highly permeable is the partially cemented sandstone of Figure 2.2. The largest interstices are about 1 mm in diameter, and the granular texture, containing many finer pore spaces, is clearly evident. Many large pores are also apparent in dolomite from the Latemar Massif, northern Italy (Figure 2.3). Calcium ions from the original calcite mineral have been replaced by magnesium; the specific volume of the dolomite produced in the reaction is less by 3 to 13 percent than that of the original calcite, so that as the reaction proceeded, the porosity increased.

Networks of small cracks or fractures allow fluid percolation even when the matrix itself is relatively impermeable. Seepage from fractures can often be discerned in roadside rock exposures. Figure 2.4 illustrates a smaller-scale network, made visible by staining, intersecting a sandstone cleavage plane. The result of fluid movement along interlacing networks of pathways on a smaller scale again is shown in Figure 2.5. Natural preferential weathering on the face of the specimen has left micropathways of dolomites standing in relief against the unaltered calcite as a dense network on the left and along what was once a single crack on the right. The stained, thin-section photograph of Figure 2.6 also shows dolomitization along micropathways in a partially dolomitized transition

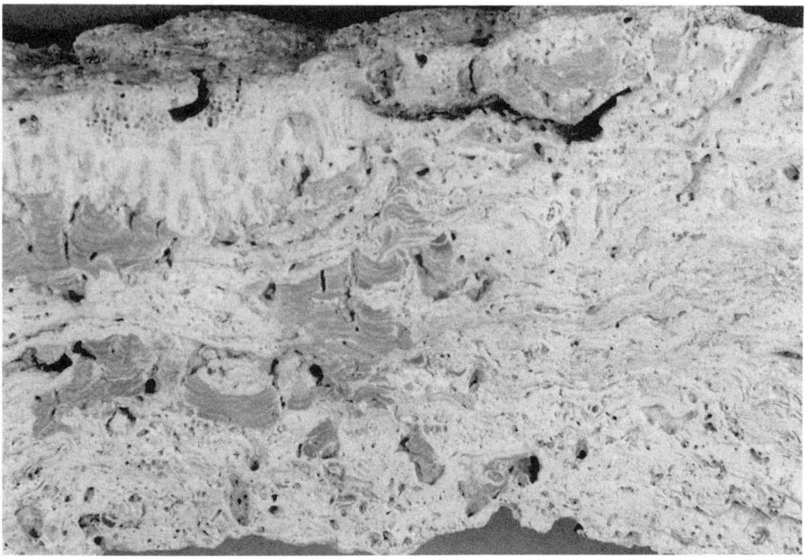

Figure 2.1. Extremely porous limestone from a Bermuda coral reef, approximately half-scale, containing many interstices with scales of up to a centimeter or so.

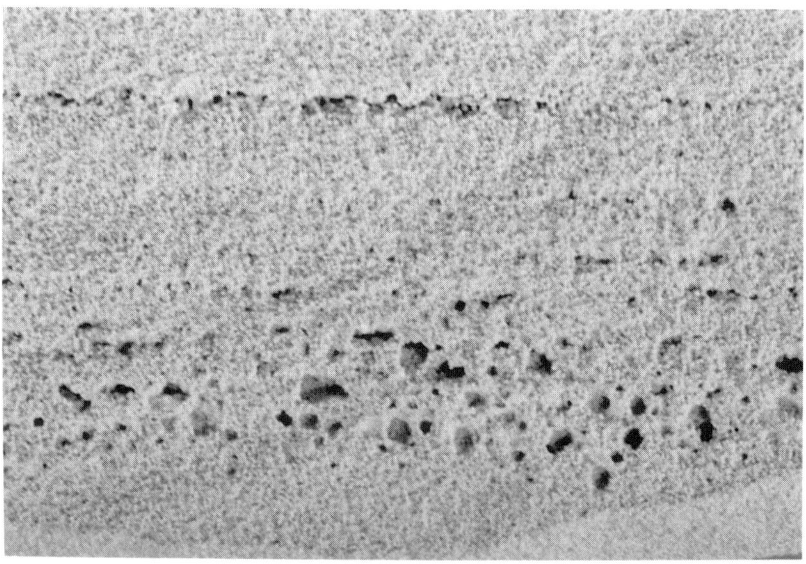

Figure 2.2. A highly porous, partially cemented sandstone, approximately full scale.

Figure 2.3. Pores in dolomite from the Latemar Massif in northern Italy. The blocks in the scale are 1 cm long. (Photograph courtesy of Dr. E. N. Wilson.)

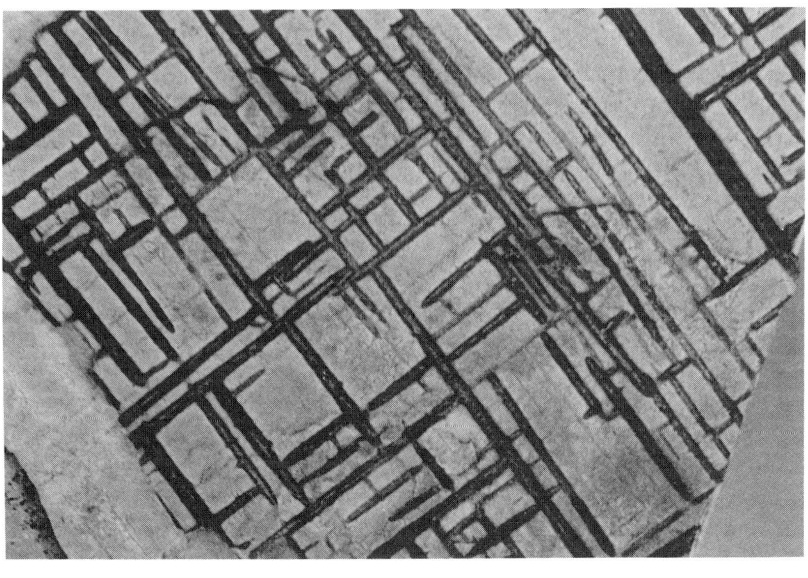

Figure 2.4. A network of cracks, made visible by staining, that provide pathways for flow along a sandstone cleavage plane. Approximately full scale.

Figure 2.5. The weathered face of a partially dolomitized calcite rock in which veinlets of dolomite produced along fluid pathways stand out in relief in a dense, fine tracery on the left and along a crack on the right. (Photograph courtesy of Dr. E. N. Wilson.)

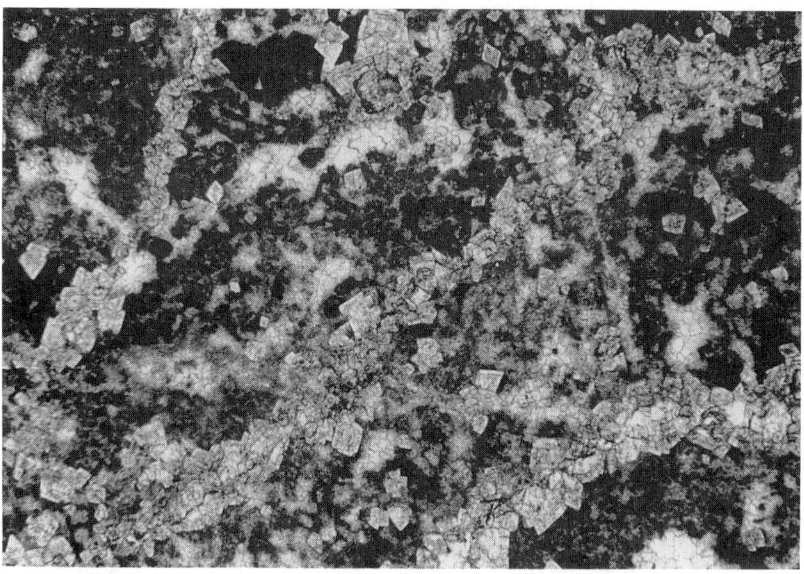

Figure 2.6. Dolomitization along linear pathways (the strings of lighter rhombic crystals) in a calcite matrix. Magnification ×20. (Photograph courtesy of Dr. E. N. Wilson.)

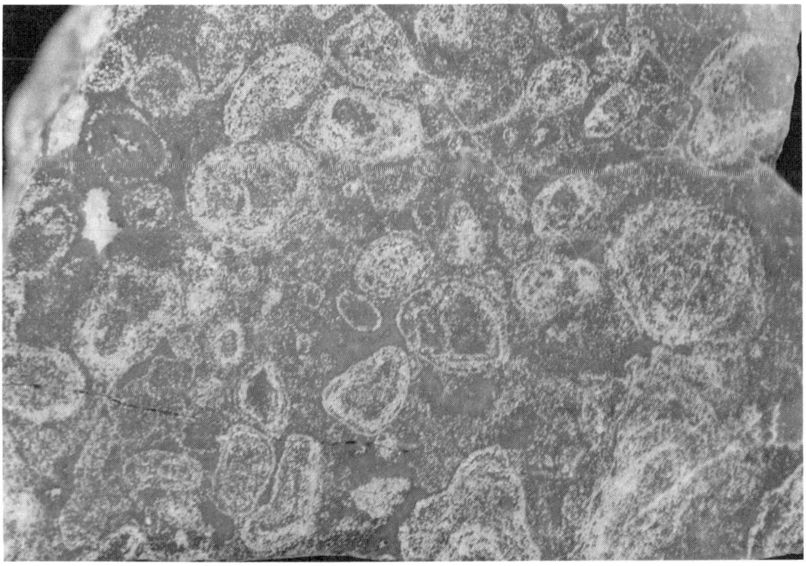

Figure 2.7. Dolomitization around the edge of grains (light) in a partially dolomitizated grainstone, indicating a generally pervasive fluid flow. Full scale. (Photograph courtesy Dr. E. N. Wilson.)

zone – the lighter rhombic dolomite crystals form a continuous network against the darker calcite background.

When the interstitial seepage is more pervasive, dolomitization (and fabric alteration in general) becomes more uniform and ultimately more complete. Figure 2.7 shows a polished and stained slab of subtidal grainstone in which only the edges of the grains have been dolomitized (again they show as light). In the photomicrograph of Figure 2.8, the lighter dolomite has replaced the darker calcite almost completely in the region on the right, though calcite central cores remain unaltered in many of the grains. Figure 2.9, at higher magnification, shows a tiny fracture, now filled with calcite cement, in a matrix of dolomitized grainstone. A pore, intersected by the fracture, has also been filled with calcite. The porosity of some of these samples is now extremely small, but evidently, during a previous time interval, it was sufficient to allow the movement of solutions from which the cements precipitated.

Fault systems also provide conduits for fluid motion. Sibson (1987, 1989) has pointed out that dilatational fault jogs in particular can open up a network of fractures that immediately draws in surrounding interstitial fluid and subsequently provides pathways for continuing fluid motion. If cementation occurs, a network of veins through the original fabric ultimately results in a fabric typically characterized, according to Sibson,

Figure 2.8. Dolomite of lighter shade has replaced calcite almost completely on the right, except in the cores of some of the grains; the reaction is only partially complete in the center. Magnification ×40. (Photograph courtesy Dr. E. N. Wilson.)

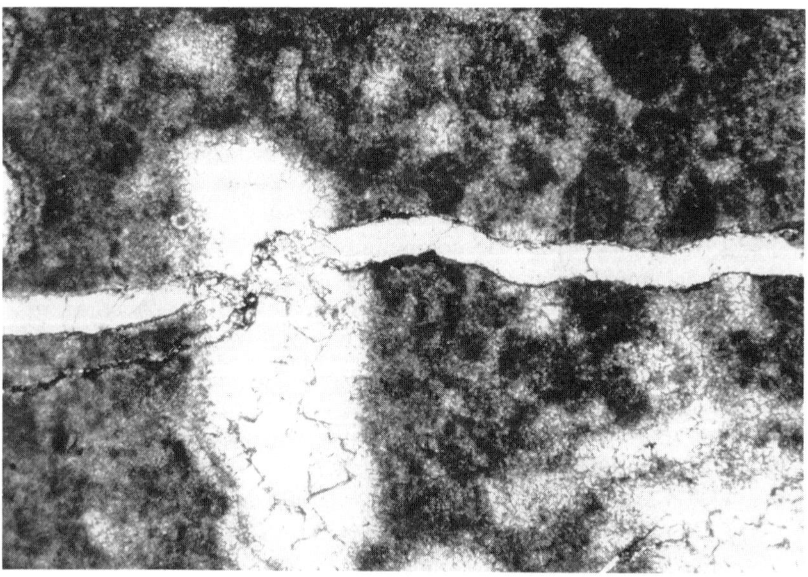

Figure 2.9. A veinlet in which calcite cement has precipitated from solution, intersecting and filling a pore. Magnification ×100. (Photograph courtesy Dr. E. N. Wilson.)

by textures that record histories of incremental opening by high-dilatation, hydrothermally cemented wall-rock breccias (angular fragments) showing multiple episodes of brecciation and recementation.

2.2 GEOMETRICAL CHARACTERISTICS

Several different geometrical length scales are pertinent to a consideration of flow through permeable rocks. The smallest is the intermolecular distance, approximately equal to the size of the molecules of liquid filling the interstices. For aqueous solutions, this is of the order 10^{-7} cm. When this intermolecular distance is small compared with the characteristic size of the interstices or width of the cracks, δ, say, then the pore fluid can be regarded as a continuum and its motion described by the equations of continuum fluid mechanics rather than those appropriate to the kinetic theory of liquids – a significant simplification! The characteristic grain size, or distance between individual fluid pathways, l_0, is usually considerably larger than δ; in unconsolidated or partly cemented sediments, $l_0 \gg \delta$. Sedimentary deposits are frequently bedded, with inhomogeneities in physical properties such as the local porosity and permeability occurring over vertical scales that are large compared with the grain size but small compared with the overall thickness of the bed. Finally, there are the macroscopic length scales h, which specify the thickness (or smallest dimension) of the porous bed as a whole, and l, its lateral extent.

Although, in order to relate the overall flow behavior to the average geometrical characteristics of the rocks, we must consider carefully certain aspects of the microscopic flow through individual pores or around individual grains, our primary concern is with flow properties, transport of heat and chemical fluxes, on the scale of inhomogeneities or on the macroscopic length scale h. Immense simplifications are possible when the microscale l_0 is sufficiently small compared with h that we can find an intermediate local averaging scale l_{av} that is large compared with the grain size l_0 yet small compared with the macroscopic scale h, that is,

$$l_0 \ll l_{av} \ll h, \tag{2.2.1}$$

where, as a rule of thumb, the inequality signs can be interpreted to mean "is less by a factor of at least 10 than." Within a given stratum, properties of the medium and characteristics of the flow through it, when averaged over the volume l_{av}^3, will vary smoothly from one averaging volume to the next. When the system is viewed on a macroscopic scale, it can again be regarded as a continuum, with the "point" properties being, in fact, local averages of this kind and being functions of three spatial coordinates (x, y, z) and possibly time t.

If a porous layer is very thin in the sense that it contains only a few fluid pathways, the averaging volume will include the entire thickness of the layer. The macroscopic geometry then becomes, in essence, two-dimensional, and the flow properties are functions of two spatial coordinates in the plane. Common examples of this geometry are found in isolated fracture planes in an otherwise impermeable rock.

When the locally averaged properties of the medium are independent of the position of the averaging volume, the medium is said to be homogeneous. If all locally averaged properties are independent of direction, the medium is described as locally isotropic. A well-mixed sand bed may be, on this averaging scale, both homogeneous and locally isotropic. As mentioned previously, however, sediments are frequently deposited in such a way that the fabric preserves a record in its layering of the vertical direction, though there may be no differences discernible in the two orthogonal horizontal directions. Seasonal variations in sedimentation rate may produce, on a microscopic scale, a stack of horizontal laminations that, on a macroscopic scale, lead to different percolation characteristics along and across the laminations. Similarly, when elongated or plate-shaped sedimenting particles tend to settle horizontally, the resulting fabric will preserve a record of the vertical direction at the time of deposition. In general, when the medium has one preferred direction but its properties are independent of rotation about that direction, it can be described as locally axisymmetric.

An important characteristic of a porous rock is the void fraction. The total void fraction ϕ_t is that fraction of the total averaging volume represented by the interstices; the solids occupy a fraction $1 - \phi_t$ of the whole. This can be measured by an examination of randomly taken thin sections; since a volumetric sample can be considered to be a stack of plane slices, the ratio of void area to total area in a typical thin section is equal to the ratio of void volume to total volume, ϕ_t. Similarly, along a sampling line the average ratio of total length of the line segments in voids to total line length is also ϕ_t. However, not all of the void spaces may be active in fluid flow through the medium. Isolated cavities or "dead end" tubes can contribute to ϕ_t but do not provide microscopic pathways to fluid motion. A more significant measure for our purposes is the connected porosity ϕ, in which only those voids that provide connections among the averaging volumes are considered. In general, this quantity cannot be estimated from thin-section examination without some additional information or assumption about the structure of the fabric, but it can be measured by comparison of the mass of saturated and dried samples or, as we shall see in the next section, by fluid observations. In this book, the term "porosity" refers to the connected porosity, since this is the property of interest in fluid motion, though in the petroleum industry it is commonly used as a synonym for "void fraction."

Clearly, $\phi \leq \phi_t < 1$. At one extreme, in a material with the geometry of swiss cheese, such as pumice, all the voids are isolated so that $\phi = 0$

while ϕ_t may be relatively large. In poorly cemented sandstone, there may be very few isolated voids and $\phi \simeq \phi_t \sim 0.2$ or 0.3.

A variety of textures are of interest in different contexts, and some simple approximate relationships are frequently useful. If, on a microscopic scale, the interstices consist of a complex of intersecting convoluted tubes, the porosity can be represented as the number of tubes per unit area found in a thin section, n, times the average area of the intersections of the tubes with the plane section, which is proportional to δ^2. Thus,

$$\phi \sim n\delta^2. \qquad (2.2.2)$$

If, however, they are in the form of intersecting cracks or fractures whose characteristic extent is large compared with the thickness δ, and if λ represents the average total length of crack per unit area in a thin section, then

$$\phi \sim \lambda\delta. \qquad (2.2.3)$$

In any case, if the medium is random, sampling lines will intersect an equal number of voids whose characteristic dimension along the line is δ and solids whose characteristic dimension is l_0, so that

$$\phi_t = \frac{\delta}{l_0 + \delta}. \qquad (2.2.4)$$

In very porous matrices such as an assemblage of well-sorted grains, $l_0 \sim \delta$, both l_0 and δ being of the order of the grain size, and ϕ is of the order 0.5, but in less porous assemblages or partly cemented granular fabrics, $\delta \ll l_0$ and

$$\phi_t \simeq \delta/l_0. \qquad (2.2.5)$$

2.3 THE MOTION OF FLUID ELEMENTS THROUGH THE MEDIUM

It is intuitively evident that as interstitial fluid moves through the convoluted passages of a porous rock or matrix of fractures, it tends to disperse. The purpose of this section is to relate the simplest dispersive characteristics to simple geometrical properties of the medium. The primary result is that the dispersion of fluid elements is asymptotically equivalent to simple diffusion, with a diffusion coefficient that is proportional to the product of the mean interstitial fluid speed and the grain size or the characteristic distance between fracture intersections. In nature, solutes are subject also to molecular diffusion, and the combined effect is considered in Section 2.4; a knowledge of the relative magnitudes of the molecular and the geometrical or, as it is usually termed, macroscopic dispersion

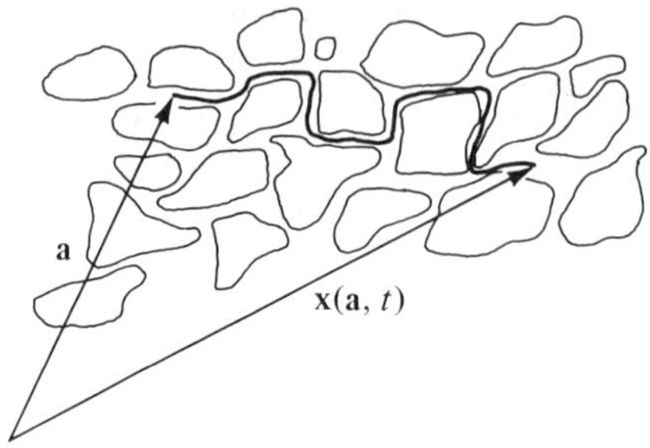

Figure 2.10. The Lagrangian displacement of fluid elements.

coefficients enables us to determine which of the two dominates in a given texture and flow.

The precise nature of the geometrical dispersion depends on the statistics of the microscopic pore structures, and this is most clearly revealed by what is known as a Lagrangian analysis in which we trace the motion of individual fluid elements and then average over an ensemble of them. Suppose that at some initial instant, a tiny region of interstitial fluid is marked with a nondiffusive tracer that enables us to follow the motion of the fluid element as it moves. If its initial position is denoted by $\mathbf{a}$, its subsequent location will depend on where it was initially, on $\mathbf{a}$, and on the elapsed time t,

$$\mathbf{x} = \mathbf{x}(\mathbf{a}, t), \tag{2.3.1}$$

as illustrated in Figure 2.10. Note that at the particular instant $t = 0$, the fluid element is still at $\mathbf{a}$, so that symbolically

$$\mathbf{x}(\mathbf{a}, 0) \equiv \mathbf{a}. \tag{2.3.2}$$

The interstitial velocity of the fluid element is defined as the time derivative of its position,

$$\boldsymbol{v}(\mathbf{a}, t) = \frac{d}{dt}\{\mathbf{x}(\mathbf{a}, t)\}, \tag{2.3.3}$$

and, again, is a function of its initial position and elapsed time. Equation (2.3.3) can be expressed equivalently in a useful integral form,

$$\mathbf{x}(\mathbf{a}, t) - \mathbf{a} = \int_0^t \boldsymbol{v}(\mathbf{a}, t')\, dt', \tag{2.3.4}$$

giving the displacement of the fluid element from its initial position as the integral over time of its velocity, as we follow the marked element of fluid along its convoluted trajectory.

In the mathematics of creeping flow such as that through these interstitices (the term will be defined more precisely in Section 2.6), there is a uniqueness theorem due to Helmholtz (see also Batchelor 1967) which guarantees that for any specified geometry of the surfaces bounding the flow, no matter how complex, and for given driving forces, there is only one possible flow pattern. Consequently, if the overall pressure gradient or temperature distribution driving the flow remains unchanged, and if we mark different fluid elements passing *precisely* the same location **a** at different times, the convoluted trajectories that the marked fluid elements trace out will be the same on each occasion. In practice, however, it is not possible to mark the fluid at a precise location within a given interstice; if we mark successive fluid elements at different points in the same averaging volume (which, on a *macroscopic* scale, is at the same point), their respective trajectories will diverge as some move one way around each grain and others another. From a given averaging volume, then, there will be an ensemble of trajectories associated with *different* starting points inside the *same* sampling volume. This ensemble is equivalent in practice to the result of marking the entire fluid in the averaging volume and following the subsequent history of the cloud.

The mean displacement of the interstitial fluid from its initial position is found by averaging equation (2.3.4) over the ensemble for a fixed time interval t. The interstitial velocity of the fluid elements can be regarded as a random function of time with mean $\bar{\mathbf{v}}$, so that the mean displacement is

$$\bar{\mathbf{x}}(\mathbf{a}, t) - \mathbf{a} = \int_0^t \overline{\mathbf{v}(\mathbf{a}, t')\, dt'} = \bar{\mathbf{v}}t. \tag{2.3.5}$$

Fluid in porous rock that is marked with a tracer, whether salt, dye, pollutant, or radioactivity, moves through the rock with, on average, the mean interstitial or pore velocity. This is a fairly obvious, but important conclusion.

Because of the multiplicity of possible trajectories from a given averaging volume, the fluid elements will also disperse about their mean position, the effect known as macroscopic dispersion. Let us subtract the mean equation (2.3.5) from (2.3.4) to obtain the location of a fluid element relative to the mean position,

$$\mathbf{x}(\mathbf{a}, t) - \bar{\mathbf{x}}(\mathbf{a}, t) = \int_0^t [\mathbf{v}(\mathbf{a}, t') - \bar{\mathbf{v}}]\, dt',$$

or, more concisely,

$$\mathbf{y}(\mathbf{a}, t) = \int_0^t \mathbf{w}(\mathbf{a}, t')\, dt', \tag{2.3.6}$$

where **y** and **w** represent the differences in the position and velocity of a fluid element from the respective mean values. The simplest measures of the dispersion of the fluid elements about the mean involve the second moments of the distribution, or the mean square values of **y**, and these can be expressed usefully in terms of the geometrical and flow characteristics. Let us consider first the dispersion in any one direction so that we are concerned with the component of the vector equation (2.3.6) in this direction, say

$$y(\mathbf{a}, t) = \int_0^t w(\mathbf{a}, t')\, dt',$$

which is equivalent to the differential form

$$\frac{d}{dt} y(\mathbf{a}, t) = w(\mathbf{a}, t).$$

When the left- and right-hand sides of these last two equations are multiplied together, there results

$$y\frac{dy}{dt} = \frac{1}{2}\frac{dy^2}{dt} = \int_0^t w(\mathbf{a}, t)w(\mathbf{a}, t')\, dt',$$

and, upon averaging,

$$\frac{d}{dt}\bar{y}^2 = 2\int_0^t \overline{w(\mathbf{a}, t)w(\mathbf{a}, t')}\, dt'. \tag{2.3.7}$$

The quantity $\overline{w(t')w(t)}$, the average product of the pore velocity fluctuations for particular fluid elements as they move along their trajectories, measured at times t and t', is called in statistics a covariance function. If the macroscopic conditions are steady in time, this covariance depends on the time interval $\tau = t - t'$, rather than on t and t' separately. When $\tau = 0$, the values of w in the covariance are taken simultaneously, and it reduces simply to the mean square of the component $\bar{w}^2$. As the time interval increases, the interstitial velocity at the end of the interval begins to differ in a random way from what it was at the beginning as each element moves through the pores or along the interstices. The velocity components become less correlated – they lose memory – and the covariance function decreases. Ultimately, when the interval τ is large, the velocity variation at the later time has lost all dependence on what it happened to be initially. When all correlation is lost, the mean value of the product of the velocities is equal to the product of the means, each of which is zero since w represents the departure from the mean. Consequently, the covariance function drops to zero as $\tau \to \infty$, in the manner shown qualitatively in Figure 2.11.

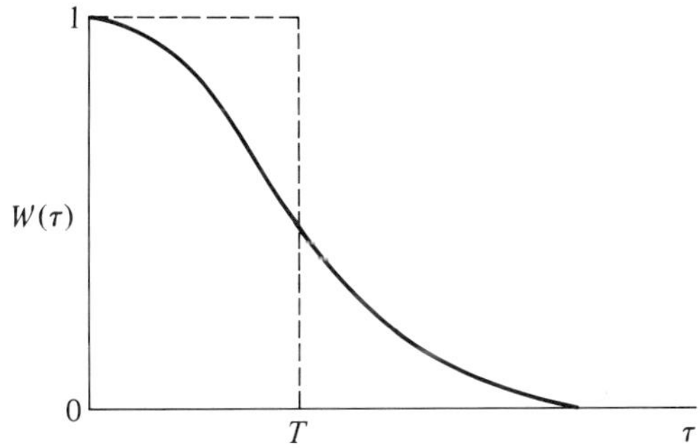

Figure 2.11. The schematic form of the Lagrangian velocity autocorrelation function.

Individual interstices are generally convoluted on a length scale of the same order as their own diameter δ, or perhaps, when they consist of intersecting linear fractures, on a scale of order l_0. The time interval over which the covariances continue to be significant is of the order of the time that a fluid element takes to traverse one or a few pore diameters or grain diameters, that is, $\delta/\bar{v}$ or $l_0/\bar{v}$, where $\bar{v}$ is the mean interstitial speed. (Note that the mean speed $\bar{v}$ and the root mean square fluctuation in velocity component are proportional, the constant of proportionality depending on the statistics of the trajectories or, equivalently, the statistical geometry of the rock structure.) When the elapsed time τ is sufficiently large compared with $\delta/\bar{v}$ or $l_0/\bar{v}$, the covariance has dropped to zero and, presumably, the integral in (2.3.7) has converged. If we write

$$\overline{w(\mathbf{a}, t)w(\mathbf{a}, t')} = \bar{w}^2 W(\tau), \tag{2.3.8}$$

where $W(0) = 1$ and $W \to 0$ as $\tau \to \infty$, then (2.3.7) becomes

$$\frac{d}{dt}\bar{y}^2 = 2\bar{w}^2 \int_0^t W(\tau)\, d\tau,$$

where the variable of integration is changed from t' to $\tau = t - t'$. When $t \gg l_0/v$,

$$\frac{d}{dt}\bar{y}^2 \to 2\bar{w}^2 \int_0^\infty W(\tau)\, d\tau = 2\bar{w}^2 T, \quad \text{say}, \tag{2.3.9}$$

where T represents the area under the curve in Figure 2.11, or the width of a box of equal area, and is of the order of the decorrelation time

$$T \simeq \delta/\bar{v} \qquad \text{or} \qquad l_0/\bar{v}. \tag{2.3.10}$$

In equation (2.3.9), for $t \gg T$, the right-hand side is constant, so that

$$\bar{y}^2 \to (2\bar{w}^2 T)t = 2Dt, \quad \text{say}, \tag{2.3.11}$$

where $D = \overline{w}^2T$, with physical dimensions $[L^2T^{-1}]$, is the macroscopic dispersion coefficient; the mean square displacement about the average increases linearly with time.

This is precisely the same result as that which emerges from the classical random-walk problem in statistics or the diffusion from a point source by molecular motion. It is pertinent in this context asymptotically when the dispersion time t is large compared with (2.3.10) or, equivalently, when the dispersion distance is large compared with l_0 (usually larger than δ), but this includes most cases of interest. Since in a medium with convoluted pathways, the variation in interstitial fluid velocity is of the same order as the mean $\bar{v}$,

$$D \sim \bar{v}^2T \sim \bar{v}\delta \qquad \text{or} \qquad \bar{v}l_0. \tag{2.3.12}$$

An important property of macroscopic dispersion is that in contrast with molecular diffusion, the dispersion coefficient (2.3.12) depends on the flow as well as the average geometry of the medium, expressed as l_0. It will be seen later (e.g., in Section 4.6) that since the advection or bodily transport of fluid elements is also proportional to $\bar{v}$, when macroscopic dispersion dominates, the *spatial pattern* of dispersion from a fixed source in a flow field is independent of the flow speed.

These results generalize simply to three dimensions. If the subscript i = 1, 2, or 3 refers to the three orthogonal directions,

$$\bar{y}_i^2 = \{2\bar{w}_i^2T\}t = 2D_it. \tag{2.3.13}$$

In an isotropic medium, (2.3.13) reduces to (2.3.11), but in layered or lenticular structures, the dispersal of fluid elements is augmented greatly by large-scale flow focusing, a process known as megadispersion (see Section 3.9).

2.4 DISPERSAL AND DIFFUSION OF DISSOLVED SPECIES

Chemical constituents in the interstitial fluid diffuse as a result of their thermal agitation as they disperse by fluid motion through the pores or along the interstices. The velocity of an individual molecule or ion can be regarded as the sum of that of its molecular agitation w_m superimposed on that of the fluid element w; the two are statistically independent. The same formulation can be used to describe the meandering of individual molecules with $w + w_m$ replacing w in (2.3.7); because of the statistical independence of w and w_m,

$$\begin{aligned}\int_0^\infty &\overline{[w(t) + w_m(t)][w(t') + w_m(t')]}\, dt' \\ &= \int_0^\infty \{\overline{[w(t)w(t')]} + \overline{[w_m(t)w_m(t')]}\}\, dt' \\ &= \bar{w}^2T + \bar{w}_m^2T_m,\end{aligned} \tag{2.4.1}$$

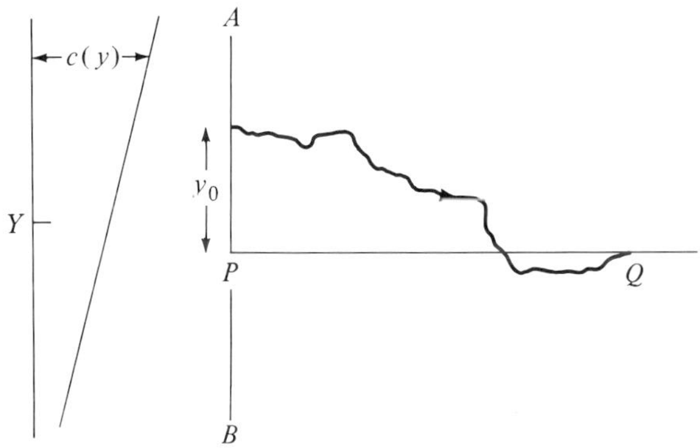

Figure 2.12. Meandering of fluid elements in a locally constant concentration gradient.

where T_m is the analogous integral time scale of molecular motion. The second term represents the dispersal described by molecular diffusivity κ_s, so that the total dispersion coefficient is simply the sum of the macroscopic and molecular coefficients, as in (2.3.11):

$$\bar{y}^2 \rightarrow 2(D + \kappa_s)t, \tag{2.4.2}$$

where $D = \bar{w}^2 T$ and T is the Lagrangian integral time scale defined in the preceding section, κ_s being the molecular diffusivity for the species concerned.

The dispersion of individual fluid elements is clearly related to the flux of the corresponding species across the surface of an averaging volume in the presence of an overall gradient in concentration c (dissolved mass per unit volume), and again it can be shown that the flux is proportional to the gradient with a diffusivity that includes the macroscopic and molecular dispersion coefficients. Consider a locally isotropic region such as that shown in Figure 2.12, in which the gradient of mean concentration of a particular species in the interstitial fluid can be regarded as constant. When the mean interstitial fluid velocity $\bar{\mathbf{v}}$ is transverse to the gradient (horizontal in the figure), the fluid elements arriving at point Q in successive realizations will have crossed the line AB at points distributed symmetrically about point P; the mean concentration c at Q is therefore equal to that at P, and the mean gradient persists. However, a fluid element that arrives at Q and that passed by AB at a distance y_0 from P has the concentration

$$c(Y + y_0) = \bar{c}(Y) + y_0 \left(\frac{\partial \bar{c}}{\partial y}\right), \tag{2.4.3}$$

with mean $\bar{c}(Y)$ together with a fluctuation about this average that is uncorrelated with its subsequent motion. The displacement y_0 can be expressed as before, as an integral following the motion before its arrival at Q at time $t = 0$:

$$y_0 = -\int_{-t_0}^{0} w(t')\, dt', \tag{2.4.4}$$

where $-t_0$ is the time at which it passed AB. When this fluid element arrives at Q, its expected concentration differs from the average value there, $\bar{c}(Y)$, by the amount

$$c' = \left\{-\int_{-t_0}^{0} w(t')\, dt'\right\} \frac{\partial \bar{c}}{\partial y} \tag{2.4.5}$$

Now, the flux in the y-direction of the dissolved species within the fluid-filled pore region is the mean product of the concentration and the fluid velocity in the y-direction, $w(0)$; since this has zero mean,

$$\begin{aligned} \text{Dispersive flux in fluid} &= \overline{cw} \\ &= \overline{c'w} \\ &= -\left\{\int_{-t_0}^{0} \overline{w(t')w(0)}\, dt'\right\} \frac{\partial \bar{c}}{\partial y} \\ &= -\bar{w}^2 T \frac{\partial \bar{c}}{\partial y}, \end{aligned} \tag{2.4.6}$$

from (2.4.5) and (2.3.9), in view of the symmetry of $W(\tau) \equiv W(-\tau)$. The mean flux of species in the fluid-filled region by macroscopic dispersion is therefore proportional to the mean gradient of concentration, and the diffusivity is equal to the macroscopic dispersion coefficient D. Since fluid occupies only the fraction ϕ of the area of the matrix across which the dispersive flux occurs, the net flux (transport across unit area of the *matrix*) is

$$-\phi D \frac{\partial \bar{c}}{\partial y}. \tag{2.4.7}$$

At the same time, molecular diffusion occurs along the pores, and on this scale the molecular fluxes and concentration gradients are roughly in the direction of the local pore axis inclined, say, at an angle θ to the direction of the overall gradient. The interstitial flux is $-\kappa_s\, \partial c/\partial s$, where s is measured along the axis; this has a component $-\kappa_s \cos\theta\, \partial c/\partial s$ in the gradient direction. Since $dy = ds \cos\theta$, the *mean* molecular diffusive flux is

$$-\phi\overline{\kappa_s \cos^2\theta\, \partial c/\partial y}, \tag{2.4.8}$$

where the average is over the ensemble of pores intersecting unit area of the fabric. One cannot expect the interstitial fluctuations in $\partial c/\partial y$ to be statistically independent of the orientation θ of the pore axis, but a useful order-of-magnitude estimate of (2.4.8) can be obtained by replacing the mean value of the product by the product of the means,

$$-\phi\kappa_s \overline{\cos^2\theta}\,\frac{\partial\bar{c}}{\partial y},$$

so that the total dispersive flux can be written

$$Fy = -\phi(D + \kappa_s \overline{\cos^2\theta})\,\frac{\partial\bar{c}}{\partial y}. \tag{2.4.9}$$

The quantity $\tau = (\overline{\cos^2\theta})^{-1/2}$ is sometimes known at the tortuosity.

The approximation made to (2.4.8) implies that (2.4.9) should not be accepted too literally, but it does suffice to demonstrate that the effective molecular diffusivity for diffusion in the interstices is less by a factor of about 2 than it is in an unconfined fluid, even when allowance is made for the porosity; the other uncertainties in many geological applications are so great that this level of precision is adequate. Consequently, the total flux can generally be expressed as

$$\mathbf{F} = -\phi(\mathbf{D} + \kappa_s\mathbf{I})\cdot\nabla\bar{c}, \tag{2.4.10}$$

where **D** is the macroscopic dispersion matrix and **I** the unit matrix; κ_s here and hereafter will refer to the effective molecular diffusivity in the matrix, of order half the unconfined solution value.

The relative importance of macroscopic and molecular dispersion of dissolved species in the interstitial fluids depends on the ratio

$$\frac{D}{\kappa_s} = \frac{\bar{w}^2 T}{\kappa_s} \sim \frac{\bar{v}l_0}{\kappa_s},$$

where l_0 is the "decorrelation distance," of the order of the grain size or, in a nonuniform medium, the scale of the inhomogeneities. Some numerical values are useful. A typical interstitial velocity in flow through sedimentary rocks might be 32 cm yr^{-1} or 10^{-6} cm s^{-1}; for small molecules or ions in water, $\kappa_s \sim 10^{-5}$ cm^2 s^{-1} at about 25° C (Harned and Owen 1958) so that if l_0 is less than about 10 cm, molecular dispersion will dominate. However, in a more porous aquifer or lens, where local flow concentrations occur as shown in Chapter 3, the interstitial velocities may be a couple of orders of magnitude greater and macroscopic dispersion will dominate even when the grain size is small.

It is worth mentioning at this point that in the case of thermal diffusion, macroscopic heat dispersion in permeable rocks rarely seems to be important, though it may be in other, more permeable media (such as insulating material), as shown by Kvernvold and Tyvand (1980). Heat is conducted through the entire fabric, not just along the fluid pathways, which occupy only the fraction ϕ of each element of area across which the dispersion occurs. Moreover, the thermal diffusion coefficient of saturated fabric, κ_m, is of the order 10^{-3} c.g.s. unit, larger by a factor of about 10^2 than the molecular diffusivity of small ions in water. Molecular heat conduction through the fabric dominates the macroscopic heat *dispersion* along the interstices when

$$\frac{\phi \bar{v} l_0}{\kappa_m} \ll 1, \tag{2.4.11}$$

a condition that is usually satisfied in geological situations. Note, however, that the *mean* transport of heat by fluid flow may or may not be large compared with the dispersive or diffusive transports, as is explained in more detail in Section 3.5.

2.5 THE TRANSPORT VELOCITY AND THE INCOMPRESSIBILITY CONDITION

Consider the fluid flow across a unit area in the fabric perpendicular to the mean pore velocity $\bar{\mathbf{v}}$. The porosity or active void fraction is ϕ, so that over the fraction ϕ of the unit cross-sectional area, the mean flow velocity is $\bar{\mathbf{v}}$; over the rest, the solids or inactive voids, it is zero. Consequently, averaged over the whole, the volume flux or the fluid volume flow per unit cross-sectional area or the transport velocity

$$\mathbf{u} = \phi \bar{\mathbf{v}}. \tag{2.5.1}$$

The velocity $\mathbf{u}$ can be interpreted as the velocity with which a fluid would be moving if it occupied the whole space and had the same volume transport. Its direction is parallel to $\bar{\mathbf{v}}$, but since for many sediments ϕ is numerically small, $\mathbf{u}$ is in magnitude substantially smaller than $\bar{\mathbf{v}}$.

The porosity of a rock sample is frequently estimated by comparing wet and dry weights of a sample. It could also be measured directly by taking advantage of equation (2.5.1), especially if the identification of active and inactive voids is difficult. If water is forced under pressure through the sample, the volume of water passing through per unit cross-sectional area per unit time is the transport velocity. If a spot of dynamically passive dye or other marker (i.e., one that does not affect the flow) is introduced, it will pass through the sample on average at the mean interstitial speed, and the ratio of the two defines ϕ.

The transport velocity $\mathbf{u}$ is one of the primary field variables used in this book, since transports, not simply of fluid volume but also of heat and chemical species, are of primary interest. Remember, however, that if we are concerned with the spreading of *particular* elements of fluid that can be marked with salinity, dye, and radioactive or other contaminants, as we are in unsteady situations, then we must consider the interstitial or pore velocity; this has some profound consequences, as we shall see later.

One physical constraint on the flow patterns through porous fabrics that can be expressed simply in terms of **u** is the conservation of total fluid mass. In a fixed arbitrary finite region of the fabric, the mass of interstitial fluid is

$$\int \phi\rho \, dV,$$

where ϕ is the porosity, ρ is the interstitial fluid density, and the integral is over the volume of the region. If there is no generation of liquid water by chemical reaction, this mass may change as a result of net fluid flow across the bounding surface; the net transport of mass out of the volume is

$$\int \rho\mathbf{u} \cdot d\mathbf{S},$$

where the integral is over the surface of the region and the element of surface area $d\mathbf{S} = \mathbf{n} \, dS$ is directed outward. The rate of decrease of fluid mass in the region is equal to the net transfer out of it,

$$\begin{aligned} \frac{\partial}{\partial t} \int \phi\rho \, dV &= -\int \rho\mathbf{u} \cdot d\mathbf{S} \\ &= -\int \nabla \cdot (\rho\mathbf{u}) dV, \end{aligned}$$

by the divergence theorem. Thus, since the volume is fixed in time,

$$\int \left\{ \frac{\partial}{\partial t} (\phi\rho) + \nabla \cdot (\rho\mathbf{u}) \right\} dV = 0,$$

and since, moreover, it is arbitrary, the integrand itself must vanish:

$$\frac{\partial}{\partial t} (\phi\rho) + \nabla \cdot (\rho\mathbf{u}) = 0. \tag{2.5.2}$$

This equation expresses the conservation of total fluid mass. It can be rewritten as

$$\rho \frac{\partial \phi}{\partial t} + \phi \frac{\partial \rho}{\partial t} + \mathbf{u} \cdot \nabla\rho + \rho\nabla \cdot \mathbf{u} = 0, \tag{2.5.3}$$

or, if

$$R_\rho = \ln(\rho/\rho_0),$$

where ρ_0 is a convenient reference value for the fluid density (the local average, say), then

$$\frac{\partial \phi}{\partial t} + \left(\phi \frac{\partial}{\partial t} + \mathbf{u} \cdot \nabla\right) R_\rho + \nabla \cdot \mathbf{u} = 0,$$

or

$$\nabla \cdot \mathbf{u} = -\frac{\partial \phi}{\partial t} - \left(\phi \frac{\partial}{\partial t} + \mathbf{u} \cdot \nabla\right) R_\rho. \tag{2.5.4}$$

In most cases of fluid flow in rocks, the two sets of terms on the right can be neglected. The first expresses the simple geometrical fact that if the porosity decreases with time, ($\partial \phi / \partial t < 0$), there is a flow divergence as fluid is expelled from the decreasing fraction of voids. If the porosity of the rock changes over geological time as cementation or dissolution occurs, the cumulative changes in ϕ may be large, but since the time scales are long, the *rates* of change are extremely small. During and following seismic events, ϕ may change rapidly, as Sibson (1987, 1989) has pointed out, and significant compaction may occur in clays or other unconsolidated sediments. When such sediments compact over a short time scale, the total volume of fluid expelled per unit volume of matrix is just the change in porosity, necessarily a small fraction of unity. Compaction is certainly important in modifying the distribution of pore fluids, whether aqueous solutions or hydrocarbons. It has sometimes been suggested as a primary mechanism for forcing fluid motion (Bethke 1985 gives a critical assessment of the process), but interesting geochemical changes require the cumulative movement through the matrix of fluid volumes that are many times the matrix volume itself. In such cases the interstitial fluid velocities associated with compaction are negligible compared with those produced by maintained hydraulic or thermal forcing.

The second set of terms on the right of (2.5.4) involving the rate of change of interstitial fluid density is also usually negligible. When variations in density occur as a result of spatial variations in temperature, $\partial R_\rho / \partial t \simeq 0$ and $\mathbf{u} \cdot \nabla R_\rho \simeq \mathbf{u} \cdot \nabla(\delta\rho/\rho_0)$, where $\delta\rho$ is the thermally induced density change. This may be important in influencing the relative buoyancy of the interstitial fluids and give rise to convective motion, but its influence on the flow divergence is minor. In geological flow fields, the individual terms in $\nabla \cdot \mathbf{u}$ are of order u/l, where u is the transport speed and l the overall flow dimension; the term $\mathbf{u} \cdot \nabla(\delta\rho/\rho_0)$ is smaller by order $\delta\rho/\rho_0$ and is therefore negligible unless boiling occurs. Another exceptional circumstance arises when the interstitial fluid pressure is suddenly altered – for example, by a natural earthquake or artificially in a pressurized well. The interstitial fluid is often much more compressible than the rock matrix, and if the pressure is suddenly reduced by a nearby fracture, the unfractured rock porosity changes little but the interstitial fluid will seek to expand, forcing a time-dependent, local flow, a process Sibson calls "seismic pumping." This situation is considered in Section 3.10.

Subject to these provisos, then, the terms on the right of (2.5.4) can be ignored, and it reduces to

$$\nabla \cdot \mathbf{u} = 0, \tag{2.5.5}$$

which, expressing the statement that the volumetric divergence of the fluid vanishes, is usually called the incompressibility condition. In Cartesian coordinates with $\mathbf{u} = (u, v, w)$,

$$\nabla \cdot \mathbf{u} = \frac{\partial u}{\partial x} + \frac{\partial v}{\partial y} + \frac{\partial w}{\partial z} = 0. \tag{2.5.6}$$

Finally, when, for example, dehydration reactions occur at considerable depth, the interstitial liquid water content may vary significantly as a result of the reaction. In these cases (considered in Section 4.8), a source term must be added to the right of (2.5.2) to represent the rate of generation of water mass per unit volume by the reaction, and this carries through as a source term on the right of (2.5.5) that depends on the reaction progress. This modification, however, will be included only in the appropriate context.

2.6 DARCY'S LAW

In a fluid at rest, the pressure increases with depth as the result of the weight of the overlying fluid; the pressure gradient is vertically downward. For a change dz in depth, the pressure itself changes by the amount $\rho g\, dz$, where ρ is the local fluid density and g the gravitational acceleration. The hydrostatic pressure gradient can therefore be expressed as

$$\frac{\partial p_{\mathrm{h}}}{\partial z} = -\rho g \mathbf{l}, \tag{2.6.1}$$

where $\mathbf{l}$ is a unit vector vertically upward. This remains true no matter what the shape of the container enclosing the fluid and no matter whether the fluid as a whole is over- or underpressurized, so that (2.6.1) holds for *single-phase* interstitial fluids at rest.

In such a fluid, if the interstitial pressure gradient is not hydrostatic, the fluid will move. Once it is in motion, viscous stresses are generated within the fluid and the fluid velocities in the pores are determined by a balance between the nonhydrostatic pressure gradients driving the motion (the difference between the actual pressure gradient and the local hydrostatic gradient) and the gradients in viscous stresses resisting it. Although the task of calculating the detailed flow patterns in the individual pores is, as a rule, both technically hopeless and of limited interest, the relation between the macroscopic transport velocity and the overall gradient of nonhydrostatic pressure can be inferred simply.

With small-molecule fluids such as water or aqueous solutions (Newtonian fluids), the viscous stresses τ_{ij} $(i, j = 1, 2, 3)$ are proportional to the rate of strain e_{ij} in the fluid:

$$\tau_{ij} = 2\mu e_{ij} = \mu\left(\frac{\partial v_i}{\partial x_j} + \frac{\partial v_j}{\partial x_i}\right), \tag{2.6.2}$$

where μ is the dynamic viscosity, a physical property of the fluid. In a pore of width δ with an interstitial fluid speed of magnitude v, the form of (2.6.2) shows that the viscous stresses are of order $\mu v/\delta$, which vary across the width of the pore to transmit the fluid drag to the pore walls. The characteristic magnitude of the viscous stress *gradient* is therefore $\mu v/\delta^2$, which balances the pressure gradient along the pore. This, in turn, is equal to the overall nonhydrostatic pressure gradient multiplied by some geometrical factors involving the shape and orientation of the pore region. Thus,

$$\frac{\mu \mathbf{v}}{\delta^2} \propto -(\nabla \hat{p} + \rho g \mathbf{l}).$$

Now, the two terms on the right side of this equation are usually very much larger than their difference, and it is this difference that drives the flow. It is therefore often convenient to define a *reduced pressure* p as the difference between the actual pressure $\hat{p}$ and the *hydrostatic pressure that would exist if the density were everywhere constant and equal to its average value* ρ_0, say

$$p = \hat{p} + \rho_0 g z,$$

where z is directed vertically upward. Thus,

$$\frac{\mu \mathbf{v}}{\delta^2} \propto -(\nabla p + (\rho - \rho_0) g \mathbf{l}).$$

When the interstitial fluid density is everywhere constant and equal to ρ_0,

$$\frac{\mu \mathbf{v}}{\delta^2} \propto \nabla p,$$

and the reduced pressure p is proportional to the potential ϕ introduced by Hubbert (1940) and widely used in the groundwater literature. When thermal effects are important in driving or modifying the flow, however, the density difference $\rho - \rho_0$ is a function of position determined by the temperature distribution and the velocity cannot be expressed in terms of the gradient of a potential, so that Hubbert's usage will not be followed in this book.

In any event, the transport velocity is obtained by averaging over a macroscopic element as described earlier; from (2.5.1),

$$\mathbf{u} \propto -\frac{\phi\delta^2}{\mu}(\nabla\hat{p} + \rho g\mathbf{l}), \tag{2.6.3}$$

where the constant of proportionality contains all the geometrical factors needed in the statistical specification of the medium. Note that, in general, the nonhydrostatic pressure gradient and the transport velocity do not necessarily coincide in direction; if the fabric contains parallel tubes inclined to the direction of the pressure gradient, a transverse velocity component (still proportional to the pressure gradient) will be induced. Accordingly, (2.6.3) can generally be written as

$$\mathbf{u} = -\mu^{-1}\mathbf{k}\cdot(\nabla\hat{p} + \rho g\mathbf{l}) = -\mu^{-1}\mathbf{k}\cdot(\nabla p + (\rho - \rho_0)g\mathbf{l}), \tag{2.6.4}$$

where the matrix (or second-order tensor) **k** has components that are proportional to $\phi\delta^2$. In a locally isotropic medium

$$\mathbf{k} = \begin{pmatrix} k & 0 & 0 \\ 0 & k & 0 \\ 0 & 0 & k \end{pmatrix},$$

and the transport velocity is parallel to the nonhydrostatic pressure gradient,

$$\mathbf{u} = -\frac{k}{\mu}(\nabla\hat{p} + \rho g\mathbf{l}) = -\frac{k}{\mu}(\nabla p + (\rho - \rho_0)g\mathbf{l}), \tag{2.6.5}$$

where $k \propto \phi\delta^2$ and incorporates the statistical geometry of the medium and is called the intrinsic permeability. The relation (2.6.5) is generally known as Darcy's law.

The intrinsic permeability k has dimensions L^2 and is measured in units of cm^2 or m^2 and so forth. A curious unit, but one widely used in hydrology, is the darcy, defined as the permeability that allows a transport velocity of 1 cm s^{-1} of a fluid with viscosity 1 centipoise (10^{-2} c.g.s. unit) under a pressure gradient of 1 *atmosphere* per centimeter! Since mean atmospheric pressure is $\sim 10^6$ c.g.s. units, a permeability of 1 darcy is about 10^{-8} cm^2, or 10^{-12} m^2. A related quantity also widely used when pressures are expressed in terms of the equivalent hydraulic head is the hydraulic conductivity

$$\mathbf{K} = \frac{\rho g\mathbf{k}}{\mu} = \frac{g\mathbf{k}}{\nu}, \tag{2.6.6}$$

where $\nu = \mu/\rho$ is called the *kinematic* viscosity. The hydraulic conductivity depends on both the permeable medium and the fluid; since its physical dimensions are $[LT^{-1}]$ (as are those of velocity), it is expressed in units of cm s^{-1}, m s^{-1}, or m yr^{-1}, and so on.

The derivation of Darcy's law given here neglects variations in interstitial pressure associated with the inertia of pore liquid as it moves around the grains or along the convoluted pathways. If, at some point, its trajectory has a radius of curvature r, the fluid inertia sets up an additional pressure gradient $\rho v^2/r$, where v is the pore velocity – this provides the centripetal acceleration associated with the curved trajectory. Darcy's law is accurate, then, only when these inertial pressure gradients are small compared with the viscous stress gradients $\mu v/\delta^2$, and since in general $r \sim \delta$, this requires

$$\frac{\rho v^2}{\delta} \ll \frac{\mu v}{\delta^2}$$

or, since $v \sim \bar{v}$,

$$R = \frac{\bar{v}\delta}{\nu} \ll 1,$$

where again $\nu = \mu/\rho$ is the kinematic viscosity. The quantity $\bar{v}\delta/\nu$, a dimensionless grouping of the pore velocity, pore width, and kinematic viscosity, is the pore Reynolds number, which, for the validity of Darcy's law, must be small. In general, flows in which $R \ll 1$ are called creeping flows. In terms of the transport speed $u = \phi\bar{v}$, the condition for validity of Darcy's law becomes

$$\frac{u\delta}{\nu} \ll \phi. \tag{2.6.7}$$

The kinematic viscosity of aqueous solutions is of the order 10^{-2} c.g.s. unit (10^{-6} SI unit); the characteristic velocities in groundwater flow are in the range 10^{-3} to 10^{-6} cm s^{-1} (320 to 0.32 m yr^{-1}) so that (2.6.7) is satisfied easily, except possibly in coarse gravel beds with $\delta > 10$ cm. It is strongly satisfied in the slower flow through porous sedimentary rocks with finer pores.

There are, however, some geological situations that do require caution. If the interstitial fluid consists of regions of two or more separate, immiscible components such as oil and water, the radius of curvature of the interface can be of order δ, and surface tension T can support a pressure difference of order T/δ across the interface without motion. If n represents the number of such interfaces per unit length along the pores, an overall pressure gradient of order nT/δ can be supported without fluid motion. When this threshold pressure gradient is exceeded, fluids will begin to move, first in the pathways with largest pores and, subsequently, as the pressure gradient increases, in the pathways with the next largest and so on. The *distribution* of pores sizes, and not simply the mean, becomes crucial in determining the flow characteristics, and the overall relation between pressure gradient and transport velocity is certainly non-linear. Interesting numerical experiments on capillary displacement and percolation of immiscible fluids, showing fingering, trapping, and incomplete displacement of one fluid by the other, have been conducted by Chandler, Koplik, Lerman, and Willemsen (1982), but this very important though specialized topic will not be pursued in this book.

In some vigorous geothermal systems where the fluid-filled interstices could better be called pipes or vents than pores and the transport velocities may be large, the interstitial Reynolds number (2.6.7) can be much larger than unity. Additional internal systematic pressure gradients proportional to and opposed (in mean) to **u** will be established, and Nield and Joseph (1985) suggest that since $k \propto \phi\delta^2$,

$$\frac{\rho v^2}{\delta} \propto \rho k^{-1/2}\phi^{1/2}v^2 = \rho(k\phi^3)^{-1/2}u^2,$$

and the Darcy equation (2.6.5) should be modified to

$$\nabla\hat{p} = -\frac{\mu}{k}\nabla\mathbf{u} - \rho g\mathbf{l} - C\rho(k\phi^3)^{-1/2}\mathbf{u}|\mathbf{u}| \qquad \text{when} \quad R \gg 1, \qquad (2.6.8)$$

where $|\mathbf{u}|$ is the magnitude of the velocity vector, the speed, and the number C, the form drag coefficient, is independent of the viscosity and other fluid properties, but depends on the geometry of the medium.

In some chemical engineering applications involving a packed bed of large porosity near a smooth impermeable wall, the packing will be modified by the presence of the wall to a distance of the order of the particle size. Moreover, the velocity of the fluid in contact with the wall must vanish in all components. In an attempt to deal with the no-slip requirement, several authors have placed an additional term, $\mu\,\nabla^2\mathbf{u}$, to the right of equation (2.6.5), and the result is sometimes known as the Brinkman equation (see, e.g., Nield 1984). In geological contexts, however, this modification is hardly pertinent, since the averaging volumes are necessarily large compared with the grain or block size.

If the porous fabric abuts an impermeable layer, there is no fluid transport across the contact surface, so that on such a surface $\mathbf{u}\cdot\mathbf{n} = 0$, where $\mathbf{n}$ is the unit normal. This boundary condition constitutes a constraint on the flow field and will be used when particular solutions to (2.6.5) are sought. Somewhat different boundary conditions are appropriate at a surface between two different porous layers; the normal component of the transport velocity must be the same on either side of the interface (a consequence, in fact, of the incompressibility condition), as must the fluid pressure (a discontinuity would imply an indefinitely large pressure gradient with consequent infinite velocities).

2.7 PERMEABILITY *k*

The actual magnitude of k depends on the statistics of the pore geometry, and an attempt to calculate it would require more detailed statistical in-

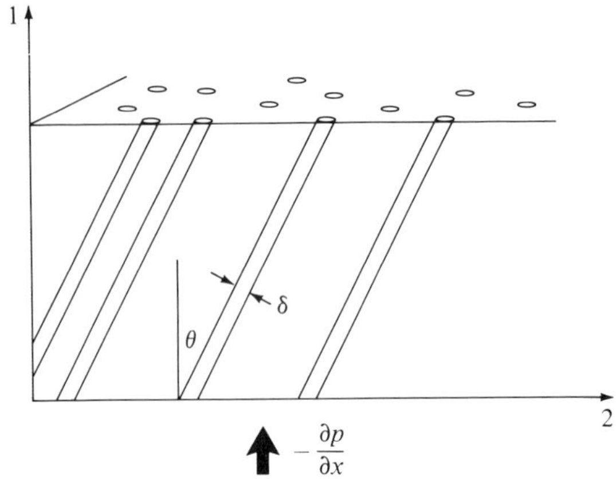

Figure 2.13. A distribution of parallel tubular interstices.

formation on the structure than is likely to be available. The permeability was shown in the preceding section to be proportional to $\phi\delta^2$; the primary conclusion of this section is that the numerical constant of proportionality is small, of order 10^{-2}, at least in simple cases that can be calculated precisely and presumably in more general cases.

For example, suppose that the fabric consists of a distribution of parallel tubes of equal diameter δ inclined at an angle θ to the (nonhydrostatic) pressure gradient (Figure 2.13). If n represents the average number of tubes per unit area in a section perpendicular to this, the number per unit area perpendicular to the pressure gradient is $n \sin \theta$. The pressure gradient along each tube is $(\partial p/\partial x) \cos \theta$, so that the volume flux along each is

$$q = -\frac{\pi r^4}{8\mu}\frac{\partial p}{\partial x}\cos\theta, \tag{2.7.1}$$

where $r = \delta/2$ is the pore radius, as shown by Poiseuille in 1840 (see, e.g., Batchelor 1967). The volume flow across unit area perpendicular to the direction of the tubes (the transport velocity) is then nq, or $\phi q/\pi r^2$, since the porosity $\phi = n\pi r^2$. Consequently,

$$\begin{aligned} u &= -\frac{\phi r^2}{8\mu}\frac{\partial p}{\partial x}\cos\theta \\ &= -\frac{\phi\delta^2}{32\mu}\frac{\partial p}{\partial x}\cos\theta \end{aligned}$$

in the direction of the tubes, which has components in the direction of, and transverse to the pressure gradient, of

$$u_1 = -\frac{\phi\delta^2}{32\mu}\frac{\partial p}{\partial x}\cos^2\theta,$$
$$u_2 = -\frac{\phi\delta^2}{32\mu}\frac{\partial p}{\partial x}\cos\theta\sin\theta. \tag{2.7.2}$$

The component in the third direction (into the page), of course, vanishes. Since we have chosen the pressure gradient in the 1-direction, by comparison with (2.6.4),

$$k_{11} = \frac{\phi\delta^2}{32}\cos^2\theta,$$
$$k_{21} = \frac{\phi\delta^2}{32}\cos\theta\sin\theta,$$
$$k_{31} = 0.$$

For a pressure gradient in the 2-direction, the problem is the same except that the labels 1 and 2 are interchanged and $\pi/2 - \theta$ replaces θ, so that

$$k_{22} = \frac{\phi\delta^2}{32}\sin^2\theta,$$
$$k_{12} = \frac{\phi\delta^2}{32}\sin\theta\cos\theta,$$
$$k_{32} = 0.$$

For a pressure gradient in the 3-direction, there is no flow at all. Consequently, for this geometry,

$$\mathbf{k} = \frac{\phi\delta^2}{32}\begin{pmatrix}\cos^2\theta & \cos\theta\sin\theta & 0\\ \cos\theta\sin\theta & \sin^2\theta & 0\\ 0 & 0 & 0\end{pmatrix}. \tag{2.7.3}$$

When the orientations of the tubes are distributed randomly in direction (the direction of each may vary along its length on the scale l_0), the overall permeability can be calculated by averaging over all directions θ, though some care is required. In the spherical polar coordinates of Figure 2.14, the number of tubes dn in the direction of the element of solid angle $d\theta\ d\varphi$ is proportional to the area of the element on a unit sphere, namely, $d\theta\ d\varphi \sin\theta$, so that

$$dn = n_0 \sin\theta\ d\theta\ d\varphi, \tag{2.7.4}$$

where n_0 is a constant independent of θ and φ. In a thin section, one measures the number per unit area in a given plane (the 2–3-plane, say), and the contribution to this from tubes in this area of solid angle is $dn \sin\theta$. The cross-sectional area of each tube observed in this plane is $\pi r^2/\sin\theta$, so that the void area per unit area ϕ is the integral of $dn \sin\theta$ times $\pi r^2/\sin\theta$ over all directions:

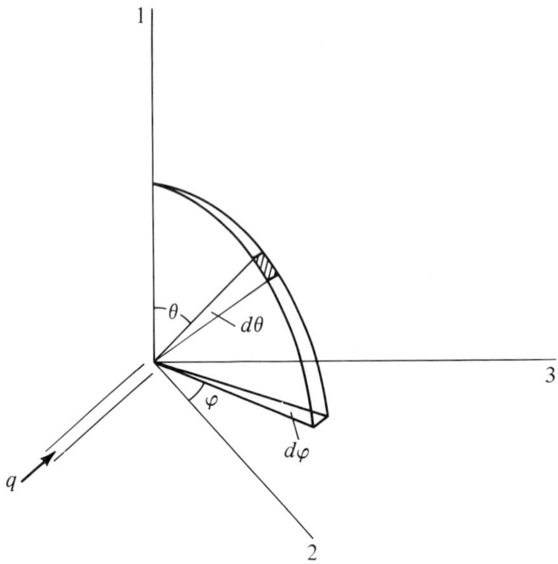

Figure 2.14. Definition sketch.

$$\phi = \pi r^2 n_0 \int_0^{2\pi} \int_0^{\pi/2} \sin\theta \, d\theta \, d\varphi$$
$$= 2\pi^2 r^2 n_0. \tag{2.7.6}$$

(Compare this with the order-of-magnitude expression (2.2.2).) The component of the transport velocity in the 1-direction associated with tubes with directions in the solid angle element is $(dn)q \cos\theta$, where q is given by (2.7.1), so that the total transport velocity in this direction is obtained by integrating over all directions:

$$u_1 = n_0 \int_0^{2\pi} \int_0^{\pi/2} q \cos\theta \sin\theta \, d\theta \, d\varphi,$$
$$= -\frac{2\pi^2 n_0 r^4}{8\mu} \frac{\partial p}{\partial x} \int_0^{\pi/2} \cos^2\theta \sin\theta \, d\theta,$$
$$= -\frac{\pi^2 n_0 r^4}{12} \frac{\partial p}{\partial x}. \tag{2.7.7}$$

The component of the transport in the lateral direction (e.g., 2) from the element of solid angle is likewise $(dn)q \sin\theta \cos\varphi$, but on integration over φ from 0 to 2π the net contribution vanishes, as it must by isotropy. The transport velocity u is in the same direction as the (negative) pressure gradient:

$$u = -\frac{\phi r^2}{24}\frac{\partial p}{\partial x},$$
$$= -\frac{\phi \delta^2}{96}\frac{\partial p}{\partial x}, \tag{2.7.7}$$

giving a scalar permeability

$$k = \frac{\phi \delta^2}{96}. \tag{2.7.8}$$

The case in which the matrix contains a distribution of fracture planes can be analyzed similarly, but the calculation will not be given in detail. For a system of parallel fractures $k = \phi\delta^2/12 = \delta^3/12l_0$, where δ is now the crack width and l_0 the crack spacing. In an isotropic assembly of plane fractures, the permeability is $\phi\delta^2/36$.

In more complex geometries, the permeability cannot be calculated so simply (if at all!), but it is of interest that while the permeability in each case is proportional to $\phi\delta^2$, the constant of proportionality is small, in the range $(1\text{–}3) \times 10^{-2}$. If, in a particular sample, the porosity $\phi = 10^{-2}$ and the pore size $\delta = 10^{-4}$ m, we would expect k to be in the range $(1\text{–}3) \times 10^{-12}$ m^2, or $(1\text{–}3) \times 10^{-8}$ cm^2. A more accurate determination would require flow measurements through the sample and the use of equation (2.6.5). Note also that the grain size per se is irrelevant; the porosity involves the size of the pores. The two are proportional only in the case of well-sorted, uncemented grains, needles, or other particles of approximately uniform shape and size. In spite of this, attempts have been made to express various sets of empirical data on permeability in terms of porosity and grain size (see, e.g., the summary in Lerman 1979), but the empirical formulas obtained should be used for media similar to those measured or, at most, only as a very general guide.

In a useful review (concerned mainly with crystalline rocks), Brace (1980) points out that laboratory measurements of permeability of small samples give values for sandstone of from 10^{-8} to 10^{-12} cm^2 and for limestone–dolomite, a wide range from 10^{-10} to as small as 10^{-18} cm^2, reflecting the variations in microstructural characteristics. Metamorphic and granitic rocks have permeabilities from 10^{-12} to 10^{-17} cm^2. In situ measurements give generally higher values, 10^{-6} to 10^{-8} cm^2 for chert and reef limestone, 10^{-8} to 10^{-14} cm^2 for granites or crystalline rocks, probably reflecting flow through fractures rather than through the rock itself. Fractures may be sealed or absent in shale. Many limestones and sandstones are already so permeable ($\sim 10^{-11}$ cm^2) that fractures add little. In the face of these variations, it is clearly rash to depend heavily on an a priori assumption or estimate of average values, but fortunately, as we shall see in subsequent chapters, flow *patterns* can often be inferred without such numerical guesses. The estimate of flow magnitudes, which are critical in many practical applications involving water or hydrocarbon extraction, does usually involve directly the numerical values of k and their spatial variability; this is often a primary uncertainty.

2.8 THE THERMAL ENERGY BALANCE

The distribution of temperature in the fabric is constrained by an equation describing the heat balance in each averaging volume – an expression of the first law of thermodynamics. Let C represent the specific heat at constant pressure; we will use subscripts m and f to refer to properties of the saturated matrix as a whole and to those of the fluid, respectively. The rate of change of the heat content in an element of unit volume is $(\rho C)_m \, \partial T/\partial t$, and this may come about as the result of heat addition (or subtraction) in advective heat transport by moving interstitial fluids, by thermal (molecular) conduction through the matrix, and by heat production (or absorption) by any chemical reactions.

The advective heat flux across unit area of the fabric is given by the product of the heat content per unit volume of the interstitial fluid $(\rho C)_f T$ times the pore velocity $\bar{\mathbf{v}}$ with which it moves, times the fraction ϕ of the area occupied by interstitial fluid, that is, by $(\rho C)_f T\mathbf{u}$. The rate of accumulation of heat in the volume element associated with this is the *net* flux inward, or minus the net flux outward across the surfaces of the element of volume,

$$-(\rho C)_f \int (T\mathbf{u}) \cdot \mathbf{n} \, dS = -(\rho C)_f \int \nabla \cdot (T\mathbf{u}) \, dV,$$

from the divergence theorem. Since we are considering an elementary unit volume, the advective heat input is

$$\begin{aligned} -(\rho C)_f \nabla \cdot (T\mathbf{u}) &= -(\rho C)_f (\mathbf{u} \cdot \nabla T + T \nabla \cdot \mathbf{u}) \\ &= -(\rho C)_f \mathbf{u} \cdot \nabla T, \end{aligned} \tag{2.8.1}$$

in virtue of the incompressibility condition (2.5.4).

When (2.4.11) is satisfied, as it almost invariably is in geological contexts, macroscopic heat dispersion is negligible compared with conduction through the matrix, and on the scale of our elementary averaging volume, the conductive heat flux down the temperature gradient is proportional to the magnitude of that gradient – this is Fourier's law of heat conduction. In a locally anisotropic medium, the constant of proportionality is a second-order tensor, but in rocks, the thermal anisotropy is much less than the anisotropy of permeability and retention of it seems to be an unnecessary refinement. Accordingly, the heat flux can be expressed as $-\kappa_m \nabla T$, where κ_m is the thermal conductivity of the saturated medium. The rate of accumulation of heat per unit volume is then minus the divergence of this, or

$$\nabla \cdot (\kappa_m \nabla T) = \kappa_m \nabla^2 T, \tag{2.8.2}$$

the simplification on the right being for a locally homogeneous medium. Finally, if chemical reactions are taking place, heat energy may be added at the rate $\hat{Q}$ per unit volume. Combining these expressions, we have the thermal energy balance,

$$(\rho C)_m \frac{\partial T}{\partial t} = -(\rho C)_f \mathbf{u} \cdot \nabla T + \kappa_m \nabla^2 T + \hat{Q}. \tag{2.8.3}$$

This can be rearranged slightly and divided throughout by $(\rho C)_f$ to give

$$\Gamma \frac{\partial T}{\partial t} + \mathbf{u} \cdot \nabla T = \kappa \nabla^2 T + Q, \tag{2.8.4}$$

where

$$\Gamma = \frac{(\rho C)_m}{(\rho C)_f} \tag{2.8.5}$$

and

$$\kappa = \frac{\kappa_m}{(\rho C)_f}. \tag{2.8.6}$$

The thermal conductivity of the *medium* divided by ρC for the *fluid* is the effective thermal diffusivity in this equation. Also,

$$Q = \frac{\hat{Q}}{(\rho C)_f}. \tag{2.8.7}$$

Note that with water as the interstitial liquid, since $\rho = 1$ and $C_f = 1$ c.g.s. unit, the numerical values of κ and κ_m are, in c.g.s. units, essentially the same. The dimensions of κ are, however, simply $[L^2T^{-1}]$.

2.9 DISSOLVED SPECIES BALANCE

The patterns of flow through sediments are of interest to the geologist largely because of the physical and chemical changes in the fabric that are produced or were produced in the distant past. These can be the result of cementation in which solid material is deposited from solution at the pore boundaries, of dissolution when it is removed and carried away, or of various chemical reactions between the solid constituents of the fabric and the solutes of the interstitial fluid.

Equations for the balances of chemical constituents in the interstitial fluid will be developed in the same way as in the preceding section. Let c represent the concentration (mass per unit volume) of a particular chemical species in the interstitial fluid. In a unit averaging volume, the total mass of that species is ϕc and its rate of change $\phi\ \partial c/\partial t$ is the net result of (i) advection of the species into the volume by movement of the interstitial fluids, (ii) diffusion and macroscopic dispersion, and (iii) addition of the species to the fluid by dissolution from the matrix or loss by crystallization or deposition of the solid species on the pore boundaries. As before, the flux of the species is $\phi\bar{\mathbf{v}}c = \mathbf{u}c$, and the net gain by advection is seen to be $-\nabla\cdot(\mathbf{u}c)$ by steps similar to those leading to (2.8.1). On an averaging volume scale, the solute diffusion occurs over only the fraction ϕ of a bounding surface, so that the diffusive flux divergence is $\phi D\ \nabla^2 c$, where D is either the molecular or macroscopic dispersion coefficient. If any source of the species in the liquid state is represented as Q_c (the mass added per unit volume of the *fluid* per unit time), then the species balance is represented by

$$\phi\frac{\partial c}{\partial t} + \nabla\cdot(\mathbf{u}c) = \phi D\nabla^2 c + \phi Q_c, \tag{2.9.1}$$

where Q_c is the rate of addition of species to the fluid per unit volume. If the rate of generation of fluid in the reaction is zero or negligible, then $\nabla\cdot\mathbf{u} = 0$ and the advection term can be written as $\mathbf{u}\cdot\nabla c$. Note that equation (2.9.1) can then be written as

$$\frac{\partial c}{\partial t} + \phi^{-1}\mathbf{u}\cdot\nabla c = D\ \nabla^2 c + Q_c,$$

which is, in form, similar to (2.8.4) *except* that the advection velocity here is $\phi^{-1}\mathbf{u}$, or the mean interstitial fluid velocity. This difference between the effective advection velocities for heat and dissolved salts, brought about by the fact that advected heat changes the temperature of the whole matrix whereas advected salts remain in the fluid phase (except for reactions), has profound consequences in the unsteady situations described in Chapter 5.

The salinity s of the interstitial fluid is defined as the total dissolved mass per unit mass of solution; when a number of dissolved species is present $s = \rho^{-1}\ \Sigma c$, summed over the concentrations c of the separate constituents. A conservation equation for s follows from (2.9.1); since this is linear in c,

$$\phi\frac{\partial s}{\partial t} + \nabla\cdot(\mathbf{u}s) = \phi D\nabla^2 s + \phi Q_s, \tag{2.9.2}$$

where $Q_s = \rho^{-1}\ \Sigma Q_c$.

The source term Q_c or Q_s may represent a variety of physical, chemical, or biological processes such as dissolution or precipitation of the solid phase, inorganic oxidation or reduction, or organic decomposition. The detailed chemical kinetics of many of these reactions is often not well known, though valuable insights are offered in the books by Lerman (1979) and Drever (1982). More recently, outstanding contributions have been made by Aagaard and Helgeson (1982) and Helgeson, Murphy, and Aagaard (1984). The rate-limiting step appears to be involved with the reaction kinetics at activated sites on the surface of the solid mineral, rather than diffusional transfer from the surface into the pores. Near equilibrium, the source Q is proportional to the active interstitial surface area per unit volume of the fabric, the rate constant for the reaction, and a function of the activities involved, and can be specified in terms of an overall rate constant γ, which lumps these factors and the temperature dependence together and consequently varies widely. Values measured in the laboratory or estimated from field observations also vary considerably. Lerman (1979) quotes values for laboratory measurements on silicate minerals that are relatively high (from 2 to 20 yr^{-1} or $(0.6\text{–}6) \times 10^{-7}\ \mathrm{sec}^{-1}$), whereas those inferred from deep ocean cores in siliceous sediments are five orders of magnitude smaller!

Near equilibrium, one might expect that

$$Q_c = \gamma(c_e - c), \tag{2.9.3}$$

where c_e is the equilibrium concentration. Reactions in which the rate of production or disappearance of a dissolved species is linearly proportional to the difference between the local concentration and the equilibrium value are described as first order.

Dissolution and precipitation may occur simultaneously when polymorphic forms of a solid phase have different solubilities. In the case of silica, for example, the solubility of amorphous silica (opal) is considerably greater than that of quartz, those of other polymorphs being intermediate (see, e.g., Lerman 1979, p. 387). The silica concentration in the interstitial fluids is then described by

$$\frac{\partial c}{\partial t} + \phi^{-1}\mathbf{u}\cdot\nabla c = D\,\nabla^2 c + \sum_n \gamma_n(c_{ne} - c), \tag{2.9.4}$$

where γ_n is the rate constant and c_{ne} the saturation concentration of the nth polymorph. Alternatively,

$$\frac{\partial c}{\partial t} + \phi^{-1}\mathbf{u}\cdot\nabla c = D\,\nabla^2 c + \gamma(C - c), \tag{2.9.5}$$

where

$$\gamma = \sum_n \gamma_n, \qquad C = \gamma^{-1}\left(\sum_n \gamma_n c_{ne}\right). \tag{2.9.6}$$

In a steady state with no flow, the interstitial fluid concentration stabilizes at $c = C$, which is the arithmetical average of the saturation concentrations of the individual polymorphs, weighted by their respective rate constants, those for which $c_{ne} > C$ are dissolving while polymorphs for which $c_{ne} < C$ are being precipitated.

In a more general way, the source term can be represented as

$$Q_c = \gamma c_e f\left(\frac{c_e - c}{c_e}\right), \tag{2.9.7}$$

where c_e represents the equilibrium concentration and $f(0) = 0$, $f > 0$ when $c_e > c$. The form of the function f is determined experimentally; it may be dependent on pH or the concentration of other ions present. Over restricted ranges of the argument, it can sometimes be represented as a power law relationship, and the power involved is called the order of the reaction. For example, the experiments of Plummer and Wigley (1976) show that calcite in water at 25° C and 1 atm partial pressure of CO_2 dissolves as a second-order reaction as the pH increases from 3.9 when $c = 0$, to 5.9 as the calcite dissolves, the partial pressure of CO_2 being maintained.

These reaction kinetics clearly present a bewildering array of possibilities, but fortunately, in the flow-controlled reactions described in Chapter 4, they become almost irrelevant. Provided that certain general inequalities are satisfied, it will be found that the reaction progress is limited by the rate at which reactants are delivered to an element of the fabric by the flow – the kinetics specify the reactions that occur, but the flow determines how rapidly they proceed through or in the fabric.

2.10 EQUATIONS OF STATE

The density of the interstitial fluid, usually an aqueous solution, depends on the temperature T, pressure p, and the concentrations c_i (mass per unit volume) of dissolved solids. The salinity s (dissolved mass per unit mass) is defined as the sum of the solute concentrations,

$$s = \frac{1}{\rho} \Sigma\, c, \tag{2.10.1}$$

where ρ is the solution density. The relationship among these quantities,

$$\rho = \rho(p, T, c) \qquad \text{or} \qquad \rho = \rho(p, T, s), \tag{2.10.2}$$

is called the equation of state. It is generally nonlinear, and over some ranges of the variables, it has not been measured with great precision.

In flow situations, variations in interstitial fluid density are most significant in their influence on the fluid buoyancy – horizontal variations in buoyancy necessarily lead to fluid motion, as will be seen in the next chapter. In many cases, the variations in temperature and salinity in the field of flow are sufficiently moderate that the equation of state can be represented with reasonable accuracy as a linear function of temperature and/or salinity. When the interstitial fluid is everywhere unsaturated with respect to all of its solutes that significantly affect its density,

$$\rho = \rho_0(1 - \alpha T + \beta s), \tag{2.10.3}$$

where $\rho_0 = \rho_0(p)$ is the reference density at the ambient pressure and

$$\alpha = -\frac{1}{\rho_0}\frac{\partial \rho}{\partial T}, \qquad \beta = \frac{1}{\rho_0}\frac{\partial \rho}{\partial s}, \tag{2.10.4}$$

the coefficient of volumetric expansion and density coefficient for salinity, respectively.

However, when significant variations in interstitial density are produced by dissolution of a major fabric component or by exsolution, then on a scale large compared with the equilibration length l_e discussed in Section 4.2 the concentration is close to saturation at the local ambient temperature: $c = c_e(T)$. The equation of state can then be expressed as

$$\rho = \rho_0(1 - \alpha_s T),$$

where

$$\alpha_s = -\frac{1}{\rho_0}\frac{\partial}{\partial T}\{\rho(T, c_e(T))\} \tag{2.10.5}$$

can be described as the thermal expansion coefficient of the continuously saturated solution. Its magnitude and *sign* depend on the chemical nature of the solute. From (2.10.5)

$$\alpha_s = -\frac{1}{\rho_0}\left\{\frac{\partial \rho}{\partial T} + \frac{\partial \rho}{\partial c_e}\frac{\partial c_e}{\partial T}\right\},$$

and if the saturated solute rather than any other that may be present dominates, the variations in density

$$\frac{\partial \rho}{\partial c_e} = -\frac{1}{\rho_0}\frac{\partial \rho}{\partial s} = \beta.$$

Consequently,

$$\alpha_s = \alpha - \frac{\beta}{\rho_0}\frac{\partial c_e(T)}{\partial T}. \tag{2.10.6}$$

Representative values of α, β, and $\partial c_e/\partial T$ can be found in the International Critical Tables (Washburn, 1929). Solutes that saturate at very low concentrations have $\alpha_s \simeq \alpha$, the ordinary coefficient of expansion for water. Those whose saturation concentration is large and increases rapidly with temperature may have $\alpha_s < 0$; as the temperature increases, so does the density, since the volumetric expansion is more than offset by the increased mass of solute in the concentrated solution.

The fluid viscosity μ is also a function of temperature and, to a lesser extent, of pressure and salinity, and a detailed collation is given by Kestin, Khalifa, and Correia (1981). For example, the viscosity of water decreases by a factor of about 3 from 20 to 100° C. However, the viscosity (appearing in Darcy's law) always occurs in combination with the permeability, and the endemic uncertainty in the latter in any natural formation is far greater. For this reason, in geological applications the fluid viscosity variations can, with convenience, be ignored – they are within the noise level of uncertainty.

3 General flow characteristics
Patterns of flow

3.1 STREAM FUNCTIONS

The principles enunciated in the preceding chapter provide the basis for our discussion in the remainder of this book. They are the individual building blocks from which our understanding of the subject is to be constructed. Though large-scale geological structures, like buildings, exist in almost infinite variety, there are a number of recurrent motifs, as there are architectural elements, that can be identified in the most disparate contexts. The recognition and understanding of these elements will constrain and illuminate the patterns of flow through the permeable rocks of a large-scale formation and the physical and chemical alterations that, over time, are produced in the geological record.

A convenient, accurate, pictorial representation of flow fields is an essential aid to an understanding of and perception of patterns, and this is offered in many instances by what is known as the stream function. In flow fields that are either two-dimensional or axially symmetrical, the stream function provides a qualitatively graphic and quantitatively revealing image of the flow and has the additional advantage of reducing the number of flow variables by one.

For two-dimensional flow, the incompressibility condition (2.5.5) becomes

$$\frac{\partial u}{\partial x} + \frac{\partial w}{\partial z} = 0, \tag{3.1.1}$$

which is satisfied by a function $\psi(x, z, t)$ such that

$$u = \frac{\partial \psi}{\partial z}, \qquad w = -\frac{\partial \psi}{\partial x}. \tag{3.1.2}$$

The function $\psi(x, z, t)$ is called the stream function.

The velocity component in any direction is thus represented as the gradient of the stream function ψ in the orthogonal direction; in particular, in the direction in which ψ = const, there is *no* transverse velocity, so that the contours ψ = const represent the directions of flow at each point. Moreover, at any given instant, if we choose local coordinates such that

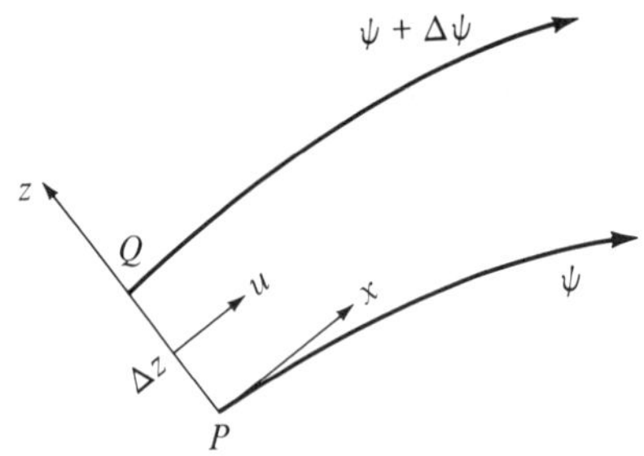

Figure 3.1. The fluid transport $u\,\Delta z = \Delta\psi$ between neighboring streamlines.

at the point P, say, in Figure 3.1, the x-axis lies along the flow direction (ψ = const), then at the neighboring point Q, the stream function is

$$\begin{aligned}\psi + \Delta\psi &= \psi + \frac{\partial\psi}{\partial z}\,\Delta z\\ &= \psi + u\,\Delta z.\end{aligned}$$

The difference in the values of the stream function at the two points is then $\Delta\psi = u\,\Delta z$, the volume flux across a line joining the two points. In a streamline pattern, represented by a distribution of curves for each of which ψ = const, the volume flux between any pair of these curves remains constant. Where the streamlines are close together, the volume transports are concentrated and the flow speed is relatively high; where they are widely separated, the flow speed is low. Streamlines cannot begin or end in the interior of a region. In recirculating flow, the streamlines are closed; if fluid enters or leaves a porous region across a bounding surface, streamlines originate or terminate at that surface, whereas at an impermeable boundary (with no flow across the surface) the stream function is constant along it. Figures 3.12 and 3.17 give examples of the use of the stream function in flow representation.

Note that the stream function is arbitrary to the extent of an additive constant, since only *differences* in its value at different points have physical significance. The value of the stream function can be assigned (usually zero) for one particular streamline; the values of ψ along the other streamlines then give the total volume transport between that streamline and the streamline $\psi = 0$. In *steady* two-dimensional flow, the streamlines correspond to the transport paths of marked fluid elements (except for diffusion), but in unsteady flow this is not generally so, since the pattern of streamlines may change in time.

In mathematical terms, use of the stream function enables us to represent the two velocity components $u(x, y, t)$, $v(x, y, t)$ in terms of the single scalar function $\psi(x, y, t)$ while satisfying the incompressibility condition automatically. We then have one variable fewer and one equation fewer – any such simplification is always welcome!

A different stream function, the Stokes stream function, can be defined for axially symmetrical flow; this has the same mathematical advantage but lacks the simple interpretation concerning relative velocities in different regions of the flow. In cylindrical polar coordinates (r, θ, z) with corresponding velocity components (u, v, w), the incompressibility condition is

$$\nabla \cdot \mathbf{u} = \frac{1}{r}\frac{\partial}{\partial r}(ru) + \frac{1}{r}\frac{\partial v}{\partial \theta} + \frac{\partial w}{\partial z} = 0. \qquad (3.1.3)$$

When the flow is axially symmetrical, $\partial/\partial\theta = 0$, and this reduces to

$$\frac{1}{r}\frac{\partial}{\partial r}(ru) + \frac{\partial w}{\partial z} = 0, \qquad (3.1.4)$$

which is satisfied by the function ψ_s, the Stokes stream function, such that

$$u = \frac{1}{r}\frac{\partial \psi_s}{\partial z}, \qquad w = -\frac{1}{r}\frac{\partial \psi_s}{\partial r}. \qquad (3.1.5)$$

As before, lines ψ_s = const indicate the direction of flow, so that their pattern gives a good visual representation of the flow field. The volume transport between the axially symmetrical *surfaces* $\psi_s = \psi_1$ and ψ_2 is $2\pi(\psi_1 - \psi_2)$, but the r^{-1} factors in (3.1.5) produce a geometrical distortion in the relationship between fluid velocities and the gradients of ψ_s at different radial positions.

3.2 NATURAL OR INTRINSIC COORDINATES

Many sedimentary strata, when they are laid down, contain horizontal or almost horizontal laminations, bedding planes, or lenses. Subsequent tectonic deformations may cause the strata to bend, tilt, fold, or fracture, but the record of the original vertical direction remains impressed upon the fabric. As will become clear later in this chapter, the flow characteristics along the laminations or bedding planes can be markedly different from those across them; the anisotropy results primarily from the geometrical patterns of permeability variation. On a scale of centimeters or tens of centimeters or more, little variation may be discernible in the fabric structure in the two original horizontal directions, whereas it may

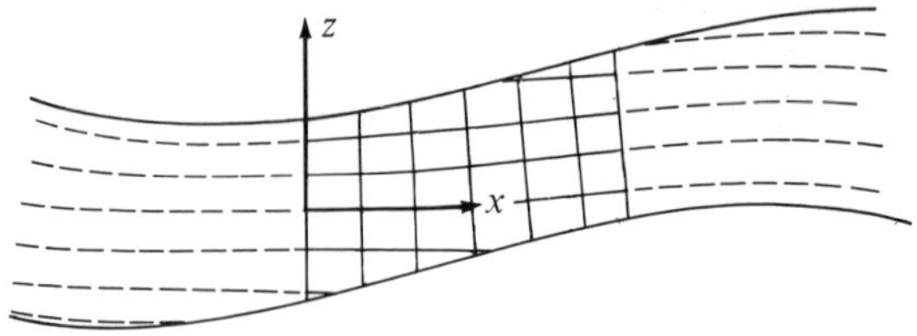

Figure 3.2. Natural or intrinsic coordinates with z normal to the bedding planes and (x, y) parallel to them.

be distinct in the vertical directions, across the laminations. For this reason, it is convenient and natural to adopt a coordinate system in which x and y specify points in planes locally parallel to the bedding and z is in the normal direction (Figure 3.2).

If the stratum is undeformed, or if the fabric is locally isotropic and homogeneous or extensively fractured in a more or less isotropic way, or, indeed, if we have no information on what the detailed fabric structure is like, the coordinates can be taken in the conventional manner, with (x, y) horizontal and z vertically upward. Where there is pronounced cross-bedding, it may be necessary to subdivide the flow domain into regions that correspond to the different bedding orientations and average over the macroscopic domains. In many instances, however, the use of these intrinsic coordinates allows for considerable simplification in analysis and representation of the flow. The penalty is, of course, that such coordinates are generally curvilinear on a macroscopic scale; the change in position vector $d\mathbf{x}$ corresponding to increments dx, dy, dz can be represented as

$$d\mathbf{x} = \mathbf{a}h_1 dx + \mathbf{b}h_2 dy + \mathbf{c}h_3\, dz, \tag{3.2.1}$$

where the direction of the unit vectors $\mathbf{a}$, $\mathbf{b}$, $\mathbf{c}$ and the magnitudes of the scale factors h_1, h_2, h_3 are functions of the coordinates. If, however, the radius of curvature R of the stratum is large compared with its thickness h, the variations in these quantities are fractionally of order (h/R), and if errors of this magnitude are acceptable, the coordinates can be regarded as locally Cartesian. If one is particularly interested in the details of flow in tight folds, local cylindrical coordinates may be preferable, or general curvilinear coordinates may be used in numerical study, though the gain in accuracy may perhaps be spurious in the presence of other, more important structural uncertainties.

With these coordinates, in Darcy's equation (2.6.5),

$$\mathbf{u} = -\mu^{-1}\mathbf{k} \cdot (\nabla\hat{p} + \rho g \mathbf{l}),$$

the permeability

$$\mathbf{k} = \begin{pmatrix} k_H & 0 & 0 \\ 0 & k_H & 0 \\ 0 & 0 & k_V \end{pmatrix}, \tag{3.2.2}$$

and the vertical unit vector

$$\mathbf{l} = (\sin\theta\cos\phi, \sin\theta\sin\phi, \cos\theta), \tag{3.2.3}$$

where k_H and k_V are the permeabilities in directions parallel to and normal to the bedding planes (the original horizontal and vertical), which now slope at the angle θ to the horizontal in the direction ϕ relative to the x-axis. Both θ and ϕ may vary slowly (over scales large compared with the layer thickness) in the x- and y-directions. In vector notation, Darcy's equation can be written as

$$\begin{aligned} \mathbf{u}_H &= -\mu^{-1}k_H(\nabla_H\hat{p} + \rho g(\mathbf{l}\cdot\boldsymbol{\nu})\boldsymbol{\nu}), \\ w &= -\mu^{-1}k_V\left(\frac{\partial\hat{p}}{\partial z} + \rho g\mathbf{l}\cdot\mathbf{n}\right), \end{aligned} \tag{3.2.4}$$

where $\mathbf{u}_H$ now is the velocity vector in the plane of the bedding and w is normal to this plane, and $\mathbf{l}$, $\mathbf{n}$, and $\boldsymbol{\nu}$ are unit vectors vertically upward, normal to the bedding planes and in the direction of $\mathbf{u}_H$, respectively. In the presence of laminations or lenses, k_H and k_V may vary rapidly (and substantially) in the z-direction while again varying only slowly in x and y. Other physical properties of the fabric such as the porosity or thermal conductivity may also vary in the z-direction over scales much smaller than those in the x- and y-directions, though the magnitudes of the fractional changes are generally less than those of the permeability and can frequently be ignored. In any event, the difference in scales sets the scene for the simplifications explored in Section 3.7.

3.3 BOUNDARY CONDITIONS FOR FLOW

The conditions that must be obeyed at the boundaries of a flow determine the flow field in the interior as much as do the governing field equations. One kind of condition is in essence kinematic, requiring that the fabric remain fluid filled.

Consider first an interface between two identifiable bodies of fluid (e.g., one fresh, one saline), its position being specified by $z = \zeta(x, y, t)$ in our intrinsic coordinates (Figure 3.3). Fluid elements at the interface move through the matrix at the interstitial fluid velocity $\mathbf{v}$, so that the component of this velocity in the z-direction is

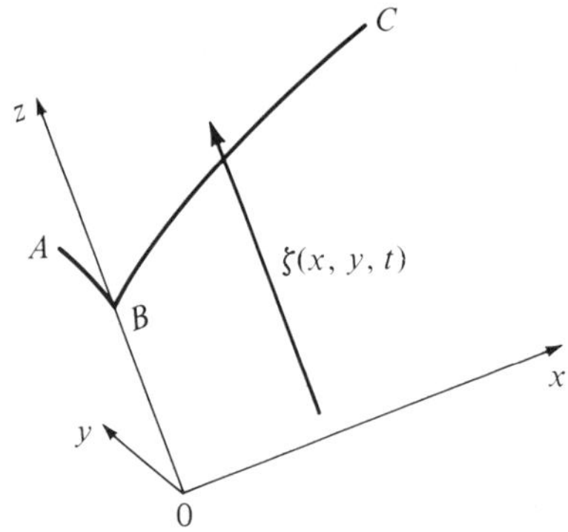

Figure 3.3. An interfacial surface *ABC* specified as $\zeta(x, y, t)$.

$$\bar{v}_z = \frac{dz}{dt} = \frac{d}{dt}\zeta(x, y, t)$$
$$= \frac{\partial \zeta}{\partial t} + \bar{\mathbf{v}} \cdot \nabla \zeta, \tag{3.3.1}$$

where $\bar{\mathbf{v}} = (dx/dt, dy/dt)$ is the interstitial fluid velocity vector in the (x, y) plane. In terms of the transport velocity (2.5.4), equation (3.3.1) can be expressed as

$$\phi \dot{\zeta} = w - \mathbf{u} \cdot \nabla \zeta. \tag{3.3.2}$$

Since this expression must hold for fluid on either side of the interface, the combination on the right-hand side must be continuous across the interface. An alternative form of this statement is that the component of the transport velocity *normal* to the interface must be continuous across it.

The interstitial fluid pressure must also be continuous across such an interface in a single-phase liquid – a discontinuity would, by Darcy's law, imply an indefinitely large velocity. In multiphase fluids such as water and oil, however, pressure differences can be maintained by capillary action, as mentioned in Section 2.6.

If the interface does not move through the matrix, or if it coincides with a fixed impermeable boundary, $\dot{\zeta} = 0$ and

$$w = \mathbf{u}_{\mathrm{H}} \cdot \nabla \zeta \qquad \text{on the boundary,} \tag{3.3.3}$$

which, in geometrical terms, asserts that at such a boundary the flow is necessarily along it – the normal velocity component vanishes.

Where the porous fabric is exposed to the atmosphere or to the ocean or a lake, fluid may enter or leave the fabric, but pressure boundary conditions can be applied directly. At the water table, differences between the interstitial fluid pressure and atmospheric pressure can be maintained by capillary action, but on a basin-wide scale, these differences can generally be ignored and the fluid pressure at the water table taken as atmospheric. If the surface of the water table is specified by $z = \zeta(x, y)$, then

$$\hat{p}(x, y, z = \zeta) = \text{const} = 0, \text{ say,}$$

whence

$$\nabla_{\mathrm{H}}\hat{p} + \frac{\partial \hat{p}}{\partial z}\nabla\zeta = 0 \qquad \text{on} \quad z = \zeta, \tag{3.3.4}$$

which is the boundary condition to be applied to the distribution of actual pressure. It is important to note that the slope $\nabla\zeta$ of the water table is measured *relative to the bedding planes,* not to the horizontal.

Finally, when the boundary is between a saturated fabric and ambient water such as the ocean or a lake, the fluid pressure at the boundary of the fabric must be equal to the ambient hydrostatic pressure. In terms of the bedding plane coordinates, the ambient hydrostatic pressure gradient is

$$\begin{aligned} \nabla_{\mathrm{H}} p_0 &= -\rho_0 g(\mathbf{l} \cdot \boldsymbol{\nu})\boldsymbol{\nu}, \\ \frac{\partial p_0}{\partial z} &= -\rho_0 g \mathbf{l} \cdot \mathbf{n}, \end{aligned}$$

so that, analogously to (3.3.4),

$$\nabla_{\mathrm{H}}(\hat{p} - p_0) + \left\{\frac{\partial}{\partial z}(\hat{p} - p_0)\right\}\nabla\zeta = 0. \tag{3.3.5}$$

This leads to a condition of the same form as (3.3.4) but with the actual pressure replaced by the reduced pressure $p = \hat{p} - p_0$ defined in Section 2.6.

3.4 VORTICITY

The vorticity is an important property of any flow field. In flow in porous media, the vorticity of the transport velocity components in vertical planes is intimately associated with the slope of the isopycnals (surfaces of constant density), and the vorticity of the components in horizontal planes will be shown to vanish. The relation between vorticity and isopycnal slope frequently enables us to make immediate inferences about the general nature of the flow without detailed calculation, while the lack of vorticity in the horizontal components of the flow field places considerable constraints on the kinds of horizontal flow patterns that are possible.

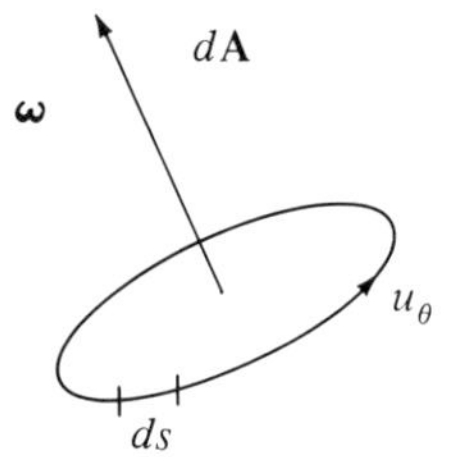

Figure 3.4. Vorticity and rotation in the transport velocity field.

The vorticity $\boldsymbol{\omega}$ is defined in terms of the transport velocity as

$$\boldsymbol{\omega} = \nabla \times \mathbf{u}, \tag{3.4.1}$$

or, in orthogonal Cartesian coordinates (x, y, z) with velocity components (u, v, w),

$$\omega_x = \frac{\partial w}{\partial y} - \frac{\partial v}{\partial z}, \qquad \omega_y = \frac{\partial u}{\partial z} - \frac{\partial w}{\partial x}, \qquad \omega_z = \frac{\partial v}{\partial x} - \frac{\partial u}{\partial y}.$$

From this definition it follows immediately that the vorticity field is divergence free:

$$\nabla \cdot \boldsymbol{\omega} = 0. \tag{3.4.2}$$

The vorticity can be interpreted most simply by making use of Stokes's theorem in vector calculus. Consider, as illustrated in Figure 3.4, a small closed circuit in the medium capped by a surface whose element of area is $d\mathbf{A}$ – magnitude dA and direction normal to the element. The integral of the vorticity component in this direction over the capping surface is, by this theorem, equal to the line integral of the transport velocity around the circuit,

$$\int \boldsymbol{\omega} \cdot d\mathbf{A} = \int \nabla \times \mathbf{u} \cdot d\mathbf{A} = \int \mathbf{u} \cdot d\mathbf{s}, \tag{3.4.3}$$

where $d\mathbf{s}$ is the line element of the contour. With a small *circular* contour of radius r, the right-hand side of (3.4.3) is equal to the mean azimuthal speed u_θ times the circumference $2\pi r$; the left-hand side is equal to the mean vorticity ω normal to the circuit times the area πr^2:

$$\pi r^2 \omega = 2\pi r u_\theta.$$

The angular velocity associated with the distribution of transport velocity is μ_θ / r, or, from the last equation, $\frac{1}{2}\omega$. The vorticity can thus be visualized as twice the local angular velocity, or rate of rotation associated with the transport velocity field. If the vorticity vanishes throughout a region, the fluid flow is said to be irrotational.

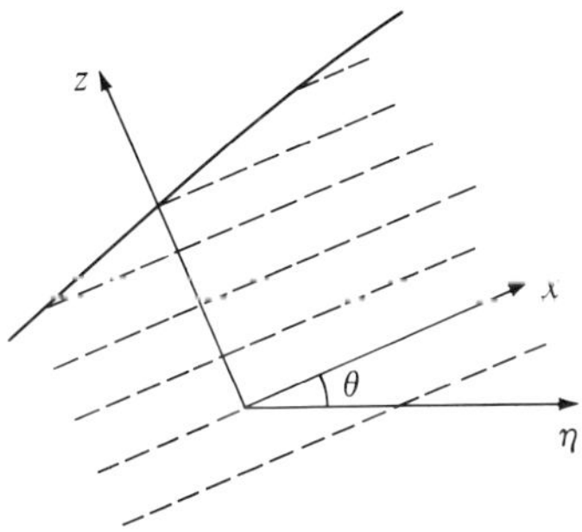

Figure 3.5. Coordinates for equation (3.4.4).

The vorticity distribution is a very useful flow characteristic both in giving a qualitative perception of flow patterns and in making precise calculations. It can be related to the distributions of permeability and interstitial fluid buoyancy whether the latter is produced by variations in temperature or in concentration of dissolved solids. Darcy's equation in natural coordinates with the x-axis taken upslope is

$$\frac{\partial \hat{p}}{\partial x} = -\frac{\mu}{k_H} u - \rho g \sin\theta, \qquad \frac{\partial \hat{p}}{\partial y} = -\frac{\mu}{k_H} v, \qquad \frac{\partial \hat{p}}{\partial z} = -\frac{\mu}{k_V} w - \rho g \cos\theta.$$

By cross-differentiation among these equations to eliminate $\hat{p}$, one obtains

$$\begin{aligned}
\frac{\partial}{\partial y}\left(\frac{w}{k_V}\right) - \frac{\partial}{\partial z}\left(\frac{v}{k_H}\right) &= -\frac{g\cos\theta}{\mu}\frac{\partial \rho}{\partial y}, \\
\frac{\partial}{\partial z}\left(\frac{u}{k_H}\right) - \frac{\partial}{\partial x}\left(\frac{w}{k_V}\right) &= \frac{g}{\mu}\left(\cos\theta\frac{\partial \rho}{\partial x} - \sin\theta\frac{\partial \rho}{\partial z}\right) = \frac{g}{\mu}\frac{\partial \rho}{\partial \eta}, \qquad (3.4.4) \\
\frac{\partial}{\partial x}\left(\frac{v}{k_H}\right) - \frac{\partial}{\partial y}\left(\frac{u}{k_H}\right) &= \frac{g\sin\theta}{\mu}\frac{\partial \rho}{\partial y},
\end{aligned}$$

where, as illustrated in Figure 3.5, η is in the *horizontal* direction; the y-axis, into the paper, is also horizontal. It is immediately evident from these equations that if the interstitial fluid density varies in any horizontal direction as a result, for example, of variations in temperature or salinity, motion will necessarily result, even in the absence of any externally imposed pressure gradient. Motions of this kind, driven by variations in fluid buoyancy, are described as convective; they are discussed in detail in Chapters 5 and 7.

In some circumstances, simplifications can be made to the set (3.4.4). When the bedding planes are horizontal, $\theta = 0$ and

$$\frac{\partial}{\partial y}\left(\frac{w}{k_V}\right) - \frac{\partial}{\partial z}\left(\frac{v}{k_H}\right) = -\frac{g}{\mu}\frac{\partial \rho}{\partial y},$$
$$\frac{\partial}{\partial z}\left(\frac{u}{k_H}\right) - \frac{\partial}{\partial x}\left(\frac{w}{k_V}\right) = \frac{g}{\mu}\frac{\partial \rho}{\partial x}, \qquad (3.4.5)$$
$$\frac{\partial}{\partial x}\left(\frac{v}{k_H}\right) - \frac{\partial}{\partial y}\left(\frac{u}{k_H}\right) = 0,$$

and if k_H is uniform in (x, y), the vertical component of the vorticity vanishes. Sloping isopycnals (surfaces of constant fluid density) induce flow with horizontal vorticity but no vertical vorticity: $\omega_z = 0$ throughout. This conclusion provides important and far-reaching constraints on possible patterns of *horizontal* components of the flow in geological basins, permeable strata, or banks, whether the flow is driven hydraulically or thermally. Since, by Stokes's theorem (3.4.4) applied to any circuit in the horizontal plane (for which dA is vertical),

$$\int \boldsymbol{\omega} \cdot d\mathbf{A} = \int \omega_z dA = \int \mathbf{u} \cdot d\mathbf{s} = 0, \qquad (3.4.6)$$

the circulation around every closed circuit in the horizontal plane must vanish. In plan form, there can be no closed streamlines or flow paths. When the horizontal permeability is uniform in a given fabric, if the flow paths are locally straight and parallel, the flow speeds along each must be the same – parallel straight streamlines are necessarily equally spaced. When they are curved, it can be seen (by taking a circuit along one streamline, cutting normally to another, back along that one, and then normally to the first to complete the circuit) that the local flow speeds along neighboring streamlines are proportional to their local curvatures. It will be seen in Chapter 6 that the dynamical flow constraints are rather different in confined or unconfined aquifers, but these general conclusions apply to both.

When the fluid density is, in effect, constant throughout, the right-hand sides of (3.4.5) are all zero, and any flow present is induced hydraulically by the reduced pressure gradient. If, further, the medium is locally homogeneous and isotropic in some region, $k_H = k_V = \text{const} = k$, all components of the vorticity $\boldsymbol{\omega} = 0$ and the motion is irrotational in this part of the domain. The reduced pressure p then constitutes a velocity potential for the motion; since

$$\mathbf{u} = -\frac{k}{\mu}\nabla p,$$

in view of the incompressibility condition $\nabla \cdot \mathbf{u} = 0$, p satisfies Laplace's equation,

$$\nabla^2 p = 0, \qquad (3.4.7)$$

for which there are a variety of powerful solution techniques. If the medium is locally homogeneous but not isotropic, $k_H \neq k_V$, but Laplace's equation for the pressure field can still be recovered by coordinate stretching. Provided that the density variations in the interstitial fluids are negligible,

$$\mathbf{u}_H = -\frac{k_H}{\mu} \nabla_H p, \qquad w = -\frac{k_V}{\mu} \frac{\partial p}{\partial z},$$

and again by use of the incompressibility condition,

$$k_H \left(\frac{\partial^2 p}{\partial x^2} + \frac{\partial^2 p}{\partial y^2} \right) + k_V \frac{\partial^2 p}{\partial z^2} = 0,$$

which, under the coordinate stretching transformation

$$x' = (k_H)^{-1/2} x, \qquad y' = (k_H)^{-1/2} y, \qquad z' = (k_V)^{-1/2} z \tag{3.4.8}$$

reduces to (3.4.7) in the primed variables. Note, however, that the shape of the flow region is also distorted by this transformation.

When density variations in the interstitial fluids are the result of temperature variations, if the isotherms slope, convective motions are set up. For the purposes of illustration, suppose that the matrix is homogeneous and istropic with $k_H = k_V = k$, so that coordinates can be chosen with z vertical and the x-direction that of the horizontal temperature gradient. From (3.4.5), then, $\partial \rho / \partial y = 0$; with $\rho = \rho_0(1 - \alpha T)$, $\partial \rho / \partial x = -\alpha \rho_0 \, \partial T / \partial x$, and the vorticity component in the y-direction is

$$\begin{aligned} \omega_y &= -\frac{k g \alpha \rho_0}{\mu} \frac{\partial T}{\partial x} \\ &= -\frac{k g \alpha}{\nu} \frac{\partial T}{\partial x}, \end{aligned} \tag{3.4.9}$$

which in a geothermal temperature gradient is proportional to the slope of the isotherms.

Some simple examples illustrating the utility of (3.4.9) are shown in Figure 3.6. The isotherms are dashed; in each sketch $T_0 > T_1$. Figure 3.6a represents a porous stratum between impermeable layers; from the point indicated, the temperature decreases along a horizontal line to the right so that $\partial T / \partial x < 0$. The vorticity in the y-direction (into the paper) is then positive; the fluid at the top of the porous layer is moving upward and to the right relative to that near the bottom. (Superimposed on this may be an arbitrary, pressure-driven flow, uniform across the width of the porous layer.) In Figure 3.6b, representing a submerged permeable escarpment, $\partial T / \partial x > 0$ at the point shown, so that the sense of rotation is reversed. This distribution of isotherms is associated with interstitial fluid being drawn into the escarpment near its base and leaving it near

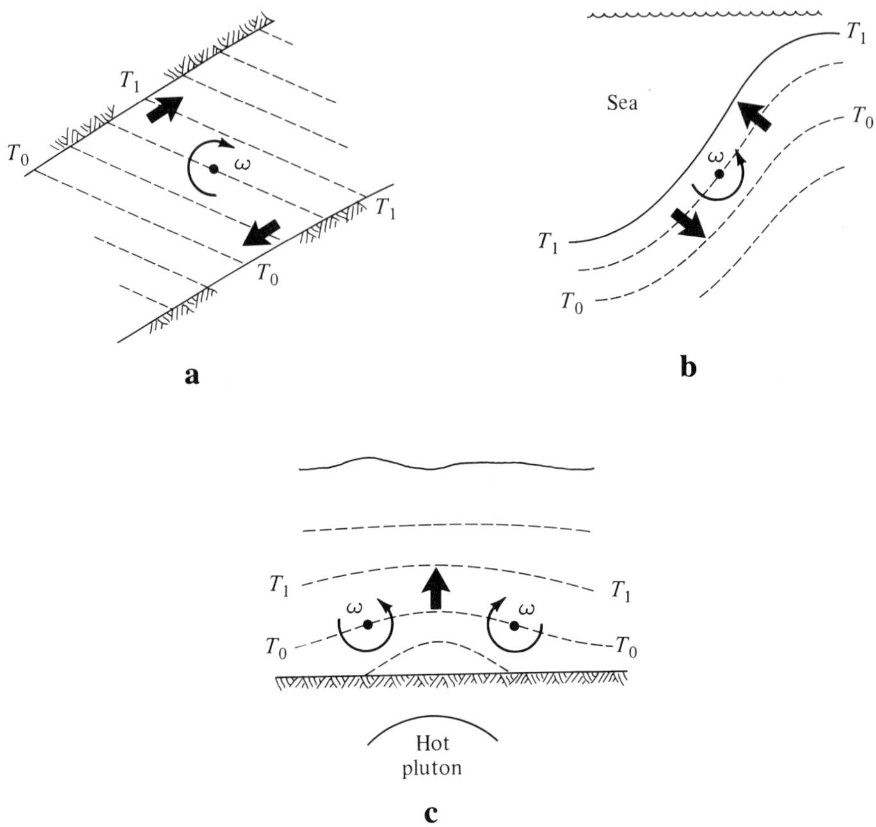

Figure 3.6. Vorticity in the distribution of transport velocity associated with sloping isotherms. In each part, $T_0 > T_1$.

the shelf break. In Figure 3.6c, the raising of the isotherms above a hot pluton produces negative vorticity on the left flank and positive vorticity on the right as the fluid rises in a plume above the heated basement region. Superficially, one might be tempted to interpret these examples in terms of "hot fluid rising," but (3.4.9) remains valid in the presence of any superimposed pressure-driven flow that can carry hot fluid in any direction.

When variations in interstitial fluid density are associated with the dissolution of major fabric constituents or with exsolution and the equilibration length l_e of Section 4.2 is small compared with flow scales of interest, the appropriate equation of state is (2.10.5),

$$\rho = \rho(c_s(T)) = \rho_0(1 - \alpha_s T),$$

where α_s is the thermal density coefficient of the continuously saturated solution. Consequently,

$$\omega_y = -\frac{kg\alpha_s}{\nu}\frac{\partial T}{\partial x}, \tag{3.4.10}$$

where α_ε may have either sign. For solutes that saturate at low concentrations or those such as NaCl whose saturation concentration varies little with temperature, $\alpha_s \simeq \alpha$, the coefficient of expansion for water, but for those with $\alpha_s < 0$ such as Na_2SO_4 at temperatures below 32° C, the senses of the rotations shown in Figure 3.6 are *reversed*.

In anisotropic layers, the senses of rotation are similar but in addition the flow circulations are distorted. When k_H is large compared with k_V, as is often the case, the flow patterns are stretched horizontally relative to the vertical, as many of the situations discussed in Chapters 6 and 7 exemplify.

3.5 DYNAMICAL SCALING

The equations governing the temperature and flow fields are coupled and nonlinear, but in many important instances, significant simplifications are possible. They appear mathematically as various limiting cases in which (usually) one physical factor is clearly dominant over another, allowing the latter to be neglected or incorporated as a small perturbation. Scaling analysis is a technique that enables us to identify these limiting cases, which are frequently of interest in themselves and applicable directly to geological formations, but which also give insight into the behavior of intermediate situations in which numerical calculation may be required for a detailed solution.

Hydraulically driven flow

One primary distinction is that between thermally and hydraulically driven flows. It is usually fairly obvious when hydrostatic pressure differences are involved (most commonly in a terrestrial aquifer or sedimentary basin with a hydrostatic head difference ζ between outcroppings), but to what extent does the geothermal temperature distribution modify the flow pattern?

The pressure difference produced by a hydrostatic head difference ζ is $\rho g\zeta$, and if the aquifer is of length l (the distance over which this pressure difference is maintained) the pressure gradient is $\rho g\zeta/l$ and the transport velocity produced has magnitude

$$u \sim \frac{k_H}{\mu}\frac{\rho g\zeta}{l} = \frac{k_H g\zeta}{\nu l}, \tag{3.5.1}$$

where k_H is the permeability for flow *along* the aquifer. If, however, the isotherms in the aquifer are not precisely horizontal, internal vorticity ω and shear in the transport velocity will be generated, with a velocity difference Δu across the depth h of the aquifer given approximately by

$$\omega \simeq \frac{\Delta u}{h} = \frac{k_H g \alpha}{\nu} \frac{\partial T}{\partial x}. \tag{3.5.2}$$

The *flow field* is hydraulically dominated if this variation in the transport velocity, Δu, is small compared with the transport velocity u itself, given by (3.5.1), or if

$$\zeta \gg \alpha h l \frac{\partial T}{\partial x}. \tag{3.5.3}$$

For example, a terrestrial aquifer of length 100 km and thickness 50 m may, it is judged, encounter temperature differences of 50° C along its length. In the condition above, $\partial T/\partial x \sim$ 50° C per 100 km or 0.5×10^{-3} °C m^{-1}. The thermal expansion coefficient α for water is about 3×10^{-4} °C^{-1}, so that $\alpha h l\ \partial T/\partial x \sim 3 \times 10^{-4} \times 50 \times 10^5 \times 0.5 \times 10^{-3}$ = 0.75 m. If the hydraulic head is large compared with this, the flow is hydraulically dominated; for example, if $\zeta \sim 75$ m, the thermally induced velocities are only about 1 percent of the hydraulically driven flow. In a given aquifer, one may not have a priori a good indication of the horizontal temperature gradient, but if an upper limit can be estimated from the variations in depth of submergence and the geothermal temperature gradient, and if (3.5.3) is found to be satisfied adequately, the procedure for solving the flow field is simplified. Any thermally induced variation in fluid density can be neglected insofar as it affects the flow field, and if $p = \hat{p} + \rho_0 g Z$ represents the nonhydrostatic pressure component (where Z is measured vertically from some convenient reference level), the flow field is determined by

$$\nabla_H p = -\frac{\mu}{k_H} \mathbf{u}_H, \qquad \frac{\partial p}{\partial z} = -\frac{\mu}{k_V} w, \tag{3.5.4}$$

where $\mathbf{u}_H$, w are the velocity components along and across the bedding, together with the incompressibility condition

$$\nabla \cdot \mathbf{u} = 0 \tag{3.5.5}$$

and the appropriate boundary conditions. As before, in a matrix whose permeability is isotropic and uniform in at least part of the region, the preceding two equations can be combined to give

$$\nabla^2 p = 0. \tag{3.5.6}$$

The distribution of temperature in the aquifer is specified by equation

(2.8.4), in which the transport velocity $\mathbf{u}$ is now determined by the solution of the preceding three equations. In a steady flow

$$\mathbf{u} \cdot \nabla T = \kappa \nabla^2 T + Q, \tag{3.5.7}$$

where Q represents any internal heat sources arising from chemical reactions or the latent heats L of crystallization or dissolution. In the latter case, over scales large compared with the equilibration length, the heat source can be estimated from the rate of dissolution (discussed in Section 4.4) as

$$Q = -\frac{L}{C_f}\left(\frac{\partial c_e}{\partial T}\right)\mathbf{u} \cdot \nabla T, \tag{3.5.8}$$

where $c_e(T)$ is the equilibrium concentration of the solute. The magnitude of Q is small compared with the advective term in the thermal balance (3.5.7), provided that

$$\frac{L}{C_f}\left(\frac{\partial c_e}{\partial T}\right) \ll 1. \tag{3.5.9}$$

This inequality is usually strongly satisfied; in the deposition of calcite cements from solution, $\partial c_e/\partial T \sim 1.5 \times 10^{-6}\ ^\circ\mathrm{C}^{-1}$ over temperatures from 10 to 100° C, while the free energy of formation of calcite is -267 kcal mol^{-1} (Bathurst 1975, p. 258), corresponding to $L = 0.17$ cal g^{-1}. The term on the left-hand side of (3.5.9) is therefore of magnitude about 3×10^{-7}, which is suitably small. These heat sources then have negligible influence on the temperature distribution, and (3.5.7) reduces to

$$\mathbf{u} \cdot \nabla T = \kappa \nabla^2 T. \tag{3.5.10}$$

Now, the groups of terms on the left- and right-hand sides of this equation represent the effects of fluid advection and thermal diffusion on the temperature distribution. Individual advective terms are of the order $u\Delta T/l$, where l, the scale of variation of the temperature field along the flow direction, is of the order of the length of the aquifer, and ΔT is the magnitude of the temperature variations. The diffusive terms on the right are at most of order $k\Delta T/h^2$, where h, the layer thickness, is the smallest geometrical length scale. Advective effects dominate when

$$\frac{u\,\Delta T}{l} \gg \frac{\kappa\,\Delta T}{h^2}$$

or, from (3.5.1), when

$$\frac{uh}{\kappa} \gg \frac{l}{h} \quad \text{or} \quad \frac{k_H g \zeta}{\nu\kappa} \gg \left(\frac{l}{h}\right)^2. \tag{3.5.11}$$

(The dimensionless ratio uh/κ is known as the Peclet number.) In this circumstance, involving highly permeable aquifers that are not too long and with a large flow rate, (3.5.10) reduces to

$$\mathbf{u} \cdot \nabla T = 0, \tag{3.5.12}$$

which asserts that the temperature is constant along the lines of flow; if the source water is of constant temperature, the temperature is constant throughout the aquifer, except perhaps near boundaries, where, as will be seen shortly, thermal conduction may be important no matter how large the overall Peclet number.

At the other extreme, when the inequality is reversed, the advective effects of heat transport on the temperature are negligible, and so within the aquifer the temperature is determined solely by thermal conduction, as it is in the underlying and overlying aquistads:

$$\nabla^2 T = 0. \tag{3.5.13}$$

In either case, then, the temperature distribution can be found simply; it is only when the groups of terms on the two sides of (3.5.11) are of the same order that a detailed solution of (3.5.10) for a given hydraulically driven flow **u** is necessary.

If the salinity or concentration of chemically inactive dissolved salts varies, its distribution can be determined similarly. In a steady flow, from (2.9.2),

$$\phi^{-1}\mathbf{u} \cdot \nabla s = D\, \nabla^2 s,$$

where D is either the molecular or macroscopic diffusivity. By reasoning analogous to that leading to (3.5.11), if

$$\frac{uh}{\phi D} \gg \frac{l}{h},$$

then $\mathbf{u} \cdot \nabla s \approx 0$; the salinity is constant along the lines of flow except near salinity boundary layers. Note that since $\phi < 1$ and the molecular diffusivity for salt $\kappa_s \ll \kappa$, this inequality can frequently be satisfied when (3.5.11) is not. When macroscopic dispersion dominates, since $D \sim \phi^{-1} l_0 u$ (where l_0 is the scale of the interstitial path irregularities) this condition is virtually always satisfied. However, if perchance $(uh/\phi D) \ll l/h$, the salinity distribution is diffusively determined by $\nabla^2 s = 0$, as apparently is the case in oil-field waters of northern Louisiana and southern Arkansas (Hanor 1984).

Thermally driven flows

Just as hydraulic pressure gradients are generally dominant in the flow through aquifers, so in isolated, fluid-saturated regions or in submarine sediments and fracture regions where there are no hydraulically imposed pressure gradients, flow will be induced unless the isopycnals happen to be precisely horizontal. In most cases, the horizontal density variations are the result of sloping isotherms, but as Evans and Nunn (1989) have pointed out, convection can be induced near salt domes by horizontal variations in salinity. The interplay between salinity, and temperature variations in driving a flow field can be quite intricate, destabilizing in some circumstances a basic state that would appear to be stable and stabilizing in other circumstances one that seems unstable. These considerations will be deferred until Chapter 5. For the present scaling purposes, let us consider flows that are thermally driven, those in which the isotherms are not horizontal. The flow and temperature fields are more intimately connected than they are in hydraulically driven flows, but again, scaling considerations allow significant simplifications in many cases.

The vorticity associated with the transport velocity field, specified by (3.4.10), provides a velocity scale for the thermally induced circulations. If θ represents the isotherm slope and $\partial T/\partial z$ the geothermal gradient in the porous matrix, then the horizontal temperature gradient

$$\frac{\partial T}{\partial x} \simeq \theta \frac{\partial T}{\partial z}$$

and a representative magnitude of the flow velocity is

$$u \sim \omega h \sim \frac{k_{\mathrm{H}} g \alpha h}{\nu} \frac{\partial T}{\partial z} \theta. \tag{3.5.14}$$

In the thermal balance (2.8.4), internal heat sources are, as we have seen, negligible, and in steady flow we have again (3.5.10), where the order of magnitude of u is now given by (3.5.14). If ΔT is a characteristic temperature difference in the flow direction in the region, the advective terms on the left are of order

$$u \frac{\Delta T}{l} \sim \frac{k_{\mathrm{H}} g \alpha h \, \Delta T}{\nu l} \frac{\partial T}{\partial z} \theta,$$

whereas the thermal diffusion terms are at most of order

$$\kappa \frac{\Delta T}{h^2},$$

where h, the thickness, is the smallest dimension of the layer. Consequently, the temperature distribution is dominated by the influence of fluid advection if the first of these is much larger than the second; since generally θ is of order h/l, this requires

$$\frac{k_{\mathrm{H}} g \alpha h^2}{\kappa \nu} \frac{\partial T}{\partial z} \gg \left(\frac{l}{h}\right)^2, \tag{3.5.15}$$

where the dimensionless combination on the left-hand side is called the

Rayleigh–Darcy number or the Rayleigh number Ra. In contrast, the influence of fluid advection on the temperature distribution is negligible when

$$\mathrm{Ra} = \frac{k_{\mathrm{H}} g \alpha h^2}{\kappa \nu} \frac{\partial T}{\partial z} \ll \left(\frac{l}{h}\right)^2. \qquad (3.5.16)$$

This latter case, that of small Rayleigh number, is by far the simplest. Because of the quadratic dependence of the Rayleigh number on h, for a given aspect ratio (h/l) it is applicable directly to relatively shallow layers; a variety of examples will be described in later chapters. Note that the term "shallow" is relative – with many permeable matrices and geometries, h may be as large as several hundred meters while still satisfying (3.5.16). When it is, the advective terms in (3.5.10) can be neglected entirely and the temperature distribution in the fluid-filled matrix is determined by thermal conduction through the matrix,

$$\nabla^2 T = 0, \qquad (3.5.17)$$

together with the appropriate boundary conditions. Once the temperature distribution is found, the density field is specified through the equation of state and the velocity field is given by the transport vorticity equation (3.4.4) with (2.8.4). These solutions for the velocity and temperature distributions together specify the distribution of cementation, dissolution, or chemical deposition, as will be described later.

The high Rayleigh number case, generally involving deep platforms, is more complex. Here, the overall balance is reversed – fluid advection generally dominates the determination of the temperature distribution and thermal diffusion is generally negligible. When thermal diffusion is neglected altogether, (3.5.10) reduces to

$$\mathbf{u} \cdot \nabla T = 0; \qquad (3.5.18)$$

the temperature is constant along the lines of flow.

But this now poses a contradiction. In a closed region or along an impermeable boundary, the fluid flow follows the boundary, yet the temperature need not be constant along the boundary – indeed, it must not be, if the motion is generated thermally. In short, (3.5.18) does not allow us to satisfy the thermal boundary conditions. This kind of problem is characteristic of equations with the structure of (3.5.10) viewed as an equation for the temperature field T in which a small parameter (κ or in dimensionless terms Ra^{-1}) multiplies the highest derivative term in the equation ($\nabla^2 T$ is second order; ∇T is first order). Equations of this type are called singular perturbation equations. If the term of highest order is dropped, the order of the differential equation is reduced and fewer boundary conditions can be satisfied. The solution to the dilemma lies in the realization that the general scaling cannot be true everywhere – that there must be regions of the flow, involving plumes or thermal boundary layers, which are sufficiently thin compared with the layer thickness that the second-derivative thermal diffusion term is *locally* comparable with the advection term, even though over most of the fluid, it does indeed remain negligible. In these large Rayleigh number flows, then, there are different domains – an interior region away from boundary surfaces or plumes in which (3.5.18) is valid and the temperature remains constant along the lines of flow, together with thin boundary regions whose width self-adjusts to produce a balance between convective and diffusive effects in the heat equation. The solution to the flow as a whole involves matching these various regions to ensure that both the temperature and velocity distributions join smoothly from one region to another, a procedure that is exemplified in Chapters 6 and 7. The temperature and flow fields must therefore be solved simultaneously rather than sequentially before patterns of fabric alteration can be inferred.

In summary, if (3.5.3) is satisfied, the flow is hydraulically determined and can be found without reference to the temperature distribution. If the Peclet number in (3.5.11) is small, the temperature field itself is decoupled from the flow and is determined conductively. If it is large, the flow distorts the isotherms substantially; the temperature is generally constant along the streamlines except for the influence of possibly growing thermal boundary layers. If, however, (3.5.3) is clearly not satisfied, the flow is thermally driven. The criterion that the flow is sufficiently feeble that it cannot disturb the isotherms to any significant extent is now the Rayleigh number condition (3.5.16). If this is strongly not satisfied, we again have a coupled thermal and flow regime involving plumes and thermal boundary layers surrounding more or less isothermal regions, more complex but much more interesting flows.

3.6 UNIQUENESS AND MINIMUM DISSIPATION THEOREMS

For the flow of a viscous fluid of constant density at very low Reynolds number, a theorem due to Helmholtz (1868; see, e.g., Batchelor 1967, p. 227) establishes that there cannot be more than one flow solution consistent with given boundary conditions – the solution is unique. This uniqueness does not extend to situations in which buoyancy forces are important; if the isotherms (or, more generally, the isopycnals) are horizontal and there is no flow across the boundaries, one solution corresponds to a state of rest, with no fluid motion. In some circumstances, however, explained in Chapter 5, this state of rest is unstable and other time-dependent or steady solutions exist and can indeed develop.

Helmholtz's uniqueness theorem, which applies to the interstitial flow, clearly implies that the solution for the transport velocity $\mathbf{u}$ in Darcy flow is also unique. This more restricted result, which covers the hydraulically driven flows considered in this book, can be established directly, as follows: Consider the flow through a locally isotropic medium in which the permeability $k(\mathbf{x})$ may be an arbitrary function of position. In the absence of buoyancy forces,

$$\mathbf{u} = -\frac{k(\mathbf{x})}{\mu}\nabla p, \tag{3.6.1}$$

where p is the reduced pressure. The incompressibility condition is

$$\nabla\cdot\mathbf{u} = 0. \tag{3.6.2}$$

Suppose, now, that two patterns of flow and pressure $(\mathbf{u}, p)$ and $(\mathbf{u}', p')$, both satisfying (3.6.1) and (3.6.2), are possible in a given region V with assigned distributions of either pressure or normal component of transport velocity at the boundaries S:

$$p = p' \qquad \text{or} \qquad \mathbf{u}\cdot d\mathbf{S} = \mathbf{u}'\cdot d\mathbf{S} \qquad \text{on} \quad S. \tag{3.6.3}$$

Consider

$$\begin{aligned}
&\int k(\mathbf{x})[\nabla p - \nabla p']^2\, dV \\
&\quad = \int k(\mathbf{x})[\nabla p - \nabla p']\cdot[\nabla p - \nabla p']\, dV \\
&\quad = -\mu \int(\mathbf{u} - \mathbf{u}')\cdot\nabla(p - p')\, dV, \qquad \text{from (3.6.1)}, \\
&\quad = -\mu \int \nabla\cdot[(\mathbf{u} - \mathbf{u}')(p - p')]\, dV, \qquad \text{from (3.6.2)}, \\
&\quad = -\mu \int(p - p')(\mathbf{u} - \mathbf{u}')\cdot d\mathbf{S}, \qquad \text{from the divergence theorem}, \\
&\quad = 0, \qquad \text{from (3.6.3)}.
\end{aligned}$$

Since $k(\mathbf{x})$ is everywhere positive, it follows that $\nabla p = \nabla p'$ everywhere – the two solutions are identical. An immediate corollary is that hydraulically driven flows are, in the absence of buoyancy forces, always stable. Since there is only one solution, it cannot evolve into another.

Another general theorem established by Helmholtz shows that the total rate of dissipation of mechanical energy in a given region when inertia and buoyancy forces are negligible (as in hydraulically driven Darcy flow) is less than that in any other conceivable incompressible flow with the same values of the velocity at all points on the boundary. In Darcy flow, the rate at which work is done on the region by the surface pressure distribution, and therefore the total rate of dissipation, is

$$\begin{aligned}
\varepsilon_0 = -\int p\mathbf{u}\cdot d\mathbf{S} &= -\int \nabla\cdot(p\mathbf{u})\, dV \\
&= -\int \mathbf{u}\cdot\nabla p\, dV \\
&= \mu \int \frac{u^2}{k(\mathbf{x})}\, dV,
\end{aligned} \tag{3.6.4}$$

which is accordingly a minimum, subject to the constraints of flow across the boundary. This theorem can be used to solve problems of hydraulically driven flow by employing the calculus of variations, though direct methods are usually simpler.

3.7 GEOMETRICAL SCALING

Many sedimentary structures, when they are formed, have horizontal length scales much greater than their thicknesses. Subsequent tectonic deformation may cause them to bend, tilt, fold, and fracture, but apart from the discontinuities associated with massive fracture displacements, the rate at which the layer thickness varies along the bed is generally small when viewed on the scale of the layer thickness. The *internal* structures of a sedimentary bed also frequently show a great disparity in geometrical scales along and across the bed; relatively impermeable layers alternate with much more permeable ones, so that the variations in the original vertical direction are far more rapid than those in the original horizontal. Geometrical scaling is a technique that takes advantage of these inequalities in scale to simplify greatly the governing flow equations; the increased simplicity correspondingly expands the range of flow situations amenable to analysis. In Section 3.9, for example, the flow patterns in and surrounding more permeable lenses or layers in an aquifer are found with only simple calculation, and in Chapter 6, it is shown how geometrical scaling in combination with numerical analysis can provide three-dimensional, basin-wide flow patterns that are beyond the reach of direct numerical simulations, presently limited, for computational reasons, to two-dimensional flows.

Consider first a local region of a laminated sedimentary bed such as shown in Figure 3.7 in which the thickness scale of the laminations, δ, is small compared with their extent in the (original) horizontal direction. On a scale of perhaps a centimeter or so, large compared with the grain size but small compared with the lamination scale δ, the rock fabric may be locally isotropic, but the permeability will vary rapidly with z over the scale δ. The fluid transport velocity in the region will vary similarly, that is, with z over the scale δ and with x over the much larger scale l, so that symbolically, for two-dimensional flow,

$$u = u(x/l,\, z/\delta),$$
$$w = w(x/l,\, z/\delta). \tag{3.7.1}$$

From the incompressibility condition

$$\frac{\partial u}{\partial x} + \frac{\partial w}{\partial z} = 0,$$

and since $\partial u/\partial x \sim u/l$ and $\partial w/\partial z \sim w/\delta$, where u and v here represent

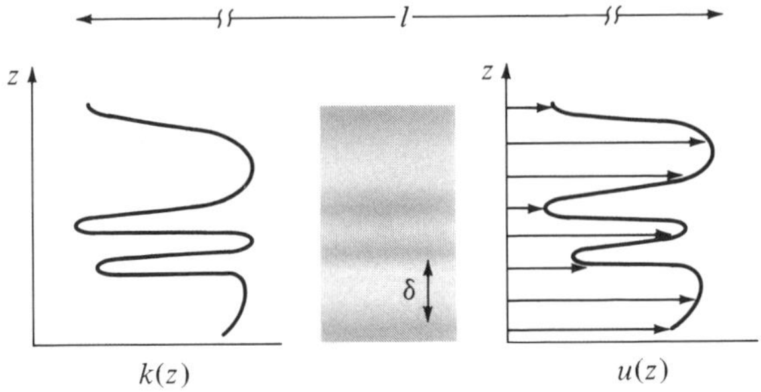

Figure 3.7. In a laminated bed in which the scale δ of the laminations is small compared with their horizontal extent, the horizontal flow velocity is proportional to the local permeability.

the general magnitudes of the transport velocity components in the x- and z-directions, it follows that

$$w \sim \frac{\delta}{l} u = \varepsilon_\delta u, \quad \text{say}, \tag{3.7.2}$$

where $\varepsilon_\delta \ll 1$. The inequality of scales forces the transport velocity along the laminations to be generally much greater than that across it, but the vorticity relation (3.4.5) provides an even stronger conclusion. On this small scale, the direct buoyancy forcing can be neglected – in convective flows it is the local pressure gradient produced by the overall, large-scale buoyancy distribution that drives the local flow. When the permeability varies on a scale δ in z,

$$k_H = k_H(x/l, z/\delta),$$
$$k_V = k_V(x/l, z/\delta),$$

where, on a local scale k_H and k_V may not be very different. From (3.4.5),

$$\frac{\partial}{\partial z}\left(\frac{u}{k_H}\right) = \frac{\partial}{\partial x}\left(\frac{w}{k_V}\right)$$
$$\sim \frac{\varepsilon_\delta u}{l k_V},$$

and the variation in u/k_H over the scale δ,

$$\text{var}\left(\frac{u}{k_H}\right) \sim \delta \frac{\partial}{\partial z}\left(\frac{u}{k_H}\right) \sim \varepsilon_\delta^2 \frac{u}{k_V} = \left(\frac{k_H}{k_V}\varepsilon_\delta^2\right)\frac{u}{k_H}, \tag{3.7.3}$$

is therefore a very small fraction of its characteristic magnitude when $\varepsilon_\delta^2(k_H/k_V) \ll 1$. Consequently, u/k_H is, in essence, constant within this region, though it may vary (slowly) from one such region to another; symbolically,

$$\frac{u(x, z)}{k_H(x, z)} = f\left(\frac{x}{l}\right) \qquad \text{when} \quad \varepsilon_\delta^2(k_H/k_V) \ll 1. \tag{3.7.4}$$

Within the region, k_H may vary substantially with z, and (3.7.4) asserts that the transport velocity along the stratum varies in proportion; flow along the strata is concentrated in layers of high permeability and avoids those of low permeability.

This result generalizes immediately to the three-dimensional case in which $\mathbf{x}$ is the position vector along the local bedding planes and z across them. The lateral velocity w is smaller than $|\mathbf{u}|$ by order ε_δ, and the ratio $\mathbf{u}/k$ is constant in magnitude and direction throughout the region.

When we are concerned with the flow on a basin-wide scale, we may not be particularly interested (at least initially) in the details of the flow distribution on a scale of tens of centimeters or of meters; our averaging volume may include many individual laminations or bedding planes. When the flow is viewed on *this* scale, striations in the distribution of local permeability with z (even though it may be isotropic) lead to a strong anisotropy in the permeability required for the specification of the macroscopic, basin-wide flow. If, locally, k is isotropic but $k = k(z)$, then for flow in the x-direction

$$\frac{\mathbf{u}(z)}{k(z)} = \text{const} = -\mu^{-1}\,\nabla_H p,$$

where ∇_H is the gradient along the bedding planes. Since from (3.7.4) $\nabla_H p$ is independent of z, the velocity along the planes, averaged over the thickness h of the averaging volume, is

$$\bar{\mathbf{u}} = \frac{1}{h}\int_0^h \mathbf{u}(z)dz = -\mu^{-1}\,\nabla_H p\left(\frac{1}{h}\int_0^h k(z)dz\right)$$
$$= \mu^{-1}\bar{k}_H\,\nabla_H p, \tag{3.7.5}$$

where

$$\bar{k}_H = \frac{1}{h}\int_0^h k(z)\,dz \tag{3.7.6}$$

is then the average permeability for the flow on the scale h. The subintervals of *high* permeability dominate the value of the integral (3.7.6); in a structure with high-permeability lenses, the flow is concentrated in the lenses and the permeability of the region as a whole approaches that of the lenses. For transverse flow, however, the focusing effect disappears; the velocity w in the z-direction is approximately constant, but the pressure gradient in this direction that is needed to drive it varies rapidly as $k(z)$ does,

$$\frac{\partial p}{\partial z} = \frac{-\mu w}{k(z)},$$

and the mean vertical pressure gradient

$$\frac{\partial \bar{p}}{\partial z} = -\mu w \frac{1}{h} \int_0^h \frac{dz}{k(z)}$$

$$= \frac{-\mu w}{\bar{k}_V}, \tag{3.7.7}$$

where

$$\bar{k}_V = \left\{ \frac{1}{h} \int_0^h \frac{dz}{k(z)} \right\}^{-1} \tag{3.7.8}$$

is now dominated by the *smallest* values of $k(z)$ in the interval. Accordingly, for flow across the bedding structures, it is the regions of least permeability that provide the greatest resistance to flow. For example, if a sedimentary deposit consists of alternating beds where the pore sizes differ by a factor of 10, the permeabilities differ by a factor of about 100. If r represents the fraction occupied by the layers of greater permeability, the permeability of the whole to mean longitudinal flow is about r times the value of the *greater* permeability. For transverse flow, the effective permeability (unless $r \simeq 1$) is close to $(1 - r)^{-1}$ times the value of the smaller permeability, or $[100(1 - r)]^{-1}$ times that of the greater permeability. The ratio k_H/k_V of the effective longitudinal to transverse permeability is then approximately $100r(1 - r)$, which is generally fairly large.

Geometrical scaling is useful also in consideration of basin-wide flows, with averaging volumes large compared with the scale of the bedding. In the geometries such as those illustrated in Figure 3.8, the common characteristic is the smallness of the *overall* thickness scale h_0 relative to the scale l along the layer over which its thickness or slope varies, so that for the layer as a whole

$$\varepsilon = h_0/l \ll 1.$$

As before, the magnitude of the transverse velocities relative to the longitudinal ones, when viewed on this scale, is smaller by order ε, but in the vorticity relation (3.4.5)

$$\frac{\partial}{\partial z}\left(\frac{u}{k_H}\right) - \frac{\partial}{\partial x}\left(\frac{w}{k_V}\right) = 0, \tag{3.7.9}$$

now usually $k_H \gg k_V$. The second term is of order

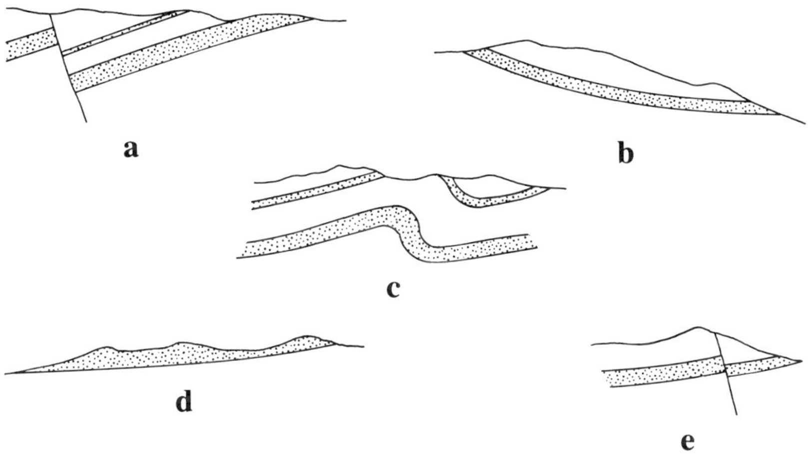

Figure 3.8. Typical configurations of permeable beds, which may form unconfined aquifers (d) or confined aquifers (b); they may be folded (c) and partially (e) or wholly (a) displaced by faults.

$$\frac{\varepsilon u}{l k_V}$$

so that the fractional variation of u/k_H across the depth of the stratum is

$$\frac{\varepsilon u}{k_V l} h_0 \bigg/ \frac{u}{k_H} = \varepsilon^2 \frac{k_H}{k_V}. \tag{3.7.10}$$

If ε is sufficiently small (10^{-2} or less, say), this fractional variation remains small even when k_H/k_V is moderately large, so that for basin-wide flow also

$$\frac{\mathbf{u}}{k_H} = \mathbf{f}\left(\frac{\mathbf{x}}{l}\right) \qquad \text{when} \quad \varepsilon^2 \frac{k_H}{k_V} \ll 1. \tag{3.7.11}$$

Note, however, that this condition on the basin-wide velocity distribution, for which k_H is usually large compared with k_V, is more restrictive than the local condition (3.7.4) in which they may be comparable. The local result (3.7.4) may hold when the larger-scale expression (3.7.11) does not. The ratio $\mathbf{u}/k_H$ does not vary much in each volume element when $\varepsilon_\delta^2 \ll 1$ according to (3.7.4), but unless $\varepsilon^2 k_H/k_V \ll 1$, it may vary from one averaging volume to the next across the whole layer. Nevertheless, the condition for (3.7.11) seems to be satisfied adequately in many terrestrial aquifers or sedimentary basins of large extent because of the smallness of ε. In atolls or coral reefs, however, layers of lithified lagoonal silts may alternate with more permeable carbonate layers, giving a very large ratio k_H/k_V while $\varepsilon = h_0/l$ may not be particularly small, so that the condition on (3.7.11) is not satisfied and a different type of approximation is necessary. Consideration of these situations is deferred until Section 7.4.

The scale ratio $S = \varepsilon^2 k_H/k_V$ has an important physical as well as geometrical interpretation. The gradient of reduced pressure across the layer is

$$\frac{\partial p}{\partial z} = -\frac{\mu}{k_V} w \sim -\frac{\mu}{k_V}\left(\frac{h_0}{l}\right) u$$

from (3.5.4), so that the difference in reduced pressure from top to bottom

$$\Delta p_z \sim \frac{\mu}{k_V}\left(\frac{h_0^2}{l}\right) u.$$

In contrast, the gradient *along* the layer is

$$\frac{\partial p}{\partial x} = -\frac{\mu}{k_H} u,$$

so that the difference in reduced pressure from one end of the stratum to the other, Δp_x, is about $\mu u l/k_H$. The ratio of differences in reduced pressure across the layer to that along its length is then

$$\frac{\Delta p_z}{\Delta p_x} \sim \left(\frac{h_0^2}{l^2}\right)\frac{k_H}{k_V} = \varepsilon^2 \frac{k_H}{k_V}, \qquad (3.7.12)$$

which, when (3.7.11) is satisfied, is small. Note that the smallness of this ratio does *not* imply that the reduced pressure gradient is primarily along the layer. The ratio of the transverse to longitudinal *gradients* is $(h_0/l)k_H/k_V$; when $k_V/k_H \sim h_0/l \ll 1$, this is of order unity while still $(h_0/l)^2 k_H/k_V \ll 1$.

It is frequently assumed in the hydrological literature, however, that when the variations along the layer are sufficiently gentle, the reduced pressure itself *is* constant across the layer and the flow is entirely along it – this is called the Dupuit assumption. This assumption, introduced more than one hundred years ago, is frequently unrealistic and, as will appear later, unnecessary. The weaker condition $\varepsilon^2 k_H/k_V \ll 1$ guarantees that the reduced pressure *gradient* along the layer is constant across it, although the pressure itself may vary (and, in general, will vary). Since the flow depends on the reduced pressure gradient, not the pressure, most of the results concerning flow that are obtained with the Dupuit assumption can be derived under this weaker and more general condition, and one's confidence in their applicability is therefore enhanced.

In summary, under the appropriate conditions, (3.7.11) is a very useful result, and many applications are given in subsequent pages. On a basin-wide scale, k_H may vary substantially over the depth of a stratum at a given location, and (3.7.11) asserts that the transport velocity, averaged over volume elements whose size is large compared with the lamination thicknesses, preserves its direction and varies in magnitude proportionately to k_H. When the overall basin, or part of it, is homogeneous with k_H independent of z, this longitudinal transport velocity is also independent of z; $\mathbf{u} = \mathbf{u}(x/l)$.

3.8 FLOW PATTERNS NEAR ISOLATED LENSES OR CRACKS

It is clear from the preceding section that when a permeable stratum contains long or continuous internal layers that are more permeable than the remainder of the stratum, the flow will be concentrated there, the ratio of the velocities being proportional to the ratio of the permeabilities. It is perhaps a little less obvious that when the lenses are of limited length in the flow direction, the same kind of flow concentration occurs; a more permeable lens in effect locally "short-circuits" the flow in the neighborhood, drawing in fluid at one end and releasing it at the other. This flow concentration then localizes in space and speeds up in time the processes of dissolution and deposition that are at the heart of mineral and ore genesis.

The detailed flow pattern is, of course, dependent on the geometry of the lens as well as the permeability ratio, but consideration of some simple cases gives a good basis for the evaluation of the magnitude of the effects in typical geological configurations. Though of possibly limited direct geological interest, the case of an isolated spherical region whose permeability is different from its surroundings does illustrate simply the variations in flow pattern that occur with variations in the permeability ratio. Suppose that a sphere of permeability k_i and radius a is embedded in a matrix of (isotropic) permeability k_0 in which the ambient gradient of reduced pressure p (far from the spherical inclusion) is ϖ. The mathematical problem can best be represented in terms of spherical polar coordinates (r, θ, ϕ) of Figure 3.9 with $\theta = 0$ in the direction down the ambient pressure gradient. Both inside and outside the sphere, the pressure distribution satisfies Laplace's equation (3.4.7), which in these coordinates reduces to

$$\frac{\partial}{\partial r}\left(r^2 \sin\theta \frac{\partial p}{\partial r}\right) + \frac{\partial}{\partial \theta}\left(\sin\theta \frac{\partial p}{\partial \theta}\right) = 0, \qquad (3.8.1)$$

since $\partial/\partial\phi = 0$.

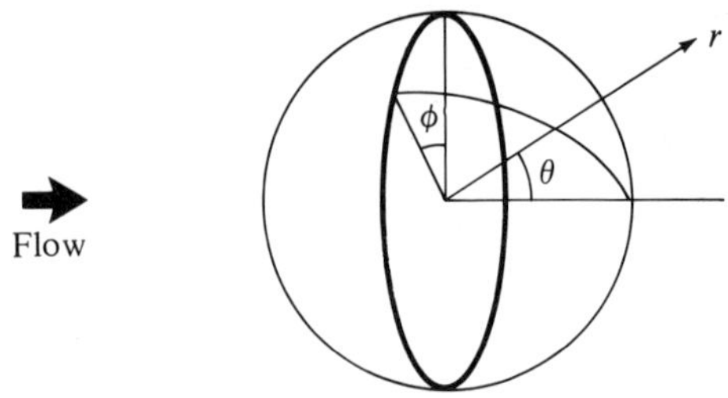

Figure 3.9. Polar coordinates for flow near a spherical region of different permeability.

The solution must obey the following boundary conditions: (i) The internal and outside fluid pressures, p_i and p_o, respectively, must be the same at all points on the surface $r = a$; (ii) the normal component of the transport velocity across the boundary, $-(k/\mu)\ \partial p/\partial r$, must also be the same on both sides; and (iii) the pressure distribution must approach the undisturbed uniform gradient $-\varpi$ far away from the sphere. Symbolically,

$$\left.\begin{aligned} p_i &= p_o, \\ k_i \frac{\partial p_i}{\partial r} &= k_o \frac{\partial p_o}{\partial r}, \end{aligned}\right\} \quad \text{at} \quad r = a,$$
$$p_o \to -\varpi x = -\varpi r \cos\theta \qquad \text{as} \quad r \to \infty. \tag{3.8.2}$$

The solutions for the internal and external pressure distributions are expressible in spherical harmonics, as described in references on potential theory:

$$p_i(r, \theta) = -\varpi \frac{3k_o}{k_i + 2k_o} r \cos\theta, \qquad \text{when} \quad r < a, \tag{3.8.3}$$

$$p_o(r, \theta) = -\varpi\left\{1 - \left(\frac{k_i - k_o}{k_i + 2k_o}\right)\frac{a^3}{r^3}\right\} r \cos\theta, \qquad \text{when} \quad r > a. \tag{3.8.4}$$

Inside the sphere, the pressure (relative to that at the center) is proportional to $x = r\cos\theta$, so that the interior flow is uniform and entirely in the x-direction; its magnitude

$$-\frac{k_i}{\mu}\frac{\partial p_i}{\partial x} = \varpi \frac{3k_ik_o}{k_i + 2k_o}$$

is greater by the factor

$$G = \frac{3k_i}{k_i + 2k_o} \tag{3.8.5}$$

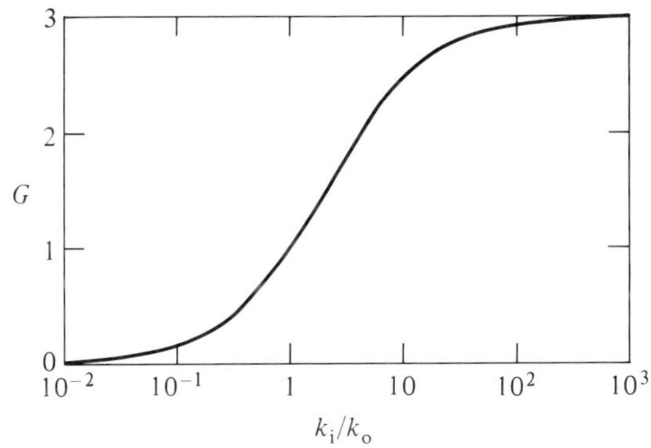

Figure 3.10. The focusing ratio or velocity ratio for a sphere of permeability k_i in a matrix of permeability k_o.

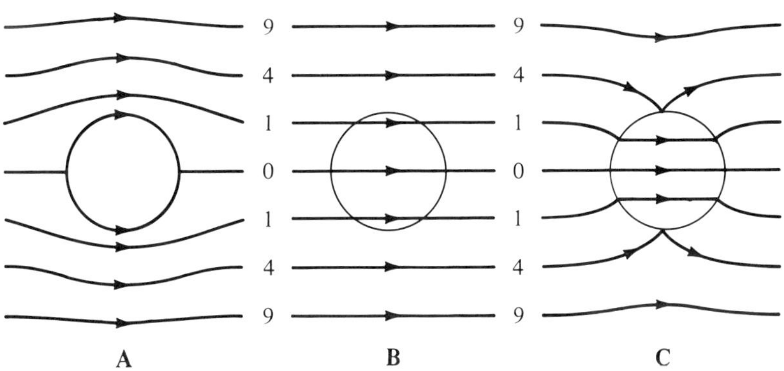

Figure 3.11. Streamline patterns for flow near a spherical region (a) when the permeability inside is much less than that outside, (b) when the two are equal, and (c) when the interior is much more permeable or liquid filled. The numerical values of the Stokes stream functions specify the relative volume transports in the contained (axisymmetrical) surfaces.

than that in the ambient fluid far from the sphere. This can be called the focusing ratio; it is illustrated in Figure 3.10 as a function of k_i/k_o. When $k_i \ll k_o$, the sphere being much *less* permeable than that outside, $G \to 0$ and the flow pattern approaches that around an impermeable isolated spherical pebble (say), as shown in Figure 3.11a. When $k_i = k_o$, $G = 1$; there is no difference in the permeabilities and the flow is undisturbed. When $k_i > k_o$, the inside now being *more* permeable, G is greater than 1, approaching the value 3 when $k_i \gg k_o$. When the interior is simply fluid filled, so that (mathematically) $k_i/k_o \to \infty$, $G = 3$ precisely; the fluid passing through the sphere comes from a cylinder of radius $\sqrt{3}a$ extending upstream. If the contrast in permeabilities is substantial (e.g., if $k_i = 50k_o$), the fractional error made by assuming that the sphere is simply fluid filled is approximately $2k_o/k_i$ (or 4 percent in this example).

Of more general geological interest are lenticular inclusions that are thin compared with their extent along the general direction of any flow. When the thickness of the lens varies slowly along its length, the scaling of the preceding section can be used for the flow *inside* the lens and shows that the reduced pressure p and the velocity u along it are uniform at any section. If $h(\mathbf{x})$ represents the thickness of the inclusion, the incompressibility condition, integrated across it, is

$$\nabla_H \cdot (\mathbf{u}h(\mathbf{x})) = w_l - w_u, \tag{3.8.6}$$

where w_l and w_u are the flow speeds across the lower and upper interfaces, respectively. This equation, in combination with Darcy's law, provides pressure boundary conditions for the solution of the flow outside. Inside the lens, and so also along its upper and lower boundaries,

$$\mathbf{u} = -\frac{k_i}{\mu}\nabla_H p,$$

whereas outside the boundaries

$$w = -\frac{k_o}{\mu}\frac{\partial p}{\partial z}.$$

Consequently, the flow outside the lens is specified by $\nabla^2 p = 0$ with undisturbed flow at infinity and along the surface,

$$\nabla_H \cdot (h(\mathbf{x})\, \nabla_H p) = \frac{k_o}{k_i}\left(\left.\frac{\partial p}{\partial z}\right|_u - \left.\frac{\partial p}{\partial z}\right|_l\right), \tag{3.8.7}$$

$$= \frac{2k_o}{k_i}\left.\frac{\partial p}{\partial z}\right|_u, \tag{3.8.8}$$

the last being relevant when the inclusion is symmetrical in vertical cross section.

Analytical solutions are not generally possible for arbitrary distributions $h(\mathbf{x})$ of lens thickness, but as in the preceding example, limiting cases are useful. The left-hand side of (3.8.8) is of order

$$\frac{h}{l}\nabla_H p = \varepsilon\, \nabla_H p.$$

When $\varepsilon \ll k_o/k_i$, the surface boundary condition (3.8.8) becomes

$$\frac{\partial p}{\partial z} = O\left(\frac{\varepsilon k_i}{k_o}\right)\frac{\partial p}{\partial x} \simeq 0 \tag{3.8.9}$$

along the lens surface. This simplification applies to lenses that are very elongated in the flow direction ($\varepsilon \ll 1$), whereas the lens permeability may be greater but not *very* much greater than that outside; under these circumstances the disturbance to the flow field is relatively small. In a long lens of this kind, or a crack extending in the direction of the ambient pressure gradient and flow, the velocity inside the lens is approximately (k_i/k_o) times that outside, so that the focusing ratio

$$G \simeq k_i/k_o. \qquad (3.8.10)$$

If the lens is indefinitely long, $\varepsilon \to 0$, and no matter what the ratio k_i/k_o, (3.8.9) is satisfied and the velocity ratio is equal to the permeability ratio, as shown previously for simple layered conditions.

If, however, the length of the lens is finite, so that ε is small but not zero, and its permeability is much greater than that outside,

$$\frac{k_o}{k_i} \ll \varepsilon \ll 1. \qquad (3.8.11)$$

This is, in the limit of $k_o/k_i \to 0$, the case of a fluid-filled crack of finite length, but note that the condition that the flow around a lens of this kind behave as if the inclusion were fluid filled, $(k_o/k_i \ll \varepsilon)$, is more restrictive than it was for a sphere $(k_o/k_i \ll 1)$. The reason for this is that the length scale for the disturbance to the outside flow is l, the lens length. Fluid can be drawn into the lens from a depth interval of order l, and since the lens thickness $h = \varepsilon l$, the interior flow speed is of order ε^{-1} times that outside. The pressure gradient along the lens is therefore $(k_o/k_i)\varepsilon^{-1}$ times that outside, and for this to be negligible, the stronger condition (3.8.11) must be satisfied. When it is, we can take

$$\nabla_H p = 0 \qquad (3.8.12)$$

along the lens surface.

Detailed solutions for the two-dimensional flow past a thin, isolated lens can be given when it is either much more permeable or much less permeable than the surrounding matrix. They are simple modifications of classical solutions for potential flow in hydrodynamics given, for example, in Lamb's *Hydrodynamics* (1932, p. 86) and are expressed in terms of elliptic coordinates,

$$\begin{aligned} x &= a \cosh \xi \cos \eta, \\ z &= a \sinh \xi \sin \eta, \end{aligned} \qquad (3.8.13)$$

which are such that

$$\frac{x^2}{a^2 \cosh^2 \xi} + \frac{z^2}{a^2 \sinh^2 \xi} = 1,$$

and

$$\frac{x^2}{a^2 \cos^2 \eta} - \frac{z^2}{a^2 \sin^2 \eta} = 1.$$

Coordinate contours ξ = const are thus ellipses that as $\xi \to 0$ shrink to the line $z = 0$, $|x| < a$ and as $\xi \to \infty$ expand to infinity; the contours η = const are a family of orthogonal hyperbolas.

When the lens or crack of length $l = 2a$ is much more permeable than the surroundings, the boundary condition (3.8.12) requires that the fluid enter perpendicular to the interface, or, in terms of the stream function, that $\partial\psi/\partial z = 0$ or $\partial\psi/\partial\xi = 0$ when $\xi = 0$. If the undisturbed stream direction is at an angle θ to the plane of the lens, the distribution of ψ is specified by

$$\begin{aligned}\psi &= Ua \cosh \xi\{\cos\theta \sin\eta - \sin\theta \cos\eta\} \\ &= Ua \cosh\xi \sin(\eta - \theta),\end{aligned} \tag{3.8.14}$$

where U is the undisturbed speed. (Note that as $\xi \to \infty$, $\psi \to U(z\cos\theta - x\sin\theta)$, since $\cosh\xi \to \sinh\xi$; the velocity components $(\partial\psi/\partial z, -\partial\psi/\partial x) \to U\cos\theta, U\sin\theta$ in the distant, undisturbed flow).

The previous discussion was concerned with an approaching flow parallel to the plane of the lens or crack, so that $\theta = 0$. The streamlines are then given by

$$\psi = Ua \cosh\xi \sin\eta, \tag{3.8.15}$$

and the pattern is illustrated in Figure 3.12a. The total volume flux across the midpoint of the region is twice the value of the stream function ψ_α, which separates the fluid entering into the lens or crack from that missing it; since this streamline passes through $x = 0$, $z = 0$, or $\xi = 0$, $\eta = \pi/2$, from (3.8.15)

$$\psi_\alpha = Ua = Ul/2.$$

Consequently, the total flux through the center of the crack is Ul; it captures fluid from the surrounding medium *over a depth range precisely equal to its own length!* If h is the thickness of the lens or crack at its midpoint, the fluid velocity there is $Ul/h = U/\varepsilon$ and the focusing ratio,

$$G = \varepsilon^{-1}, \tag{3.8.16}$$

which is very large, since $\varepsilon \ll 1$. The variation in G over the whole range of conditions for isolated lenses or cracks parallel to the incident stream is therefore as shown in Figure 3.13; when k_i/k_o is small compared with ε^{-1}, G increases linearly with k_i/k_o in accordance with (3.8.10) until $k_i/k_o \sim \varepsilon^{-1}$, after which G asymptotes to ε^{-1} as in (3.8.16).

When the plane of a highly permeable or fluid-filled region is inclined at the angle θ to the incident flow direction, the solution is given by (3.8.14). When $\theta = \pi/2$,

$$\psi = -Ua\cosh\xi\cos\eta = -Ux$$

everywhere; the flow passes undisturbed normally through the thin lens. This is illustrated in Figure 3.12c, which has been rotated through 90° to

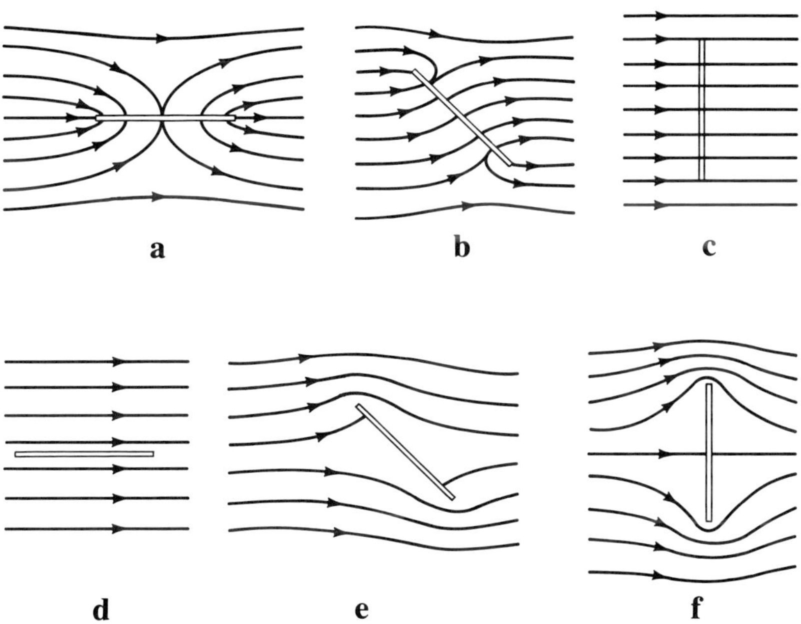

Figure 3.12. Streamline patterns near thin two-dimensional isolated lenses. In (a), (b), and (c), the lens is much more permeable than the matrix (or is a liquid-filled crack); in (d), (e), and (f), it is highly impermeable.

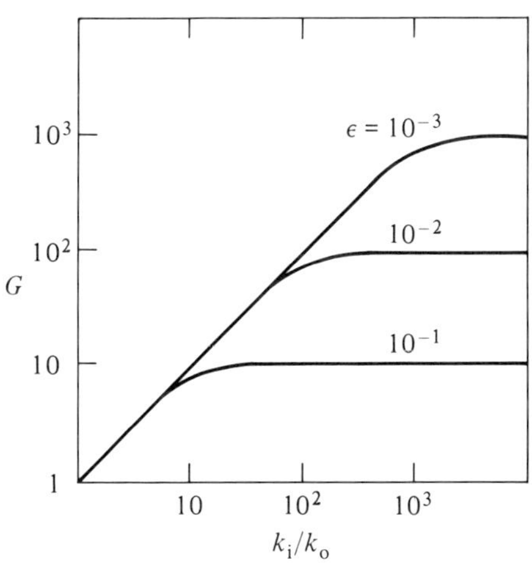

Figure 3.13. The focusing ratio G for two-dimensional lenses in the stream direction as a function of permeability ratio and ε, the thickness-to-length ratio.

keep the incident flow direction the same. The solution for $\theta = 45°$ is shown in Figure 3.12b.

When the lens is *impermeable*, no flow can enter it and $\psi = \text{const} = 0$, say, on the surface. The solution for the stream function is then

$$\begin{aligned}\psi &= Ua \sinh \xi(\cos \theta \sin \eta - \sin \theta \cos \eta) \\ &= Ua \sinh \xi \sin(\eta - \theta).\end{aligned} \tag{3.8.17}$$

When $\theta = 0$,

$$\psi = Ua \sinh \xi \sin \eta = Uz,$$

so that the velocity field $(u, w) = (\partial\psi/\partial z, -\partial\psi/\partial x) = (U, 0)$ is again undisturbed by a thin impermeable lens parallel to the flow. When $\theta \neq 0$, the flow must pass around the lens and the streamline patterns are as shown in Figures 3.12e and f when $\theta = 45°$ and $90°$, respectively. In the latter case particularly, fluid in the matrix can be seen to move faster around the ends of the impermeable region than it does elsewhere.

3.9 LONG AND MULTIPLE LENTICULAR LAYERS

Flow focusing

When the length of a lens is much greater than the total thickness of the permeable layer, the presence of the boundaries significantly modifies the overall flow pattern, but the inequality of scales allows the geometrical scaling of Section 3.7 to be applied to the flow outside the lens as well as that inside. The greatest interest attaches to lenses or internal layers that are more permeable than the fabric as a whole, since in these, flow focusing occurs which increases the transport of dissolved solutes and produces enhanced horizontal dispersion and lateral mixing.

If h represents the total depth of the layer and l the characteristic length of a lens in the flow direction, when the scale ratio $S = (h/l)^2 k_H/k_V = \varepsilon^2 k_H/k_V \ll 1$, the relation (3.7.4) assures us that the ratio $\mathbf{u}(z)/k_H(z)$ is constant throughout the thickness of the layer at any location (x, y). This relation, in combination with overall pressure or continuity constraints, allows the local flow patterns to be found very simply for a variety of realistic configurations.

As an illustration of the kinds of flow pattern produced, consider the two-dimensional flow in a confined permeable stratum such as that shown in Figure 3.14 containing one or more lenses or layers in which the thickness d may vary along its length. For the sake of definiteness, the fabric upstream of the inclusions will be supposed homogeneous, with k_H independent of z; the approaching flow velocity U is therefore also independent of z. If k_i and k_o represent the permeabilities for flow in the bedding planes inside and outside the lens, then at any location x

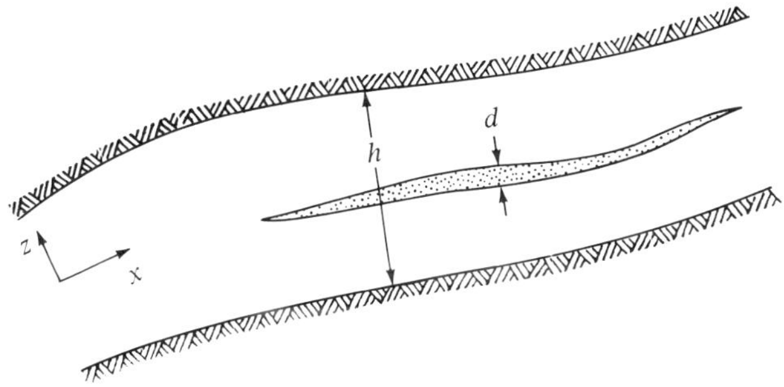

Figure 3.14. A permeable lens in a confined aquifer.

$$\frac{u_i}{k_i} = \frac{u_o}{k_o}, \tag{3.9.1}$$

whereas by incompressibility, the volume flux of fluid along the whole layer is constant,

$$u_i d + u_o(h - d) = \text{const} = UH, \tag{3.9.2}$$

where $h = h(x/l)$ is the total thickness of the permeable region, H its thickness where the velocity is U, and $d = d(x/l)$ represents the thickness of the lens at any section, or the total thickness of multiple lenses if they overlap. Note that insofar as the velocities u_i and u_o are concerned, the location of the lens within the layer is immaterial; if it is in the interior, the velocity outside is the same above and below as long as k_o is also the same on either side of the lens.

The pair of equations (3.9.1) and (3.9.2) can be solved for u_i and u_o,

$$u_o = \frac{UH}{h} f(x/l), \qquad u_i = \frac{k_i}{k_o}\frac{UH}{h} f(x/l),$$

where (3.9.3)

$$f(x/l) = \left\{1 + \left(\frac{k_i}{k_o} - 1\right)\frac{d}{h}\right\}^{-1}.$$

The focusing ratio

$$G = \frac{u_i}{U} = \frac{k_i}{k_o}\frac{H}{h} f\left(\frac{x}{l}\right)$$
$$= \frac{H}{h}\left\{\frac{d}{h} + \frac{k_o}{k_i}\left(1 - \frac{d}{h}\right)\right\}^{-1}. \tag{3.9.4}$$

In these expressions, the factor H/h merely represents the variations in flow velocity produced by changes in the overall layer thickness. The

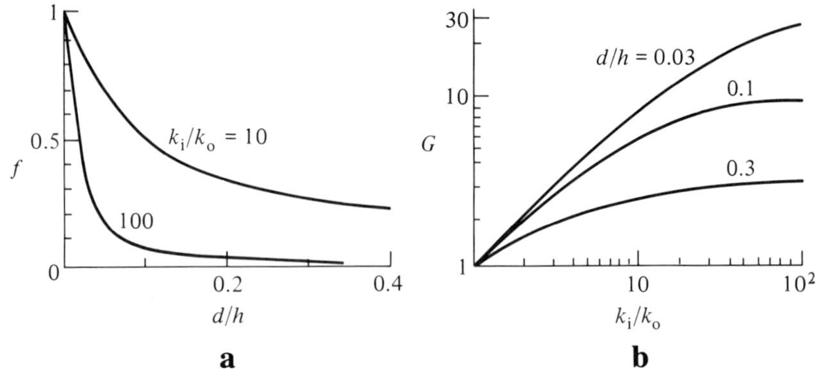

Figure 3.15. (a) The function f of equation (3.9.3). (b) The focusing ratio in a confined aquifer as a function of the permeability ratio and the thickness of a permeable lens.

more important factor $f(x/l)$ is illustrated in Figure 3.15. When $k_i > k_o$, the function f *decreases* as the thickness d of the lens increases from zero at its leading edge, so that the fluid velocity in the lens decreases from its initial value of $(k_i/k_o)U$ as the lens thickens. The reason for this is that in a confined aquifer, as the lens begins to "short-circuit" the volume flux, the longitudinal pressure gradient outside decreases and consequently the longitudinal velocity decreases in proportion. The volume flux $u_i d$ in the lens (or lenses) does increase, however, as long as the lens thickness d increases, the maximum volume flux and minimum fluid velocity in the lens occurring when the lens thickness is a maximum, d_m, say. The fraction of the total volume flux moving along the lens or lenses at this point is

$$\frac{u_i d_m}{UH} = \frac{k_i}{k_o}\frac{d_m}{h}\left\{1 + \left(\frac{k_i}{k_o} - 1\right)\frac{d_m}{h}\right\}^{-1}$$

$$\simeq 1 - \frac{k_o}{k_i}\frac{h}{d_m} \qquad \text{when} \quad k_i \gg k_o. \qquad (3.9.5)$$

Clearly, when k_i/k_o is sufficiently large, almost all of the fluid transport passes through the lenses even if the fractional depth d_m/h is rather small. The focusing ratio

$$G \to \frac{h}{d} \qquad \text{when} \quad k_i \gg k_o, \qquad (3.9.6)$$

as illustrated in Figure 3.15b.

For a given lens shape $d(x/l)$, flow distributions can be calculated simply from (3.9.3). For example, when the lens at a given section occupies

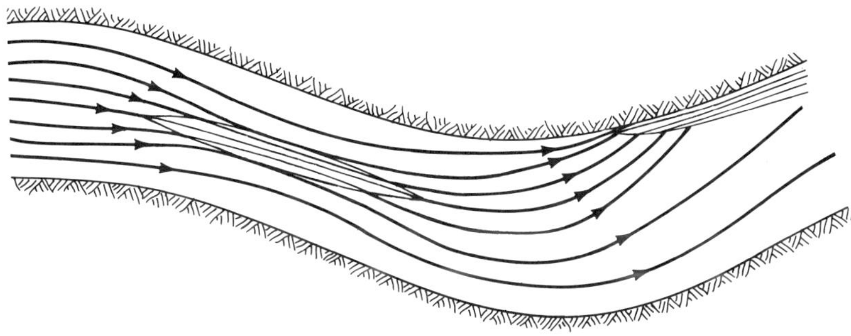

Figure 3.16. Streamline patterns in more permeable lenticular layers – the permeability ratio here is 10. Note that above and below the layers, the fluid velocity is less than in the incoming flow (the streamlines are more widely separated), so that the longitudinal pressure gradient is reduced as the lens "short-circuits" the flow.

the interval $(z_0, z_0 + d)$ of the cross section, the streamlines in the surrounding fabric are given by

$$\begin{aligned} \psi &= Uzf(x/l), \qquad 0 < z < z_0 \\ &= U\{h - (h - z)f(x/l)\}, \qquad z_0 + d < z < h. \end{aligned} \tag{3.9.7}$$

Figure 3.16 is an example of a streamline pattern calculated in this way.

Striations in the distribution of local permeability (which may be locally isotropic on a grain-size scale) lead to the strong anisotropy in mean permeability measured on, say, a scale of meters; multiple layering or lenses augment the anisotropy in a similar way on a basin scale. The overall permeability for longitudinal flow of a basin with multiple, more permeable lenses is approximately that of the lenses themselves times the fraction of the total section that they occupy, as in equation (3.7.6).

The reduction in pressure gradient in a confined aquifer as the flow is focused into a lens does not occur in an unconfined aquifer, since in these elongated geometries, the pressure gradient is maintained throughout the aquifer by the water-table slope θ. The fluid velocity outside a lens is $(k_o g/\nu) \sin\theta$ downslope and that inside is $(k_i g/\nu) \sin\theta$. The total transport

$$\{k_o(h - d) + k_i d\} \frac{g \sin\theta}{\nu}$$

increases as the lens thickness d does; water is drawn down from the water table above the upstream end of the lens and discharged across it at the downstream end (Figure 3.17). Note that fluid is drawn into the inclusion only from above, and that in a water table of constant slope, the streamlines outside the lens retain equal vertical intervals as the downslope velocity is constant.

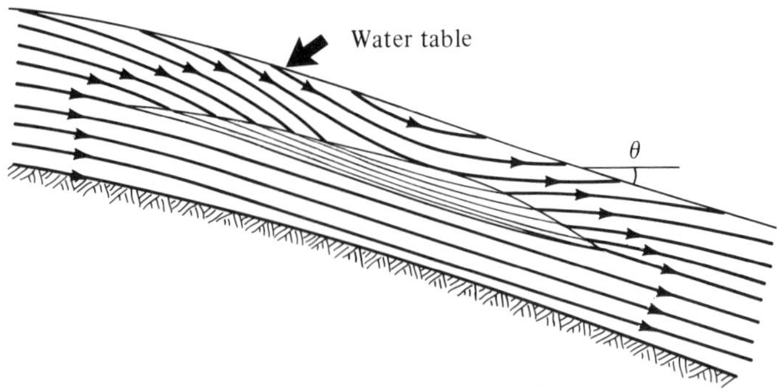

Figure 3.17. Streamlines in a more permeable lenticular inclusion in an unconfined aquifer $k_i = 4k_o$. Note the contrast with Figure 3.16. Here the flow beneath the lens is undisturbed, but above it, water is drawn in across the water table at the upper end and discharged across it at the lower.

Dispersion and mixing in lenses

Variations of permeability associated with bedding or lenses also have a strong influence on the overall dispersal characteristics of marked fluid elements, such as dissolved chemical species or contaminants in a groundwater aquifer. Freeze (1975) has noted that in many field formations, $J = \ln(k/\bar{k})$ has an approximately Gaussian distribution with horizontal correlation length scales l that are 10 or 20 times larger than those in the vertical. The consequent variations in percolation velocity, closely proportional to those in permeability, then persist over horizontal scales of order l, rather than over the pore scale l_0 as in the "macroscopic" dispersion discussed in Section 2.3. If the variations in k are supposed random on a basin-wide scale, the mathematics of the dispersal is essentially the same as discussed in that section, though the interpretation is somewhat different. Dagan (1984) has shown explicitly that if the covariance of J decreases exponentially with increasing separation, the longitudinal dispersal of marked particles from their mean position is given asymptotically by

$$\overline{(X - X_0)^2} \to 2\bar{J}^2\phi^{-1}Ult \qquad (3.9.8)$$

when $Ut/l \gg 1$, where U is the mean transport velocity. This can be compared with (2.3.11) and (2.3.13); the effective longitudinal diffusivity is then $\bar{J}^2\phi^{-1}Ul$, where l is the *lens* length scale rather than Ul_0/ϕ, where l_0 is the *grain* size as it is in macroscopic dispersion in a uniform matrix. If k varies by a factor of, say, 10^2 between lenses and surrounding matrix, J varies by a factor of about 5 and $\bar{J}^2 \sim 4$, so that the dispersion on a

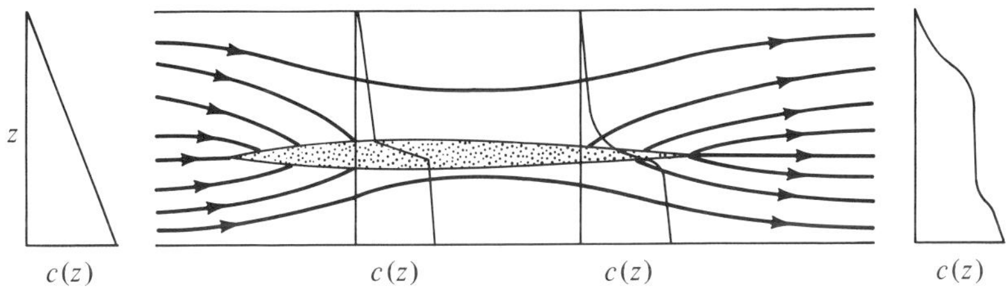

Figure 3.18. Mixing produced by flow focusing in a lens. When oncoming interstitial fluid has a uniform concentration gradient as shown on the left, the concentration tends to remain constant along the streamlines, provided that the fluid is not drawn into the lens. The streamlines that do enter the lens are crowded within it, so that inside, the vertical concentration gradient is magnified by the focusing ratio. Internal dispersion smooths out this gradient as the fluid moves along the lens, so that when it is reinjected into the matrix and spreads out, the vertical concentration gradient in this fluid is greatly reduced.

basin-wide scale, described by (3.9.8), is many orders of magnitude larger than that discussed in Section 2.3 for a uniform matrix. The term "megadispersion" has been coined to describe this large-scale effect associated with variations in k in lenses and layers.

Finally, the concentration of flow in lenses also enhances the lateral mixing of solutes. As fluid enters a lens, any existing lateral gradient of an unsaturated solute (or one whose equilibration length – see Section 4.2 – is large compared with the lens size) is amplified by the convergence by a factor equal to the focusing ratio, if diffusion is neglected during this convergence. Inside the lens, the lateral concentration gradient is then very large and lateral diffusion is greatly enhanced. This large internal gradient is smoothed out in a time d^2/D, where d is the lens thickness and D the dispersion coefficient in the lens; if this is comparable with the passage time through the structure l/u_i, or if

$$l \gtrsim \frac{u_i d^2}{D},$$

the fluid in the lens has become mixed before it emerges at the other end and spreads out laterally. The final lateral gradient in concentration of the solute in the fluid that has passed through the lens is then very much less than it was initially (Figure 3.18).

If, before entering the lens, the interstitial fluid contains near the top and the base of the layer different solutes that can react chemically, the convergence and intermixing inside the lens allow the reaction to occur, even though the solutes were initially separated in the vertical. If the equilibration length is short compared with the length of the lens, the reaction will occur predominantly within the lens; if it is large, it will occur predominantly downstream, over the depth interval of the mixed fluid.

Sedimentary structures that contain highly permeable continuous layers of approximately uniform thickness throughout the whole length may not exhibit this effect to any marked extent, since the fluid remains essentially confined to these layers. Only lenses of long but finite extent act as efficient fluid mixers.

3.10 FLOW TRANSIENTS

Diffusion of pressure

An important characteristic of any dynamical system is its intrinsic response time, the time that it takes to adjust to changes in the boundary conditions or the driving forces. If, for example, an earthquake suddenly changes the pressure distribution along a fracture surface, a seismic wave packet radiates outward and is gone, but the altered pressure remaining along the fracture alters the flow, at first locally but over larger distances as time proceeds. When the elapsed time is equal to the response time of the system, the flow is approaching its new steady state throughout. Changes in most other geological driving forces for interstitial rock flow – changes in configuration associated with erosion or sedimentation or changes in the internal temperature – occur much more slowly, and these might be considered a sequence of infinitesimal changes distributed over a long time. If the time scale over which the changes in driving forces occur is very long compared with the response time of the system, the flow pattern is always infinitesimally close to a steady state under the gradually evolving thermal or hydraulic forcing, and transient effects are negligible. It is clearly important to be able to assess simply the order of magnitude of this reponse time for fluid flow in a given geological structure and, in particular, how it varies with the size of the structure, its permeability, and other characteristics.

Following an impulsive disturbance, both the stress field in the rock and the pressure in the interstitial fluid are altered. The compressibility of most rock material is only about 5 percent of that of water, a useful tabulation having been given by Daly (1951). An approximate estimate of the ratio can be made by comparing the speeds of propagation c_p and c_s of seismic P- and S-waves, respectively, in rocks and sound in water. In rocks, the compressibility is $[\rho(c_p^2 - \frac{4}{3}c_s^2)]^{-1}$, where $c_p \sim 6$ km s^{-1} and $c_s \sim 3$ to 4 km s^{-1} in sandstone or limestone. The compressibility of water is $(\rho_w c^2)^{-1}$, where $c \sim 1.5$ km s^{-1}; these figures give a ratio of some 4 percent. Care must be taken, particularly in the case of saturated clay minerals, to distinguish between the compressibility of the matrix as a whole and that of the clay mineral itself; when the porosity is relatively large and the rigidity small, the compressibility of the overall matrix is dominated by that of the interstitial water. With this proviso, then, to a reasonable accuracy the rock material can be regarded as incompressible, whereas the interstitial fluid seeks to expand or contract following the pressure change. The porosity ϕ can then be considered constant, and the mass conservation equation (2.5.3) becomes

$$\phi \frac{\partial \rho}{\partial t} + \mathbf{u} \cdot \nabla \rho = -\rho \, \nabla \cdot \mathbf{u}. \qquad (3.10.1)$$

The adjustment process in the interstitial fluid is close to isothermal, so that density and pressure changes are related by the isothermal sound speed c_i:

$$dp = c_i^2 \, d\rho.$$

Consequently,

$$\phi \frac{\partial p}{\partial t} + \mathbf{u} \cdot \nabla p = -\rho c_i^2 \, \nabla \cdot \mathbf{u}. \qquad (3.10.2)$$

Suppose, for simplicity, that the rock is locally isotropic (the influence of anisotropy is readily included by use of the transformation (3.4.8)), so that

$$\nabla p = -\frac{\mu}{k}\mathbf{u}$$

and

$$\nabla \cdot \mathbf{u} = -\frac{k}{\mu} \nabla^2 p.$$

Equation (3.10.2) then becomes

$$\frac{\partial p}{\partial t} + \phi^{-1}\mathbf{u} \cdot \nabla p = \frac{k c_i^2}{\nu \phi} \nabla^2 p, \qquad (3.10.3)$$

which is a classical unsteady heat conduction or diffusion equation of the form (2.8.4) or (2.9.1) with no source term, but with a "pressure diffusivity"

$$\kappa_p = \frac{k c_i^2}{\nu \phi}. \qquad (3.10.4)$$

For water $c_i \sim 1.4 \times 10^3$ m s^{-1} and $\nu \sim 10^{-6}$ m^2 s^{-1}; with $\phi = 0.2$ and $k = 10^{-13}$ m^2, it is found that $\kappa_p = 1$ m^2 s^{-1}. For higher or lower permeabilities, the pressure diffusivity varies proportionately.

The advection term in (3.10.3) is almost always negligible. It is of the order $\phi^{-1}u\,\Delta p/l$, where Δp is the magnitude of the pressure disturbance over the length l, while the diffusion term is of order $\kappa_p\,\Delta p/l^2$. The diffusion term dominates if

$$\frac{ul}{\phi\kappa_p} \ll 1. \tag{3.10.5}$$

For example, with a flow speed of 30 cm $\text{yr}^{-1} = 10^{-8}$ m s^{-1} in a basin of scale l = 100 km and the figures above, the ratio on the left-hand side has the value 5×10^{-2} and (3.10.5) is satisfied quite strongly. In such cases (3.10.3) reduces to the simple diffusion equation

$$\frac{\partial p}{\partial t} = \kappa_p\,\nabla^2 p, \tag{3.10.6}$$

appropriate to a medium at rest. The analogy with heat conduction is very useful; the way that the internal pressure field changes in the interstitial fluid in response to an impulsive change is exactly analogous to the way that the internal temperature field would change in response to an impulsive temperature change at the boundary, with the thermal conductivity equal to κ_p. With a uniform medium, a pressure disturbance diffuses through a characteristic distance d in time t given by

$$d = (2\kappa_p t)^{1/2}, \tag{3.10.7}$$

as does heat, with the appropriate thermal diffusivity. Conversely, if d represents the dimension of the region, the time taken for a pressure disturbance to diffuse throughout, the response time, is characterized by

$$\tau = \frac{d^2}{2\kappa_p}. \tag{3.10.8}$$

When d = 1 m with the above value for κ_p, the response time is about half a second; when d = 1 km, it is about 6 days, and when d = 100 km, it is about 200 years.

As an example of estimating the response time of a permeable basin, consider the geometry of Figure 3.19 modeled by Ge and Garven (1988) in a detailed numerical study. In it, a highly conductive sandstone layer or carbonate aquifer is supposed to lie between much less conductive layers, and an impulsive pressure disturbance is applied along the left vertical margin. If one keeps in mind the thermal analogy, it is apparent that the pressure disturbance diffuses from the end rapidly along the conductive layer but only slowly in the insulating (or less conductive) layers. With d = 500 km and $\kappa_p = 0.3\ \text{m}^2\ \text{s}^{-1}$ appropriate for diffusion along the layer, the response time for diffusion from one end to the other is found from (3.10.8) to be about 4×10^{11} sec, or about 13,000 years. For diffusion of pressure vertically through the upper layer and release

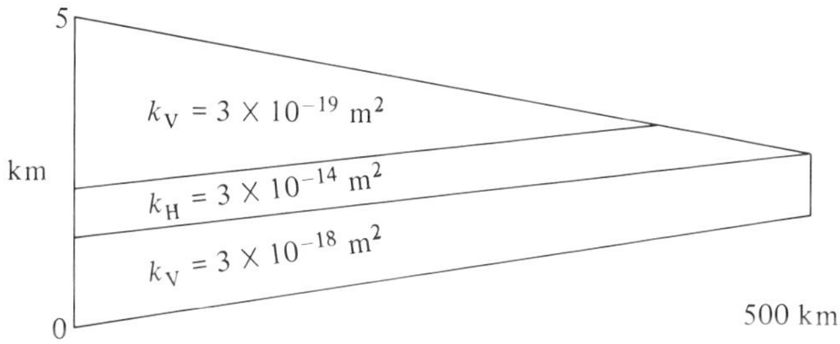

Figure 3.19. Basin geometry modeled by Ge and Garven (1988).

at the surface, the diffusion distance averages about 1 km; with $k_V = 3 \times 10^{-19}$ m^2, the pressure diffusivity is 3×10^{-6} m^2 s^{-1} and the response time is a little shorter, 1.6×10^{11} sec, or about 5,000 years. Ge and Garven's detailed calculation leads to an estimate for essentially *complete* relaxation of about 8,000 years, which agrees well with the simple order of magnitude estimates. Note, incidentally, that if the upper permeability were an order of magnitude greater than assumed here, the response time for vertical diffusion across this layer would be half an order of magnitude smaller. The pressure disturbance would diffuse upward to the surface before it reached the end of the more highly conductive, but long layer. The response time of the basin as a whole, then, would be in essence that of the upper layer alone.

The maintenance of overpressure in deeply buried strata over geological time scales requires confinement by deep layers of very low permeability, possibly combined with the formation of new water by chemical reaction. The time scale for pressure diffusion varies, according to (3.10.8), as the square of the thickness of the confining overburden; for a layer of depth 10 km with $k_V = 10^{-19}$ m^2, the relaxation time is some 2 million years.

Water-table oscillations

Another type of time-dependent flow can be generated by oscillations in the water table driven, for example, by the rise and fall of the tides on a beach or by seasonal changes in water level of a reservoir or lake. Close to the water's edge, the water table in the permeable region will rise and fall in response, with an amplitude that decreases with increasing distance from the shoreline, and it is of interest to estimate the distance shoreward over which these oscillations persist. In nature, the problem is complicated by the influence of capillarity at the interface between the water-saturated and unsaturated parts of the medium and these tend to immobilize it. Pressure differences of order $T/\rho\delta$ can be supported without

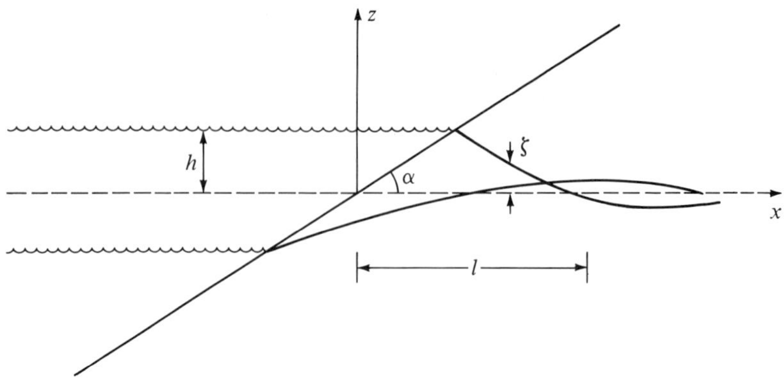

Figure 3.20. Water-table transients near a shoreline.

motion, where T is the surface tension and δ the pore diameter. An upper limit to the extent of the oscillations can be found by neglecting this effect – actual penetration distances will be less. Our primary concern here is with very permeable shoreline materials (e.g., sand) whose pressure diffusivity is of the order 10^3 to 10^4 m^2 s^{-1}, and for these, with even a high-frequency diurnal oscillation, the pressure diffusion scale is some 10 km at least. It is anticipated that the oscillations will die out over distances much smaller than this, and over these distances, diffusion of pressure is practically instantaneous – the compressibility effects of the previous section can be neglected. In the water-saturated region, then, $\nabla^2 p = 0$, whereas at the saturated–unsaturated interface $z = \zeta$, the pressure $\hat{p}(\zeta)$ can be taken as zero. In terms of values along the mean level $z = 0$ of Figure 3.20,

$$\hat{p}(\zeta) = \hat{p}(0) + \zeta \left.\frac{\partial \hat{p}}{\partial z}\right|_0 + \ldots = 0,$$

so that to the first order

$$\hat{p}(0) = -\zeta \left.\frac{\partial \hat{p}}{\partial z}\right|_0 = \zeta\left(\frac{\mu}{k}w + \rho g\right) \qquad \text{at} \quad z = 0, \qquad (3.10.9)$$

where

$$w_0 = \phi \frac{\partial \zeta}{\partial t} \qquad \text{on} \quad z = 0 \qquad (3.10.10)$$

is the vertical transport velocity across $z = 0$; the kinematic condition (3.3.2) has been approximated to the first order. On the right-hand side of (3.10.9) the first term in parentheses is generally small compared with the second (the vertical pressure gradient is always hydrostatic to zeroth order) and decreases with distance from the shore as the amplitude of oscillation decreases, so that this equation can be approximated by

$$\hat{p}(0) = \rho g \zeta \qquad \text{on} \quad z = 0. \tag{3.10.11}$$

Along the boundary edge AB, the water pressure

$$\begin{aligned} \hat{p} &= \rho g(h(t) - z) \qquad && \text{when} \quad z < h(t), \\ &= 0 && \text{when} \quad z > h(t). \end{aligned} \tag{3.10.12}$$

The two boundary conditions at $z = 0$ can conveniently be combined. From (3.10.11), (3.10.10), and the vertical component of Darcy's equation,

$$\frac{\partial \hat{p}(0)}{\partial t} = \rho g \frac{\partial \zeta}{\partial t} = \frac{\rho g}{\phi} w_0 = -\frac{\rho g k}{\phi \mu}\left(\frac{\partial \hat{p}}{\partial z} + \rho g\right). \tag{3.10.13}$$

In terms of the reduced pressure $p = \hat{p} + \rho g z$, the governing equations become

$$\nabla^2 p = 0,$$

$$\frac{\partial p}{\partial t} + \frac{gk}{\phi \nu}\frac{\partial p}{\partial z} = 0 \qquad \text{on} \quad z = 0, \tag{3.10.14}$$

and on AB,

$$\begin{aligned} p &= \rho g h(t) \qquad && \text{when} \quad z < h(t) \\ &= 0 && \text{when} \quad z > h(t). \end{aligned}$$

Even without a detailed solution (which is, in fact, untidy), these equations determine the length scale of penetration of the oscillation for a given radian frequency σ of water level oscillation. The problem is linear, and the last condition shows that the pressure perturbations near the shoreline scale in magnitude with $\rho g h_0$, where h_0 is the tidal (or seasonal) amplitude. The boundary condition on $z = 0$ defines the only intrinsic length scale in the problem, l_p, which therefore specifies the penetration distance,

$$l_p = \frac{gk}{\sigma \phi \nu}. \tag{3.10.15}$$

For diurnal tides on a sandy shore, $\sigma = 2\pi/(1 \text{ day}) = 7 \times 10^{-5}\ \text{s}^{-1}$, $k \sim 10^{-10}\ \text{m}^2$, so that $1 \sim 100$ m or so. For seasonal oscillations of water level in a lake adjoining an aquifer with $k \sim 10^{-14}\ \text{m}^2$, the corresponding distance is only a few meters.

If the upper region (above the x-axis in Figure 3.20) is saturated with fresh water, the dynamical problem is exactly the same, except that g is replaced by the reduced gravitational acceleration $(\nabla\rho/\rho)g$, where $\nabla\rho$ is the density difference between the underlying (perhaps saline) water and

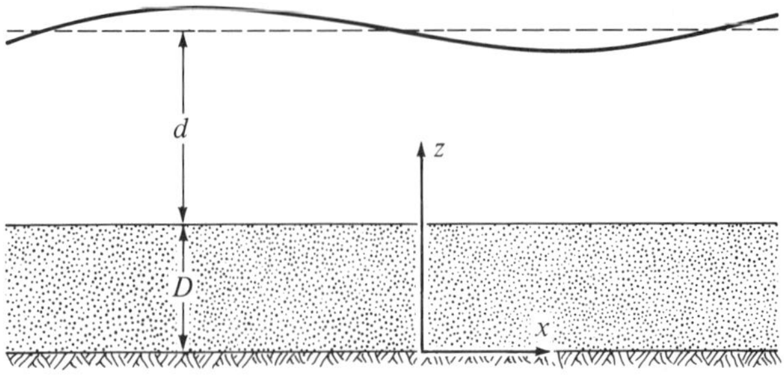

Figure 3.21. Water waves over a permeable bed.

that above. The penetration distance is then also reduced by the factor $\nabla\rho/\rho$ and is therefore generally small. Enhanced dispersion in the permeable medium, associated with tidal oscillations, is then limited to a very narrow strip along the shoreline.

Wave-induced flow in shallow banks

When long ocean waves or swell travel across a shallow sandy bed, the variations in pressure at the sand–water interface induce an oscillating flow in the sand that augments the diffusion of oxygen and chemical nutrients for biota inhabiting the substrate. If as illustrated in Figure 3.21 the water depth is d, the wave amplitude a, wavelength $\lambda = 2\pi/m$, and radiant frequency σ, the pressure at the top of the sand bed is (Phillips 1977)

$$p = \rho g a \frac{\cos(mx - \sigma t)}{\cosh md}. \qquad (3.10.16)$$

When md is large, that is, when the water depth is large compared with the wavelength, these pressure variations are exponentially small, but in shallow water, when md is of the order of unity or less, $\cosh md \sim 1$ and the variations in pressure are approximately the variations in hydrostatic head between the wave crests and troughs.

If the sand layer is of uniform depth D and overlies impermeable bedrock, then $\partial p/\partial z = 0$ at the bedrock surface $z = 0$. The pressure field inside the sand layer is readily found to be

$$p = \rho g a \frac{\cosh mz}{\cosh mD \cosh md} \cos(mx - \sigma t), \qquad (3.10.17)$$

consistent with the boundary condition (3.10.16) at $z = D$. The induced transport velocity components are then

$$u = -\frac{k}{\mu}\frac{\partial p}{\partial x} = \frac{gamk}{\nu}\frac{\cosh mz}{\cosh mD \cosh md}\sin(mx - \sigma t),$$

$$w = -\frac{k}{\mu}\frac{\partial p}{\partial z} = -\frac{gamk}{\nu}\frac{\sinh mz}{\cosh mD \cosh md}\cos(mx - \sigma t). \qquad (3.10.18)$$

Near the sand surface $z = D$, the root mean square vertical transport velocity, pumping in and out of the sand bed, is

$$(\bar{w}^2)^{1/2} = \frac{gamk}{2\nu}\frac{\tanh mD}{\cosh md}, \qquad (3.10.19)$$

and the root mean square interstitial fluid velocity is ϕ^{-1} times this. Note that with a given geometry, the interstitial speeds are proportional to the product of the hydraulic conductivity gk/ν and the maximum wave slope $am = 2\pi a/\lambda$. The macroscopic dispersion coefficient, or the effective diffusivity for dissolved species, is, from Section 2.4, of the order of the characteristic interstitial fluid velocity times the grain size l_0, so that

$$D \sim \frac{(am)}{2}\left(\frac{gk}{\nu}\right)\frac{l_0}{\phi}\frac{\tanh mD}{\cosh md}. \qquad (3.10.20)$$

Consider, for example, shallow water $md \ll 1$, $\cosh md \simeq 1$ and a deep sand bank $mD > 1$, $\tanh mD \simeq 1$. If the maximum wave slope $am \sim 0.1$ over sand with $\phi = 0.4$, $l_0 \sim 10^{-1}$ cm, and $k \sim 10^{-6}$ cm^2, then $K = gk/\nu = 10^{-1}$ cm s^{-1}, and the effective diffusivity is about 10^{-3} cm^2 s^{-1}, some two orders of magnitude greater than the molecular value.

3.11 DIFFUSION-INDUCED FLOW IN FISSURES OR FRACTURES

Hanor (1984) has pointed out that a noteworthy characteristic of oil-field waters in northern Louisiana and southern Arkansas is an almost linear increase in dissolved Na and Cl concentrations with depth. Associated with these gradients are diffusive fluxes, but in addition, if a density gradient resulting from variations in the concentration of dissolved salts exists along a fissure or fracture in otherwise impermeable rock, a slow internal circulation develops that can greatly augment the flow of solute along it, *even when there is no overall pressure or temperature gradient.* The reason for this circulation is evident from Figure 3.22. Suppose that the salinity and therefore density decrease upward along the fissure. If there is no flux of solute into the walls, the concentration gradient normal to the walls must vanish so that the isopycnals, the surfaces of constant density, are bent upward at the upper wall and downward at the lower one. At any horizontal level, then, the denser fluid will drift downward

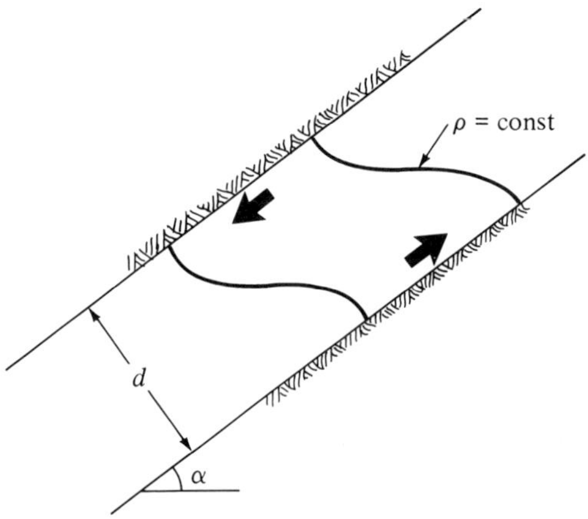

Figure 3.22. Diffusively induced flow in a fissure. Since the isopycnals at the walls are perpendicular to them, fluid at the upper wall is denser than that in the interior and at the lower wall is less dense. Fluid then rises along the lower wall and sinks along the upper one.

near the upper surface and the lighter fluid upward along the lower one until a balance is reached between longitudinal convection of solute and lateral diffusion. If the fissure is precisely vertical, the isopycnals remain horizontal and the effect disappears, but in all other cases, this kind of diffusively driven flow will occur.

A detailed analysis of the flow for a two-dimensional fissure inclined at an angle α to the horizontal has been given by Phillips (1970). The transport characteristics depend critically on the parameter

$$\gamma d = \left(\frac{1}{\nu \kappa_s} \frac{g}{\rho} \frac{\partial \rho}{\partial s} \right)^{1/4} d, \qquad (3.11.1)$$

where d is the fissure width, $\partial\rho/\partial s$ the density gradient along it, and κ_s the solute molecular diffusivity. The convective transport F_c of solute along the fissure is negligible when $\gamma d < (\sin\alpha)^{-1/4}$, but when γd is large,

$$F_c \simeq \kappa_s \frac{\partial \rho}{\partial s} d \cot^2 \alpha. \qquad (3.11.2)$$

The purely diffusive transport is just $\kappa_s(\partial\rho/\partial s)d$, so that the ratio of convective to diffusive transport is simply $\cot^2\alpha$ provided that $\sin\alpha > (\gamma d)^{-4}$. The *vertical* transports involve an additional $\sin\alpha$ term in both, but somewhat surprisingly, it is the gently sloping fissures that provide the greatest vertical transports! (Completely horizontal or vertical fissures have, of course, no convective transport.) For example, if a fractured region is saturated with a brine solution below and fresh water above, with a density gradient between them of 0.1 per 100 m, $(g/\rho)\, \partial\rho/\partial s = 10^{-2}\ \mathrm{s}^{-2}$; with $\nu \sim 10^{-2}$, $\kappa_s \sim 10^{-5}\ \mathrm{cm}^2\ \mathrm{s}^{-1}$, and $\gamma \sim 20\ \mathrm{cm}^{-1}$. For moderately small angles α, salt transport associated with diffusively induced flow dominates direct diffusion when the fissure width is greater than about 5×10^{-2} cm; when $\alpha = 5°$, it is 130 times as large and when $\alpha = 20°$, it is still larger by a factor of about 8.

4 General flow characteristics
Patterns of Reactions

4.1 SIMPLE REACTION TYPES

As water, carrying dissolved chemical species, moves through a permeable matrix or a network of fissures or fractures, a variety of chemical reactions can occur. The resulting patterns of dissolution, precipitation, and fabric alteration depend on the reaction kinetics and the influence on them of temperature, pressure, and other constituents, but in many geological situations, the controlling factor is the rate at which reactants in solution can be delivered to the reaction site by advection and diffusion in the flow. In these flow-controlled reactions, then, there are two sets of balances to be considered: the chemical balances that specify the nature of the reactions and the physical balances that specify the transports, sources, and sinks. Our primary concern in this book is with the physical balances, but their detailed form must reflect the chemistry, so that it is necessary to review briefly a few of the simplest and most common reaction types. In any discussion of chemical reactions, it is convenient to express concentration in terms of molarity, moles per unit volume, a mole being a mass unit equal to the molecular weight of the species involved, so that the ratios of moles produced or consumed per unit volume are simple integral multiples. In precise geochemical calculations, the molality (moles of solute per unit mass of *water*) is a more convenient quantity, but unless the temperature and pressure are near critical, the difference is minor. A very clear discussion of these matters is given in Garrels and Christ's classic book (1965).

The simplest type of chemical reaction is pure dissolution or precipitation, in which a solid mineral S dissociates reversibly into dissolved constituents or ions D_1 and D_2. This can be expressed as

$$S \rightleftharpoons D_1 + D_2 \tag{4.1.1}$$

and is exemplified by the dissolution or precipitation of halite:

$$NaCl \rightleftharpoons Na^+ + Cl^-.$$

At equilibrium, where saturated solution and solid can coexist indefinitely, the saturation or equilibrium concentration c_{e1} of ion D_1 is a function of temperature, pressure, and also the concentration of ion D_2 – an

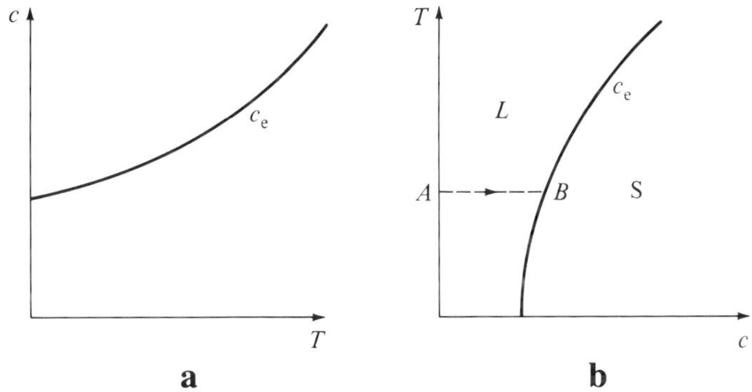

Figure 4.1. (a) A conventional solubility diagram and (b) the corresponding phase diagram, with the solid phase to the right and below the curve c_e, and liquid solution to the left and above.

excess of chloride ions drives the reaction above to the left and reduces the equilibrium concentration of sodium ions. With the other quantities held constant, the variation of saturation concentration with temperature can be represented by a solubility diagram, as in Figure 4.1a, or by an equivalent phase diagram, as in Figure 4.1b, in which c represents the molar concentration of one of the dissolved ions (say, D_1) in the interstitial fluid. At a given temperature, when the ion concentration $c = c_e$, solid and liquid phases can coexist in equilibrium; when $c < c_e$ the liquid (dissolved) state is in equilibrium only when no solid is present. Water may enter the soluble matrix with an initially negligible concentration of the ions involved, and its state is therefore represented by point A of Figure 4.1b; in the presence of the solute, it is no longer in equilibrium and the point representing its state moves to the right toward B, toward saturation, as simultaneously the physical processes of advection and diffusion occur.

The *rate* at which dissolution occurs depends, as described in Section 2.9, on (i) the proportion of active interstitial area per unit volume of the fabric and so on the mass fraction r of dissolving mineral present in the fabric and on (ii) the degree of disequilibrium, which can be expressed as the dimensionless ratio

$$b = \frac{c_e - c}{c_e} \tag{4.1.2}$$

where $c_e = c_e(T, p, c_1, c_2, \ldots)$ is the saturation or equilibrium concentration and c_1, c_2, and so on represent the molar concentration of the other dissolved species influencing the reaction per unit volume of the interstitial fluid (including possibly the hydrogen ion concentration, the pH).

Thus, the source term per unit volume of fluid for either ionic constituent in the species balance equation can be written generally as

$$Q_c = \gamma c_e f\left(r, \frac{c_e - c}{c_e}\right), \tag{4.1.3}$$

where the "rate constant" γ has dimensions $(\text{time})^{-1}$ and $f = 0$ when the interstitial fluid is in equilibrium, $c = c_e$, or when there is no dissolving mineral left, $r = 0$. Without loss of generality, we can take $\partial f/\partial b = 1$ when $b = 0$; this merely fixes the numerical value of γ, which is otherwise arbitrary. In pure dissolution, then, the balance equation for either dissolved ion (assuming that it is not involved in another simultaneous reaction) is then

$$\frac{\partial c}{\partial t} + \phi^{-1}\mathbf{u}\cdot\nabla c - D\,\nabla^2 c = \gamma c_e f\left(r, \frac{c_e - c}{c_e}\right), \tag{4.1.4}$$

or in terms of the degree of undersaturation b of (4.1.2),

$$\frac{\partial b}{\partial t} + \phi^{-1}\mathbf{u}\cdot\nabla b - D\,\nabla^2 b = -\gamma f(r, b). \tag{4.1.5}$$

An interesting asymmetry can be expected when precipitation occurs from a locally supersaturated solution ($b > 1$) since the *rate* of precipitation does not necesssarily vanish when the mass fraction r of mineral density already precipitated does–a finite density of active nucleation sites for precipitation must be expected to exist even when $r = 0$, so that in this case, $f(0, b) \neq 0$.

A second type of reaction involves the dissolution of two solid minerals S_1 and S_2 into a third S_3 and a single dissolved species D:

$$S_1 + S_2 \rightleftharpoons S_3 + D. \tag{4.1.6}$$

This is exemplified by the formation of wollastonite (calcium pyroxene) by the reaction between calcite and silica:

$$\underset{\text{Calcite}}{CaCO_3} + \underset{\text{Silica}}{SiO_2} \rightleftharpoons \underset{\text{Wollastonite}}{CaSiO_3} + CO_{2(aq.)} \tag{4.1.7}$$

The silica may exist as impurities of quartz silt or chert in limestone, or the calcite may be an impurity in sandstone. In either case, the phase diagram can be represented in terms of the molar concentration c of the dissolved solute (CO_2 in the example). At a given pressure and concentration of other ions influencing the reaction, the phase diagram for the reaction is as illustrated in Figure 4.2, where the equilibrium boundary $T = T_e(c)$ separates the (temperature–concentration) region below, where $S_1 + S_2$ is stable, from that above, where $D + S_3$, is stable. Along the

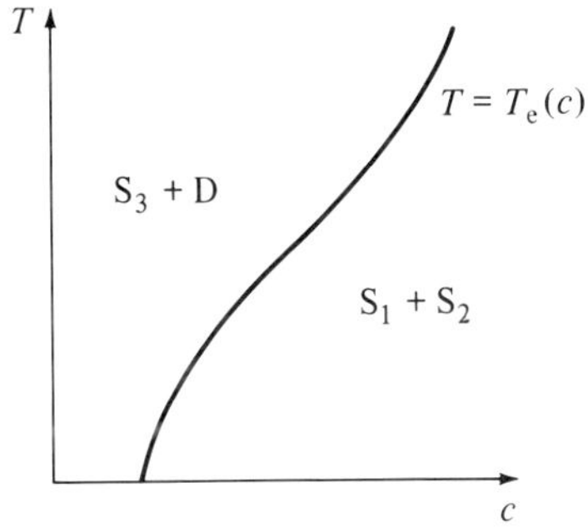

Figure 4.2. Phase diagram for the reaction $S_1 + S_2 \rightleftharpoons S_3 + D$; c represents the concentration of the dissolved solute D. All four species can coexist in equilibrium along the curve $T = T_e(c)$.

line $T = T_e(c)$ the four can coexist in equilibrium with aqueous interstitial fluid. The equilibrium is stable: At a fixed c a small increase in temperature above the line causes the balance in the reaction (4.1.6) to move to the right as long as both S_1 and S_2 are present, generating additional dissolved solute. The point representing the state of the system in Figure 4.2 then also moves to the right, back to the equilibrium curve $T_e(c)$.

Here again, the rate of generation of the dissolved species D can be represented by the expression (4.1.3), where r now represents the molar fraction of the *less abundant* mineral S_1 or S_2. If $r = 0$, only one of the two minerals on the left of (4.1.6) or (4.1.7) is present, the reaction cannot proceed, and no dissolved solute is generated. Consequently, the molar concentration of solute D is specified by

$$\frac{\partial c}{\partial t} + \phi^{-1}\mathbf{u}\cdot\nabla c - D\,\nabla^2 c = Q_c$$
$$= \gamma c_e f(r, b), \qquad (4.1.8)$$

where, again, $f = 0$ when either $b = 0$ or $r = 0$. In this case, however, as each mole of the solute is generated in the reaction and carried away by the moving fluid, 1 mole of the mineral S_3 is being deposited and left behind. In (4.1.8), Q_c represents moles of solute generated per unit volume of the fluid, so that ϕQ_c is the number per unit volume of the fabric, and if s represents the moles of the immobile mineral S_3 (e.g., wollastonite) per unit volume of matrix,

$$\frac{\partial s}{\partial t} = \phi Q_c = \phi\gamma c_e f(r, b), \qquad (4.1.9)$$

where the functions $f(r, b)$ in (4.1.8) and (4.1.9) are the same. We also have equations for the rate of disappearance of minerals 1 and 2 in the

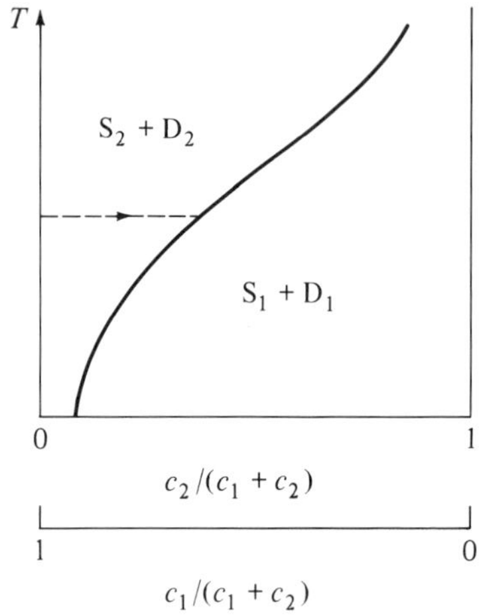

Figure 4.3. Phase diagram for the reaction $S_1 + D_1 \rightleftharpoons S_2 + D_2$, where c_1 and c_2 represent the molar concentrations of D_1 and D_2 in solution.

reaction. For the reaction-limiting, less abundant species, with r moles per unit volume of fabric,

$$\frac{\partial r}{\partial t} = -\phi Q_c = -\phi \gamma c_e f(r, b). \tag{4.1.10}$$

If the variation of r is significant, (4.1.10) and (4.1.8) are a pair of coupled equations for c (or, equivalently, b) and r whose solution in turn specifies the rate of deposition of S_3 by means of equation (4.1.9).

A third type of reaction is replacement, in which aqueous ions D_1 entering the matrix replace another in the solid mineral S_1, the new dissolved species D_2 being advected away. This can be represented as

$$S_1 + D_1 \rightleftharpoons S_2 + D_2, \tag{4.1.11}$$

an example being the dolomitization replacement reaction

$$2\,CaCO_3 + Mg^{2+} \rightleftharpoons CaMg(CO_3)_2 + Ca^{2+}, \tag{4.1.12a}$$

or albitization of potassium feldspar,

$$KAlSi_3O_8 + Na^+ \rightleftharpoons NaAlSi_3O_8 + K^+. \tag{4.1.12b}$$

The phase diagram for this type of reaction is illustrated in Figure 4.3, the abscissa representing the molar concentration of ion 2 as a fraction of the total molar concentration $c_1 + c_2$. When the concentration c_1 of entering fluid is zero, S_1 is stable at all temperatures since there is nothing to react with, so that the equilibrium curve slopes upward to the right. If, however, seawater containing Mg^{2+} but little Ca^{2+} in this example enters a calcite bed, the reaction proceeds to the right of (4.1.11) or (4.1.12), magnesium is lost from solution as calcium is added, and the reaction point moves horizontally to the right in Figure 4.3 until equilibrium is attained at that particular temperature. Again, as in the case of pure solution, the equilibrium concentration of solute D_1, c_{e1}, is a function not only of temperature and pressure but also of the concentration of D_2.

The flow equations for the dissolved species are similar to those already described, but now the source terms for the rates of change of the molar concentrations of D_1 and D_2 in the interstitial fluid are equal and opposite. If the respective concentrations are designated by c_1 and c_2,

$$\begin{aligned} \frac{\partial c_1}{\partial t} + \phi^{-1}\mathbf{u}\cdot\nabla c_1 - D\,\nabla^2 c_1 &= -\,Q_c, \\ \frac{\partial c_2}{\partial t} + \phi^{-1}\mathbf{u}\cdot\nabla c_2 - D\,\nabla^2 c_2 &= Q_c. \end{aligned} \tag{4.1.13}$$

When the reaction proceeds to the right (as it usually does when magnesium-rich seawater enters calcite), the sink of ions from solution is dependent on the amount of mineral S_1 with which they can react and on the extent of disequilibrium, so that again

$$Q_c = \gamma c_{e1} f_1(r_1, b), \tag{4.1.14}$$

where r_1 now represents (twice, in the case of (4.1.12), since two molecules of $CaCO_3$ are involved in the reaction) the moles per unit volume of S_1 as a fraction of the total solid moles per unit volume and b_1 is the degree of disequilibrium of D_1 (>0 if $c_1 < c_{e1}$). The rate of deposition of S_2 per unit volume of the matrix is again given by an equation such as (4.1.9).

In more complex reactions, similar balance equations can be written for the concentrations of dissolved species, the main elaboration occurring in the specification of the equilibrium concentration c_e of each, which depends on the concentration of the others in the reaction. Note that in the species balance equations, the left-hand sides represent *rates of change* of concentration, so that the addition of an excess of any particular species to the whole system affects only the source term through the equilibrium concentration c_e.

When water is produced or consumed by the reaction, the fluid flow divergence balances the rate of fluid generation,

$$\nabla\cdot\mathbf{u} = \phi n_w V Q_c, \tag{4.1.15}$$

where n_w is the number of moles of fluid generated per mole of solute, V is the fluid molar volume, and the full advection term $\nabla\cdot(\mathbf{u}c)$ must be used in the species balance equation. When fluid is continually being driven through the matrix by either hydraulic or thermal forces, the incremental water addition may be insignificant, and in such cases, the modification to the flow velocity is slight, but there are important situations to be discussed later, involving dehydration reactions in only slightly permeable rocks, in which this represents the major water source.

In these expressions for the source terms Q_c, the reaction rates and the forms of the functions f for particular reactions are often not known to any degree of accuracy, but fortunately, even in this state of ignorance, useful results can be obtained, as demonstrated in Section 4.4. Some limiting cases, however, do provide valuable simplification. When r is small, that is, when the reacting or *dissolving* mineral constitutes only a small fraction of the whole matrix, f can be expressed as the first (linear) term in a Taylor series expansion about $r = 0$ (for which $f = 0$), so that

$$f(r, b) \simeq rg(b). \tag{4.1.16}$$

Near equilibrium, again, b is small, so that

$$Q_c \simeq \gamma c_e rb = \gamma r(c_e - c). \tag{4.1.17}$$

In *precipitation,* however, $f(r, b)$ does not generally vanish when $r = 0$, so that

$$f(r, b) \simeq f(0, b) = f(b), \text{ say.} \tag{4.1.18}$$

When r is small and if the solution is near equilibrium,

$$Q_c \simeq \gamma(c_e - c). \tag{4.1.19}$$

Finally, when r is large (nearly unity), the reacting solid constitutes almost the whole matrix, and the dependence of $f(r, b)$ on r can probably be neglected.

4.2 THE APPROACH TO EQUILIBRIUM AT DISSOLUTION OR REACTION BOUNDARIES

As Palciauskas and Domenico (1976) have pointed out, when interstitial fluid enters the matrix from an adjacent water body or moves across a mineralogical boundary, the concentration of dissolved species moves toward a local equilibrium with the solid phase, but the extent of the region of disequilibrium and the spatial distribution of the mineralogical changes that can result depend on the speed of flow, the reaction rates, and the solute diffusivity. The relationships among these quantities can, in fact, be used in field situations to constrain possible values of the least certain of them (frequently the reaction rate) as well as to provide criteria under which local equilibrium between the fluid and the matrix would or would not be expected.

Let us first consider the dissolution or reaction of a major constituent of a rock fabric (say, calcite in a limestone bed) when water, unsaturated with respect to the dissolving species, enters the fabric across a sharp interface. There is an important distinction between this situation and the dissolution of minor constituents, which may become depleted and disappear without much alteration of the *physical* characteristics of the matrix–this case is discussed in Section 4.3. We shall assume here that the time over which the reaction or dissolution process has continued is not so large that the cumulative effect has caused substantial changes in either the porosity or mass per unit volume of the solid undergoing reaction or dissolving, so that the influence of the mass fraction r in the equations of the preceding section can be ignored. Quantitative upper limits for this time scale are given later in this section.

In a frame of reference with the x-direction normal to the water–matrix interface (or the mineralogical boundary) and directed into the matrix, the concentration c of ions in the interstitial fluid generated by reaction or solution is specified by equation (4.1.4) or in terms of $b = (c_e - c)/c_e$ by (4.1.5), which in unidirectional flow and with r = const reduces to

$$\frac{\partial b}{\partial t} + \phi^{-1}u\frac{\partial b}{\partial x} - D\frac{\partial^2 b}{\partial x^2} = -\gamma f(b), \qquad (4.2.1)$$

where u is the velocity component normal to the interface. A steady-state distribution of concentration of the interstitial fluid is reached in a time of order γ^{-1} (short on a geological time scale if the reactions are to be of interest), so that the first term in (4.2.1) can be ignored and the distribution $b = b(x)$ is specified by

$$\phi^{-1}u\frac{\partial b}{\partial x} - D\frac{\partial^2 b}{\partial x^2} = -\gamma f(b). \qquad (4.2.2)$$

At the interface $x = 0$, $b = b_0$, say, determined by the ionic concentration of the incoming fluid; if this is zero, $b_0 = 1$. Far from the interface, the interstitial fluid eventually attains equilibrium, so that $b \to 0$ as $x \to \infty$. In general, $1 \geqslant b_0 \geqslant b \geqslant 0$.

Now, the rate of generation of ions by dissolution in the reaction is in (4.2.2) balanced by the rate at which they are swept away by advection (represented by the first term) or by the rate of diffusion (the second) or by both at comparable rates. Note that $f(b)$ is not generally equal to but is of the order of b, so that relative to the dissolution term, the respective orders of magnitude of the other terms are

$$\frac{u}{\phi\gamma l_e}, \qquad \frac{D}{\gamma l_e^2}, \qquad 1, \qquad (4.2.3)$$

where l_e is the differential length scale in the equation, or the length scale over which equilibrium is approached. Now, if

$$\frac{u}{\phi\gamma l_e} \gg \frac{D}{\gamma l_e^2},$$

or

$$u \gg \frac{\phi D}{l_e}, \tag{4.2.4}$$

the diffusion term is minor and the balance must be found between dissolution and advection, so that $u/\phi\gamma l_e \sim 1$ and

$$l_e \sim \frac{u}{\phi\gamma}. \tag{4.2.5}$$

The equilibration length l_e is then the distance the solute moves at the interstitial fluid velocity u/ϕ through the matrix in the reaction time γ^{-1}. The condition under which this interpretation is valid is found by substitution of (4.2.5) into (4.2.4), namely,

$$u \gg \phi(D\gamma)^{1/2}. \tag{4.2.6}$$

If, however, $u \ll \phi D/l_e$, the balance in equation (4.2.2) is between the last two terms and in (4.2.3) $D/\gamma l_e^2 \sim 1$; advection is negligible and the dissolving ions are removed by diffusion. The equilibration length is then

$$l_e \sim (D/\gamma)^{1/2}, \tag{4.2.7}$$

when $u \ll \phi D/l_e = \phi(D\gamma)^{1/2}$. In this circumstance, the equilibration length can be interpreted as the diffusion distance $(Dt)^{1/2}$ in the reaction time $t = \gamma^{-1}$. Finally, when $u \simeq \phi(D\gamma)^{1/2}$, both advection and diffusion are important and the expressions (4.2.5) and (4.2.7) become identical in order of magnitude.

The significance of the equilibration length is akin to that of the response time to pressure changes discussed in Section 3.10. It expresses the characteristic distance in the flow direction over which the interstitial fluid remains in disequilibrium with respect to a particular constituent. At distances from the interface large compared with l_e, equilibrium is in essence established. More generally, if the temperature, pressure, or mineralogy varies smoothly in a formation over length scales that are large compared with l_e, given by (4.2.5) or (4.2.7) as appropriate, then the interstitial fluid is everywhere close to *local* equilibrium with respect to the constituent concerned. If the changes occur over distances in the flow direction that are small compared with the equilibration length, the disequilibrium persists over a distance of order l_e.

This length also, of course, expresses the distance from the interface over which the mineral alterations have occurred as a result of the reaction. As long as the reacting mineral remains a major constituent of the fabric, the rate of increase of any solid product of the reaction is given by (4.1.9):

$$\frac{\partial s}{\partial t} = \phi \gamma c_e f(b), \tag{4.2.8}$$

which is greatest at the interface when $f(b)$ is order unity and decreases to zero beyond the equilibration distance since $b \to 0$ and $f(b) \to 0$. In this equation, s represents the molar concentration of reaction product; after a time t_r this concentration near the interface is of order $\phi\gamma c_e t_r$. If r represents the moles per unit fabric volume of the initial mineral undergoing reaction, the condition that r has not changed much is that

$$\phi \gamma c_e t_r \ll r,$$

or that the time of reaction

$$t_r \ll (\phi\gamma)^{-1}\left(\frac{r}{c_e}\right). \tag{4.2.9}$$

When the original mineral constituted a substantial fraction of the matrix, the solid molar fraction r is generally larger than the molar interstitial fluid concentration c_e by many powers of 10, so that these distributions persist for times enormously longer than the reaction time γ^{-1}. When, finally, the original mineral approaches depletion, a reaction front develops, as described in the next section. Another pertinent time scale, that of alteration of the porosity of the fabric, is discussed in Section 4.7.

Numerical values of the equilibration length can vary widely as a result of the wide range of reaction and flow rates encountered in different situations. For example, suppose that the fluid velocity $u \sim 10^{-6}$ cm s^{-1} = 30 cm yr^{-1} and the diffusivity $D = 10^{-5}$ cm^2 s^{-1} in a rock with porosity $\phi = 0.2$. For a fairly slow reaction, $\gamma \sim 10^{-10}$ s^{-1}, say, and $\phi(D\gamma)^{1/2} \sim 1.4 \times 10^{-8}$ cm s^{-1}, which is very much less than u, so that (4.2.6) is satisfied easily. The equilibration length is advectively determined and given by (4.2.5):

$$l_e = \frac{u}{\phi\gamma} = 5.10^4 \text{ cm} = 500 \text{ m}.$$

In a faster reaction, $\gamma \sim 10^{-6}$ s^{-1}, say, and with the same flow speed, $\phi(D\gamma)^{1/2} \sim 1.4 \times 10^{-6}$ cm s^{-1} $\sim u$, so that diffusion and advection effects are comparable. On substitution, the equilibration length is given by (4.2.7) as 2 cm and by (4.2.5) as 5 cm. Note that when $u \ll \phi(D\gamma)^{1/2}$, the equilibration length is independent of the flow speed, so that in this last example, if the flow speed decreased almost to zero, the equilibration length would remain the same, a few centimeters.

The same formulation applies to the concentration of dissolved species in the ocean bed when, as a result of sedimentation, the interface is rising at speed u. If the interstitial fluid is not moving through the matrix, then in a frame of reference attached to the interface the distribution of concentration is again represented by (4.2.2) but with $\phi^{-1}u$ replaced by the sedimentation rate u. The above expressions can then be used, but with $\phi = 1$. Reactions more complex than simple dissolution may involve more than one equilibration length – steps in the overall reaction that are more rapid produce concentration distributions that change rapidly near the surface, whereas slower reactions have an influence distributed over greater depths. Detailed profiles are given by Toth and Lerman (1977) and Lerman (1979).

It may have been noticed that these considerations have required only very general information on the form of the source function f in equation (4.2.1), so that the concept of equilibration length and the relations defining it are independent of the detailed kinetics. In the case of a first-order reaction in which $f \equiv b$, detailed solutions can be found easily. The degree of undersaturation b as a function of distance x from the interface is given from (4.2.2) as

$$b = b_0 \exp(-x/l_e),$$

where

$$l_e = \left\{ \frac{u}{2\phi D} - \left[\left(\frac{u}{2\phi D} \right)^2 + \frac{\gamma}{\phi D} \right]^{1/2} \right\}^{-1}, \qquad (4.2.10)$$

which reduces to (4.2.5) and (4.2.7) in the appropriate limits.

When the flow is across a distinct mineralogical boundary, solute dissolved from the downstream assemblage may be diffused back across the interface against the advection. On the upstream side, the concentration of solute in the interstitial fluid can be shown (from (4.2.1) with $f = 0$; no dissolution on this side) to decrease exponentially with distance upstream over the length scale

$$l_d = \frac{\phi D}{u}.$$

For the more rapid fluid velocities, when $u > \phi\kappa_s/l_0$ or so, where l_0 is the grain size, macroscopic dispersion is dominant, $D \sim \bar{v} l_0$ and $l_d \sim l_0$: The back diffusion distance is only of the order of the grain size. When $u \ll \phi\kappa_s/l_0$ and molecular diffusion dominates, $D \sim \kappa_s$ and $l_d = \phi\kappa_s/u$. With the generic numbers $\phi = 0.2$, $\kappa_s = 10^{-5}$ cm^2 s^{-1}, $u = 10^{-6}$ cm s^{-1}, and $l_d = 2$ cm.

4.3 ISOTHERMAL REACTION FRONTS

When, as considered in the preceding section, fluid enters a permeable matrix or crosses a mineralogical boundary and undergoes reaction, reaction fronts, as described by Lichtner (1985) and others, may originate at the boundary and propagate away from it. In a reaction of a type similar to (4.1.6) or (4.1.11) the original mineral near the interface becomes replaced by the reaction product after a time $(\phi\gamma)^{-1}(r/c_e)$, where r is the original number of moles in the solid mineral per unit volume of the fabric or, in the case of (4.1.6), that of the less abundant solid. When the original mineral is depleted, the reaction cannot, of course, continue and the fluid concentration of D remains at its entering value until it reaches a position farther downstream where S_1 in the case of (4.1.11) or both S_1 and S_2 in the case of (4.1.6) are present and reaction occurs. As a result, the zone of reaction gradually moves downstream, leaving behind the solid product. The solute resulting from the reaction is carried away downstream, so that if the temperature of the region subsequently changes or the flow ceases, the reaction does not reverse and the pattern remains.

Although, as we shall see, many important aspects of these reaction fronts are independent of the detailed reaction kinetics, their initial formation and propagation can be illustrated by a numerical calculation in which the reaction kinetics is (necessarily) prescribed. Suppose for the purpose of this illustration that no water is *produced* by the reaction and that the source of solute per unit volume of fluid is linearly proportional to both the degree of disequilibrium b and the moles s per unit volume of the reacting solid at any point as a fraction of the moles initially present, s_0, say. Thus,

$$Q_c = \gamma(c - c_e)\left(\frac{s}{s_0}\right) = \gamma c_e bS,$$

where $S = s/s_0$ and $b = (c_e - c)/c_e$. When the x-direction is chosen normal to the interface at $x = 0$ and u is the component of the transport velocity in this direction,

$$\frac{\partial c}{\partial t} + \phi^{-1}u\frac{\partial c}{\partial x} - D\frac{\partial^2 c}{\partial x^2} = \gamma c_e bS,$$

or, as in (4.1.5),

$$\frac{\partial b}{\partial t} + \phi^{-1}u\frac{\partial b}{\partial x} - D\frac{\partial^2 b}{\partial x^2} = -\gamma bS. \tag{4.3.1}$$

For the solid reactant, if n_s is the moles of solid consumed per mole of solute produced,

$$\frac{\partial s}{\partial t} = -n_s\phi\gamma c_e bS,$$

or

$$\frac{\partial S}{\partial t} = -\left(\frac{n_s\phi\gamma c_e}{s_0}\right)bS. \tag{4.3.2}$$

Imagine that the fluid starts moving through the matrix at time $t = 0$, so that initially $S = s/s_0 = 1$ throughout. The incoming fluid has $c = 0$ at $x = 0$ (so that $b = 1$), and far from the interface it is in equilibrium with the matrix, so that $c = c_e$ and $b = 0$.

The simultaneous solution of the coupled nonlinear equations (4.3.1) and (4.3.2) specifies the evolution of the distribution $s(x, t) = s_0 S(x, t)$ of the solid reactant and that of the concentration $c(x, t)$ or the degree of disequilibrium of the interstitial fluid. The equations can be rewritten in a form suitable for computation by taking as a time scale the quantity $(n_s\phi\gamma)^{-1}(s_0/c_e)$ (c.f. equation (4.2.9)) and as a length scale the advective equilibration length $l_e = u/\phi\gamma$. With

$$t = (n_s\phi\gamma)^{-1}(s_0/c_e)\tau,$$
$$x = l_e\xi = (u/\phi\gamma)\xi,$$

the equations reduce to

$$\left(\frac{n_s\phi c_e}{s_0}\right)\frac{\partial b}{\partial \tau} + \frac{\partial b}{\partial \xi} - \left(\frac{D\phi^2\gamma}{u^2}\right)\frac{\partial^2 b}{\partial \xi^2} = -bS,$$
$$\frac{\partial S}{\partial \tau} = -bS. \qquad (4.3.3)$$

In this pair of dimensionless equations, the terms without coefficients are numerically of order unity. The diffusion term is small under the condition (4.2.6); the quantity $\phi c_e/s_0$, representing the ratio of the number of moles in solution at equilibrium in the interstitial fluid to the number of moles of solid reactant initially present, is also tiny. Neither of these terms has any perceptible influence on the numerical solutions.

The evolving patterns, calculated from (4.3.3), are shown in Figure 4.4; fluid crosses the interface at $x = 0$ from the left, and the curves represent the profiles of s/s_0 and c/c_e after equal time intervals $(s_0/n_s\phi\gamma c_e)$. Initially, as shown by the upper line and the curve marked 0, $s = s_0$ throughout and the interstitial fluid concentration c attains its equilibrium value c_e within a distance of order l_e, as described in the preceding section. As the solid reactant is depleted near the interface, the distribution of c flattens; by the time $\tau = 3$ or 4, the solid reactant is in essence zero at the interface, and both profiles have assumed an elongated S-shape, which then moves through the matrix to the right without further change in shape. The thickness of the front is a few times the equilibration length l_e; it moves a distance l_e in the time interval $(s_0/n_s\phi\gamma c_e)$, so that its speed

$$U = \frac{u}{\phi\gamma}\cdot\frac{n_s\phi\gamma c_e}{s_0} = un_s\left(\frac{c_e}{s_0}\right), \qquad (4.3.4)$$

which is independent of the reaction rate γ. The steadily propagating profiles of s/s_0 and c/c_e can, in fact, be found analytically from (4.3.3) with the terms in parentheses neglected; it is found that

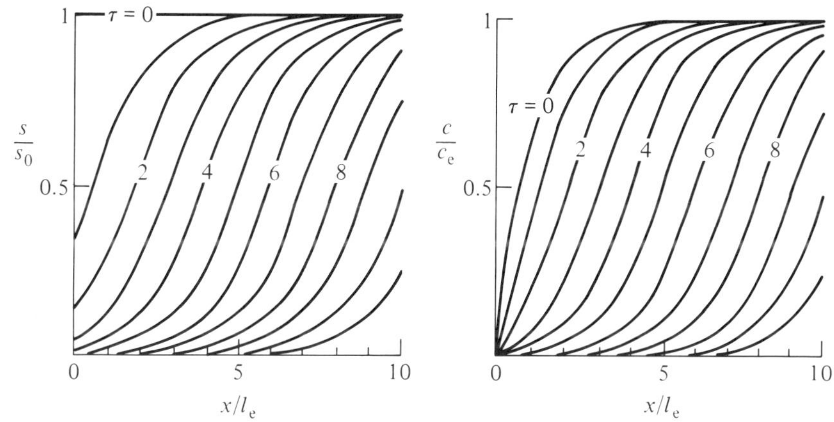

Figure 4.4. The development of an isothermal reaction front. Initially, at time $\tau = 0$ the concentration of mineral s, is uniform and the interstitial fluid approaches equilibrium with the mineral over the equilibration length l_e. Profiles are shown for equal time intervals $s_0/n_s\phi\gamma c_e$; by the time $\tau = 2$ the solid mineral is becoming seriously depleted near the interface, and by $\tau = 4$ the depletion is virtually complete. The profiles of c/c_e and s/s_0 have become essentially identical, and they subsequently move away to the right without further change in shape as the reaction front propagates through the fabric.

$$\frac{c}{c_e} = \frac{s}{s_0} = \left\{1 - \exp\left[\frac{-\phi\gamma(x - Ut)}{u}\right]\right\}^{-1}, \tag{4.3.5}$$

where U is given by (4.3.4). These are the profiles in Figure 4.4 when τ is greater than about 4.

The propagation of a reaction zone or front can be described more generally without the specification of the reaction kinetics involved in this example. In many reactions, water is produced or consumed, so that (4.1.15) must be used in place of the incompressibility condition,

$$\nabla \cdot \mathbf{u} = \frac{\partial u}{\partial x} = \phi n_w V Q_c, \tag{4.3.6}$$

where n_w is the moles of fluid generated per mole of solute and V the fluid molar volume. The concentration of solute (often CO_2) produced by the reaction is specified by

$$\frac{\partial c}{\partial t} + \phi^{-1}\frac{\partial}{\partial x}(uc) - D\frac{\partial^2 c}{\partial x^2} = Q_c, \tag{4.3.7}$$

where again the x-direction is normal to the zone and u the transport velocity component in this direction. The rate of change of molar concentration of the solid mineral consumed by the reaction is, from (4.1.9),

$$\frac{\partial s}{\partial t} = -n_s\phi Q_c, \tag{4.3.8}$$

where n_s is the moles of solid consumed per mole of solute produced in the reaction. Since Q_c is a function only of c and s at any particular temperature and pressure, these equations allow solutions of the form

$$c = c(x - Ut), \qquad s = s(x - Ut), \tag{4.3.9}$$

representing distributions that move through the matrix with speed U to be determined. Substitution of (4.3.9) into (4.3.7) and (4.3.8) gives

$$\begin{aligned} u' &= \phi n_w V Q_c, \\ -Uc' + \phi^{-1}(uc)' - Dc'' &= Q_c, \\ Us' &= n_s\phi Q_c, \end{aligned} \tag{4.3.10}$$

where the primes denote differentiation with respect to the variable $(x - Ut)$. Now far upstream of the reaction zone, the concentration c of the dissolved solute is whatever it happened to be, c_0, in the entering fluid and the (less abundant) mineral has already been consumed, so that $c \to c_0$, $s \to 0$, $u \to u_0$ when $\xi = (x - Ut) \to -\infty$. Also, ahead or downstream of the reaction zone, the solute is in equilibrium and the mineral is unconsumed so that $c \to c_e$, $s \to s_0$, its initial value, and $u \to u_1$, the inflow velocity augmented by any fluid generated in the reaction.

The equations (4.3.10) can be integrated across the reaction zone to obtain overall balances, and with the use of these conditions, it appears that

$$\begin{aligned} u_1 - u_0 &= \phi n_w V \mathcal{Q}, \\ -U(c_e - c_0) + \phi^{-1}(u_1c_e - u_0c_0) &= \mathcal{Q}, \\ s_0 &= \phi(n_s/U)\mathcal{Q}, \end{aligned} \tag{4.3.11}$$

where

$$\mathcal{Q} = \int Q_c \, d\xi, \tag{4.3.12}$$

the total rate of production of solute per unit area of the reaction zone normal to the flow. The elimination of $\mathcal{Q}$ and u_1 from the set (4.3.11) gives for the speed of front propagation

$$\frac{U}{u_0} = \frac{c_e - c_0}{\phi(c_e - c_0) + (s_0/n_s)(1 - n_w V c_e)}, \tag{4.3.13}$$

or, when $s_0 \gg n_s\phi c_e$,

$$\frac{U}{u_0} = \frac{n_s(c_e - c_0)}{s_0(1 - n_w V c_e)}. \tag{4.3.14}$$

This simplifies further when the rate of generation of fluid in the reaction is zero or can be neglected. Then $n_w = 0$ and we recover (4.3.4),

$$\frac{U}{u_0} = \frac{n_s \Delta c}{s_0}; \qquad (4.3.15)$$

the propagation speed of the reaction zone is smaller than the flow velocity by the ratio of the change in molar concentration of the interstitial fluid to the initial concentration of reacting mineral, then multiplied by the moles of solid consumed per mole of solute produced.

The fluid–rock ratio r in these reactions is defined as the volume of fluid with which unit volume of rock reacted. In a time interval dt the reaction zone moves ahead a distance $U\,dt$ and the volume of fluid moving into it is $(u_0 - U)\,dt$, so that

$$r = \frac{u_0}{U} - 1. \qquad (4.3.16)$$

Equivalently,

$$\frac{U}{u_0} = \frac{1}{r + 1}.$$

The fluid–rock ratio can be expressed from (4.3.14) as

$$\begin{aligned} r &= \frac{s_0(1 - n_w V c_e)}{n_s(c_e - c_0)} - 1 \\ &= \left(\frac{\phi \mathcal{Q}}{U}\right) \frac{1 - n_w V c_e}{c_e - c_0} - 1. \end{aligned}$$

Now $\phi \mathcal{Q}\, dt$ represents the total production of volatile solute in the zone in the time interval dt and the volume of reaction is $U\,dt$ so that $\phi \mathcal{Q}/U$ represents the moles n_v of volatile species released per unit volume of rock during metamorphosis and

$$r = \frac{n_v(1 - n_w V c_e)}{c_e - c_0} - 1, \qquad (4.3.17)$$

which is equivalent to the expression given by Ferry (1987). If during a certain time interval t, the reaction front moves a distance $d = Ut$, the total volume of interstitial fluid that has crossed unit area of the fabric over this time interval is $u_0 t = (r + l)d$. If the time over which the reaction has proceeded can be estimated and the distance of propagation measured, the fluid–rock ratio provides a local flow meter for the interstitial flow.

The fluid–rock ratio is sometimes calculated assuming that the incoming fluid is pure water, so that $c_0 = 0$. This is often geologically unreasonable, and the resulting values (generally of order unity or less) cannot be interpreted in terms of the volume of fluid that has passed through the sample. If r_0 represents the number calculated in this way, then the true fluid–rock ratio r is such that

$$r + 1 = \left(\frac{c_e}{\Delta c}\right)(r_0 + 1),$$

where $\Delta c = c_e - c_0$ is the concentration change across the front, and the total volume of fluid that has crossed unit area of the fabric is $(c_e/\Delta c)(r_0 + 1)d$, where d is the distance that the front has progressed.

It has been asserted (Lichtner 1988) that traveling wave solutions are not possible when diffusion is taken into account, but this does not seem to be so. Diffusion merely tends to spread the reaction zone, and the specific shape and thickness of the profiles in the reaction zone or front depend on both the reaction kinetics and the influence of solute diffusion. The role of the latter can be evaluated by consideration of the second equation of (4.3.10). The diffusion term Dc'' can at most be of the same order as the advection term $\phi^{-1}(uc)'$, and this occurs when the thickness l of the zone is

$$l \sim \frac{\phi D}{u_0}.$$

If macroscopic dispersion dominates molecular diffusion, $D \sim \phi^{-1}ul_0$ from Section 2.4, where l_0 is the grain size, so that dispersion becomes important only when the thickness of the front approaches this scale – to the continuum approximation the reaction region shrinks to a discontinuity, a true front. If the actual thickness is at all greater than this, diffusion can be neglected. Note that if Q_c is represented as $\gamma c_e f(s, b)$, where, as before, the dimensionless function is of order unity, then

$$\frac{\phi l \gamma}{u} = 0(1),$$

and the zone thickness is given by the advective equilibration length, whatever the detailed reaction kinetics may be. The faster the reaction, then, the thinner the reaction zone; for relatively rapid reactions and slow transport velocities it approaches the grain size.

These results apply also to dissolution fronts of a distributed, relatively sparse mineral in a less soluble permeable matrix or to the reaction fronts between mineral assemblages in which, say,

$$S_1 + S_2 + S_3 \rightleftharpoons S_2 + S_3 + S_4 + D,$$

where the interstitial fluid, initially buffered by and in equilibrium with the minerals 1, 2, and 3, exhausts one of them at a given temperature and moves to a new equilibrium with 2, 3, and 4 at a different concentration.

4.4 GRADIENT REACTIONS

The reaction fronts described in the preceding section arise at mineralogical boundaries or interfaces and propagate in the flow direction; they separate an unaltered region ahead from that behind where the reaction has proceeded to a different equilibrium between the altered host rock and the interstitial fluid. Flow-controlled reactions of another type, termed gradient reactions, can occur simultaneously throughout the whole matrix or along systems of cracks at rates proportional to the flow speed and the gradients of temperature and pressure in the flow direction.

As fluid moves into a fabric or from one mineral assemblage to another, beyond the equilibration length l_e it is close to equilibrium with the new assemblage at the ambient pressure and temperature. If, however, these variables change along the flow path, the equilibrium concentration also changes; thus, even though the composition of interstitial fluid may remain the same in time at each point, it varies along the flow trajectory so that solute is continually being added to or taken from solution. As a consequence, the mineral composition is gradually altered along the flow path. In general, the equilibrium concentration of the interstitial fluid $c_e = c_e(T, p, c_1, c_2, \ldots)$, where $c_1, c_2, \ldots$ express the dissolved concentrations of other species (including hydrogen ions) that may influence the reaction, and is specified by the chemistry, not the flow. Wood and Hewett (1982) realized that the fluid concentration balance equation then becomes a prescription for the source terms, the rate of addition of species to the solution (moles per unit volume of the fabric per unit time) needed to maintain the equilibrium at all points in the moving fluid:

$$\phi Q_c = \mathbf{u} \cdot \nabla c_e - \phi D \, \nabla^2 c_e. \tag{4.4.1}$$

As was shown earlier, the macroscopic dispersion coefficient is of the order $\phi^{-1} u l_0$, where l_0 is the grain size, which is *very* much smaller than the distances l over which the temperature and the pressure vary significantly, so that in these situations, dispersive effects are quite negligible. Accordingly,

$$\begin{aligned} \phi Q_c &= \mathbf{u} \cdot \nabla c_e(T, p, c_1, c_2, \ldots) \\ &= \mathbf{u} \cdot \left\{ \frac{\partial c_e}{\partial T} \nabla T + \frac{\partial c_e}{\partial p} \nabla p + \sum_r \frac{\partial c_e}{\partial c_r} \nabla c_r \right\}, \end{aligned} \tag{4.4.2}$$

from the chain rule for partial differentiation. The summation is over whatever other species $r = 1, 2, \ldots$ influence the equilibrium. Even though the total fluid pressure may be close to lithostatic, the local *gradient* in total pressure in a connected region is primarily hydrostatic, so that $\mathbf{u} \cdot \nabla p = -\rho g \mathbf{l} \cdot \mathbf{u} = -\rho g w$, where $\mathbf{l}$ is a vertical unit vector and w the vertical component of transport velocity. Equation (4.4.2) becomes

$$\phi Q_c = \left(\frac{\partial c_e}{\partial T}\right) \mathbf{u} \cdot \nabla T - \left(\frac{\partial c_e}{\partial p}\right) \rho g w + \sum_r \left(\frac{\partial c_e}{\partial c_r}\right) \mathbf{u} \cdot \nabla c_r \qquad (4.4.3)$$

The last term, involving the concentration gradients of other solutes, can be simplified considerably. If they do not enter the reaction but may be purely inhibitory, their concentration remains the same along the flow path and this term vanishes. If any of them do, a species balance can be written for this species with a source term equal to $-\phi Q_c$ times the moles consumed per mole of the solute (4.4.1) generated:

$$\mathbf{u} \cdot \nabla c_r = -m_r \phi Q_c,$$

and the rate of mineralization $Q_m = -\phi Q_c$ becomes

$$Q_m = -\left\{1 + \sum_r m_r \left(\frac{\partial c_e}{\partial c_r}\right)\right\}^{-1} \left\{\left(\frac{\partial c_e}{\partial T}\right) \mathbf{u} \cdot \nabla T - \left(\frac{\partial c_e}{\partial p}\right) \rho g w\right\}. \qquad (4.4.4)$$

Several comments should be made concerning these expressions. At first sight it is surprising that the rate at which reaction or dissolution or precipitation occurs is independent of the chemical kinetics, provided only that the reaction rate is sufficiently rapid to make the equilibration length small compared with the size of the region. The determining factor is physical—the rate at which fluid moves through the system, the variation in fluxes of species into and out of each volume element at the matrix. Moreover, Q_c represents the rate of addition of the species to the solution (mass per unit volume of the fluid per unit time) so that in dissolution, $Q_c > 0$. In precipitation, $Q_c < 0$. In the calcite dissolution reaction

$$CaCO_3 + CO_2 + H_2O \rightleftharpoons Ca^{2+} + 2HCO_3^-,$$

for example, (4.4.4) specifies the rate of generation of Ca^{2+} ions or the rate of dissolution of $CaCO_3$; CO_2 in solution is consumed by the reaction and the term $\partial c_e/\partial c_1 = (\partial c_1/\partial c_e)^{-1}$ represents the variations in equilibrium concentration of Ca^{2+} ions with concentration of dissolved CO_2. The equilibrium concentration of Ca^{2+} is smaller when CO_2 is depleted, so that $\partial c_e/\partial c_1 > 0$. Many experimental values for the variation of saturation solubilities with temperature and total pressure are summarized by Barnes (1979). For example, in the case of calcite with a partial pressure of CO_2 of 1 atm, $(\partial c_e/\partial T) \sim -4 \times 10^{-6}$ °C^{-1} at about 100°C; $(\partial c_e/\partial p)$ is positive, but its magnitude is insufficient to overcome the increase in solubility of a rising fluid moving down the temperature gradient. The sum of the two flow terms in (4.4.4) is therefore positive, and dissolution will tend to occur. In the case of quartz, both $(\partial c_e/\partial T)$ and $(\partial c_e/\partial p)$ are positive, so that in a rising fluid both terms in the second factor of (4.4.4) are negative and precipitation occurs as described, for example, by Haszeldine, Samson, and Cornford (1984) and McBride (1989). The relative magnitude of the temperature and pressure effects depends not only on these coefficients, but also on the direction of flow. In purely horizontal flow, the pressure effect disappears and Q_c is proportional to the horizontal temperature gradient. The rate of reaction is in any event proportional to the flow speed, so that the rates of fabric alteration are greatest when the flow is concentrated by lenses or layers of relatively high permeability, whatever their orientation.

In many reactions, the pressure term in (4.4.4) appears to be subordinate to the temperature term, and the *pattern* of mineral alteration within a given structure is determined primarily by the distribution of $\mathbf{u} \cdot \nabla T$* associated with the flow. As suggested to me by Dr. James Wood, this quantity can conveniently be called the *rock alteration index,* or RAI. Distributions of this index are found for various flow patterns in later chapters of this book, regions where it is relatively high indicating sites of more rapid alteration by gradient reactions and, over a given time interval, sites of denser mineral accumulation or depletion.

Fault zones, particularly dilatational fault jogs connecting offset planes along which shear displacement has occurred, provide especially favorable sites for gradient reactions. Newly opened dilatational cracks attract fluid through flow focusing and allow concentrated, relatively high fluid velocities across isotherms. Sibson (1987) has recognized the importance of these fault structures in mineral deposition, citing examples from the Camp Bird vein system, Colorado, and the Chuquicamata copper deposit in Chile. Sibson, Robert, and Poulsen (1988) likewise associate the formation of many mesothermal gold–quartz deposits in high-angle fault veins to fluid discharges following fracture.

A few simplifications of (4.4.4) are appropriate in particular circumstances. When the multiplicative factor is simply unity and the isotherms are horizontal, $\mathbf{u} \cdot \nabla T = -wG$, where w is the vertical component of the transport velocity and $G = -\partial T/\partial z$, the geothermal gradient. Equation (4.4.4) then becomes

$$Q_m = w\left\{G\left(\frac{\partial c_e}{\partial T}\right) + \rho g\left(\frac{\partial c_e}{\partial p}\right)\right\}. \qquad (4.4.5)$$

Over a given time interval, the total moles of mineral deposited by the reaction per unit volume of the matrix is therefore

* This is read "*u*-dot-grad-*T*."

$$M = V\left\{G\left(\frac{\partial c_e}{\partial T}\right) + \rho g\left(\frac{\partial c_e}{\partial p}\right)\right\}, \tag{4.4.6}$$

where V is the *total volume of fluid per unit area* that has moved vertically through the matrix in this time interval, whether it is brief or long on a geological time scale. If two different reactions are proceeding simultaneously, the number of moles of each is the ratio of the respective factors on the right of (4.4.6).

It is evident that observations of mineral deposition M allow us to find the total volume of fluid per unit area that has moved through the fabric (vertically if the isotherms are horizontal) by means of equation (4.4.6). Note that this is *not* the fluid–rock ratio discussed earlier, having dimensions of length rather than being a pure number, but it is an important measure of reaction progress. If ϕ represents the porosity, then $\phi^{-1}V$ is the distance that fluid elements have moved since the reaction began; if this is large compared with the dimensions of the flow, either the reaction must have been sustained by a continuous supply of fluid from outside the region or the flow must involve convective recirculation.

Fluid–rock ratios as defined earlier can, of course, be calculated in mineral assemblages that have been altered by gradient reactions, but the results are difficult to interpret accurately in this case. For example, if we neglect any generation of water in the reaction and if $n_v(\propto M)$ represents the moles of volatiles generated per unit volume in the alteration, we define

$$r_0 = n_v/c_e$$

(as is usual, supposing the incoming fluid to be pure water). Then, in a temperature gradient

$$n_v = \left(\frac{\partial c_e}{\partial T}\right) \int \mathbf{u} \cdot \nabla T \, dt,$$

the integration being over past time. If the flow and temperature fields are unchanged over the time t_0,

$$r_0 = n_v/c_e = \frac{1}{c_e}\left(\frac{\partial c_e}{\partial T}\right)(\mathbf{u} \cdot \nabla T)t_0,$$

which is proportional to the time interval over which the reactions took place. Although it is difficult to ascribe significance to the numerical values of r_0 in gradient reactions, the *ratios* of values of r_0 at different points in a given lithography and temperature field do represent ratios of mean fluid velocities across isotherms, a useful conclusion exploited in Section 6.7.

In summary, in gradient reactions the interstitial fluid remains close to equilibrium with the suite of minerals in the surrounding fabric, but as time goes by, the relative abundance of these minerals changes, some increasing and some decreasing as the reaction continues. The process is ubiquitous, occurring throughout those parts of the fabric where the fluid flow crosses isotherms or isobars. As long as all the minerals are present, the interstitial solution is buffered at this equilibrium value for the ambient temperature and pressure. However, if, locally, one of the minerals involved in the reaction, say S_1, disappears entirely, the interstitial fluid there moves to a new equilibrium among the remainder of the minerals present. As this fluid moves into an adjacent region where S_1 is not yet depleted, a reaction front will develop between the now distinct mineralogical regions and start to propagate as described in the preceding section (see also Hewett, 1986).

When the Peclet number of a hydraulically driven flow or the Rayleigh number of a convective flow is large, however, the isotherm pattern is substantially modified by the flow. In interior regions, as described in Section 3.5, the temperature becomes almost constant along the streamlines and $\mathbf{u} \cdot \nabla T = 0$, so that gradient reactions are produced predominantly by the variation of c_e with pressure p as fluid moves across isobars and are correspondingly slower. Distributions of the rock alteration index, associated with the variation of c_e with temperature T, then tend to be confined to regions near the boundaries of a permeable domain and particularly near the surface, where fluid may rise through the water table down a geothermal gradient that is locally large. Some examples of this are given in Chapters 6 and 7.

When the reaction rates are so low or the flow speeds so high that the equilibration length l_e is large compared with the length scale l of the flow, the concentration of that species does not change much from the initial value (c_0) in the fluid entering the region and will remain either unsaturated or supersaturated throughout. The reaction rate

$$Q_c = \gamma c_e f\left(\frac{c_e - c_0}{c_e}\right) \tag{4.4.7}$$

is then independent of the flow speed but may vary throughout the matrix as the reaction rate γ and the local equilibrium concentration c_e vary, largely with temperature. Since, however, the temperature fields are characteristically much more diffuse than flow fields (which may be focused by lenses, inclusions, or fractures), the patterns of fabric alteration in this case are also much more diffuse. The rates of alteration are independent of the distribution and magnitude of the flow, provided only that it is sufficient to ensure that $l_e \gg l$.

4.5 MIXING ZONES

Alterations in mineral composition or precipitation or dissolution can also occur by the mixing of different interstitial waters. One common example is afforded by a coastal salt wedge, where seawater infiltrates beneath the freshwater outflow of a coastal aquifer and mixing occurs along its upper edge. Fluid diffusion or dispersion is necessarily involved, and this is greatly augmented by the presence of lenses or more permeable layers, as described in Section 3.9. If one is concerned with the structural details of the reaction patterns, it is necessary to resolve the flow details on the scale of the lenses; if one is content with a description of only the general characteristics of the region, it may be adequate to represent the mixing in terms of a megadispersion coefficient

$$D \sim \phi^{-1} U l$$

consistent with (3.9.8), where l is the length scale of the lenses. It should be remembered, though, that this is a fairly coarse approximation. The overall mixing behavior can be represented only asymptotically in terms of a dispersion coefficient when the lens scale is small compared with the scale of the overall flow, and this may not be so in many applications. Usually, however, the lens structures or the distributions of permeability inhomogeneities are not known in detail, and this approximation may be the best available.

In any event, when the equilibration length l_e is small, the source term is again specified by (4.4.1), but now the length scale over which the fluid composition varies may be small (the lens thickness when D is the usual macroscopic dispersivity or the mixing zone thickness with D, the megadispersion coefficient, when we do not choose to resolve the individual lens flows). The equilibrium concentration c_e of any particular species may depend on the local mixing ratio of the two fluids, and it will be seen that reactions occur when c_e is other than a linear function of this ratio.

For the sake of definiteness, consider the dissolution or precipitation of calcium carbonate in a coastal freshwater, saltwater mixing zone. The equilibrium concentration of dissolved $CaCO_3$ in the fresh water of the aquifer is generally less than that in seawater, and although the concentration of dissolved NaCl may not enter the dissolution process directly, the salinity s is a useful index of the mixing ratio. The equilibrium concentration c_e is therefore a function of s (Figure 4.5). When seawater and fresh water, each initially saturated with respect to $CaCO_3$, are mixed, the resulting concentration of $CaCO_3$ lies along the straight line joining $(c_e)_s$ and $(c_e)_f$ in this figure, and unless c_e is precisely linear in s, the mixture is unsaturated over a certain range or ranges of the mixing ratio s and supersaturated over others, or possibly undersaturated or supersa-

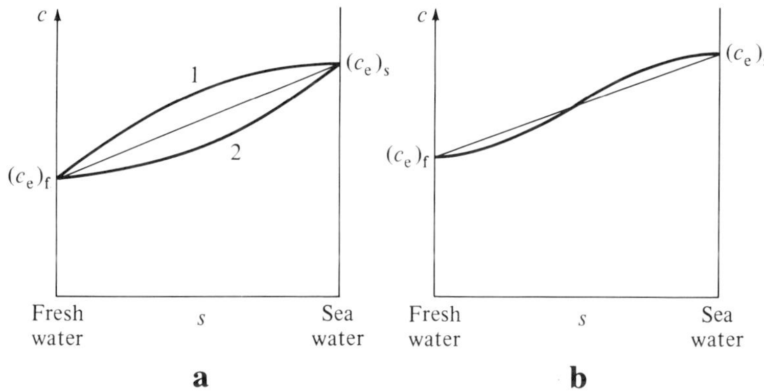

Figure 4.5. When seawater and fresh water, both saturated with respect to $CaCO_3$ (say), are mixed, the resulting concentration lies along the straight line joining $(c_e)_s$ and $(c_e)_f$. If $c_e(s)$ lies above the line ($\partial^2 c_e/\partial s^2 < 0$, curve 1 in the left-hand figure), the solution is unsaturated with respect to $CaCO_3$ and dissolution occurs. When it is below, $\partial^2 c_e/\partial s^2 > 0$, as in curve 2; it is then supersaturated and precipitation occurs. If the curvature of $c(s)$ changes as in part (b), dissolution occurs at high salinities and precipitation at low.

turated over the whole range if the curvature of $c_e(s)$ is everywhere concave downward or upward (Figure 4.5a). When the solution is undersaturated, we would expect local dissolution; when supersaturated, precipitation.

The rate of dissolution of $CaCO_3$ per unit volume of fabric is specified by (4.4.1),

$$Q_s = \mathbf{u} \cdot \nabla c_e - \phi D \, \nabla^2 c_e, \tag{4.5.1}$$

where now $c_e = c_e(s)$ and D is either the macroscopic or megadispersion coefficient as described earlier. The distribution of salinity s is governed by a similar advection–diffusion balance but without a source term:

$$0 = \mathbf{u} \cdot \nabla s - \phi D \, \nabla^2 s. \tag{4.5.2}$$

Now, since $c_e = c_e(s)$,

$$\nabla c_e = \left(\frac{\partial c_e}{\partial s}\right) \nabla s,$$

and

$$\nabla^2 c_e = \left(\frac{\partial^2 c_e}{\partial s^2}\right) (\nabla s)^2 + \left(\frac{\partial c_e}{\partial s}\right) \nabla^2 s$$

from the chain rule for partial differentiation. Consequently,

$$
\begin{aligned}
Q_s &= \left(\frac{\partial c_e}{\partial s}\right)\{\mathbf{u}\cdot\nabla s - \phi D\,\nabla^2 s\} - \phi D\left(\frac{\partial^2 c_e}{\partial s^2}\right)(\nabla s)^2 \\
&= -\phi D\left(\frac{\partial^2 c_e}{\partial s^2}\right)(\nabla s)^2, \qquad (4.5.3)
\end{aligned}
$$

in virtue of (4.5.2).

This very simple and important result shows clearly (a) that the rate of dissolution when $(\partial^2 c_e/\partial s^2) < 0$ or precipitation, $(\partial^2 c_e/\partial s^2) > 0$, is proportional to the curvature of the function $c_e(s)$, which is determined by the solution kinetics; (b) that the rates are proportional to the *square* of the salinity gradient and are therefore greatest when this gradient is largest, at the sea-level outflow in the salt wedge situation; and (c) that they increase linearly with the flow velocity, through the linear dependence of D on U. The formation of sea-level caverns in the limestone Yucatan aquifer can be interpreted in these terms, the time required for their formation being estimated in Section 4.7.

More generally, when fluid in an aquifer is drawn into a more permeable lenticular inclusion as described in Section 3.9, any vertical gradient in interstitial fluid concentration is amplified by the focusing ratio G, as is the flow speed. The reaction rate is, then, from (4.5.3) increased by the factor G^3, which can be very large. If $\partial^2 c_e/\partial s^2 \neq 0$, *mixing zone reactions are especially concentrated* in more permeable layers – the rate of gradient reactions and the speed of advance of isothermal reaction fronts increase only linearly with the flow velocity.

A similar kind of expression describes the rate of precipitation of sulfides from black smokers, sea-floor vents from tectonic divergence zones. In this case, $c_e = c_e(s, T)$, and the conservation equations for sulfide concentration, salinity, and temperature in the mixing region in the turbulent flow above the vent are the usual fluid equations,

$$
\begin{aligned}
Q_s &= \mathbf{u}\cdot\nabla c_e - D\,\nabla^2 c_e, \\
0 &= \mathbf{u}\cdot\nabla s - D\,\nabla^2 s, \\
0 &= \mathbf{u}\cdot\nabla T - D\,\nabla^2 T, \qquad (4.5.4)
\end{aligned}
$$

where $\mathbf{u}(\mathbf{x})$ is the mean velocity distribution in the water above the vent and D the *turbulent* diffusivity, of order ul, where l is the width of the jet. The rate of precipitation per unit volume of water is found by similar steps to be

$$
Q_s = -D\left\{\left(\frac{\partial^2 c_e}{\partial s^2}\right)(\nabla s)^2 + \left(\frac{\partial^2 c_e}{\partial T^2}\right)(\nabla T)^2\right\}. \qquad (4.5.5)
$$

Measurements sufficiently detailed to test this result apparently are not yet available.

4.6 MIXING AND REACTION PATTERNS FROM FAULTS AND FRACTURES

Fault and fracture zones provide pathways for fluid transport through otherwise impermeable layers and, in already permeable zones, produce planes or surfaces of high permeability that attract focusing of fluid flow in a manner similar to that described in Section 3.8. This local flow enhancement in turn may accelerate the processes of deposition or dissolution that become part of the geological record; deposition leads to veins of mineralization and dissolution to enlargement of the interstices and possibly increased flow rates.

The rate of deposition in gradient reactions (say) within an individual fracture can be established by a slight modification of the analysis of Section 4.4. When the equilibration length is small, the source term for dissolved minerals in the interstitial fluid Q_c (negative for deposition) is $\mathbf{v} \cdot \nabla c_e$, where $\mathbf{v}$ is the interstitial fluid velocity, or $v\ \partial c_e/\partial x$, with x measured along the fracture in the flow direction. The rate of deposition of solid mineral per unit area of the fracture is then $-q\ \partial c_e/\partial x$, where q is the volume flux along it per unit distance transverse to the flow $[L^2T^{-1}]$. Consequently, the expressions of Section 4.4 can be used in this context also, but with slightly different interpretations: $-Q_s$ now represents the rate of deposition per unit area of the fracture and $\mathbf{u}$ the volume flux per unit transverse distance or, in (4.4.5) w the vertical volume flux per unit distance. In (4.4.6), M is moles deposited per unit area of the fracture and V the total volume of fluid per unit transverse distance that has passed through.

If one averages over a network of fractures, the original interpretations of Section 4.4 are recovered. The overall rate of mineralization, for a given transport velocity or fluid volume flux per unit area, is independent of the number of individual fractures, but the fewer there are, the greater the volume transport along each and, consequently, the greater the rate of mineralization along each.

Venting from fractures into permeable layers

When a fracture traverses a relatively impermeable layer between two more permeable ones, it opens a fluid pathway that allows mixing of interstitial fluids from one region to the other and provides the opportunity for chemical reactions between the entering fluid and its new environment. Burley, Mullis, and Matter (1989) suggest that in the Tartan reservoir of the UK North Sea oil field, faults were primary conduits for migrating hot fluids that mixed with sulfate brine in the structures above. In the fracture, the flow is, in the main, simply along it; in the permeable layers above and below, it is a combination of whatever layer flow there

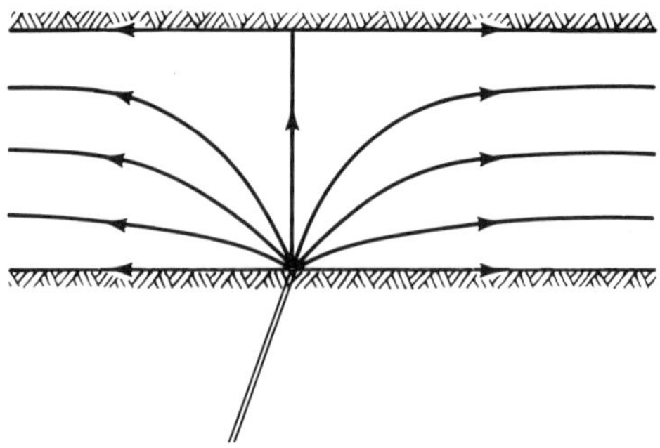

Figure 4.6. Streamlines of flow issuing from a linear fracture into a permeable layer with no superimposed ambient pressure gradient.

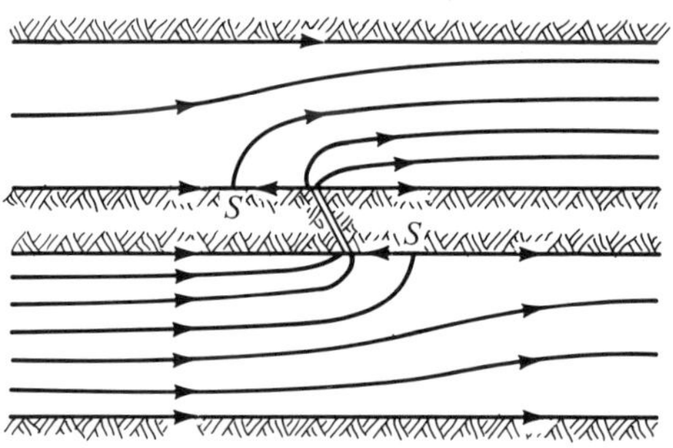

Figure 4.7. Streamlines of flow when fluid moves through a linear fracture from one permeable layer to another.

may be approaching the fracture, together with a fluid source (at the discharge end of the fracture) or a fluid sink (at the entry end). With no such flow (or if it is parallel to the fracture), the streamlines from a linear fracture are as illustrated in Figure 4.6 when the layer permeability is isotropic. (If $k_H \neq k_V$, the horizontal scale is stretched by the ratio $(k_H/k_V)^{1/2}$, as described in Section 3.4.) At a sink, the streamline pattern is the same, but the flow reversed. Near a discharge line, the velocity $q/2\pi r$ is radially outward, where r is the distance from the orifice. If, however, the fluid in the permeable layer has a velocity component U normal to the fracture, the entering fluid is swept downstream. When the flux per unit width q issuing from the fault is less than the volume flux per unit width approaching, Uh, stagnation points S develop upstream of the exit line and downstream of the entry line where the two flow components balance (Figure 4.7), the distance between the stagnation point and the fracture being approximately $q/2\pi U$. The streamline dividing the discharged fluid from the oncoming stream rises asymptotically to occupy

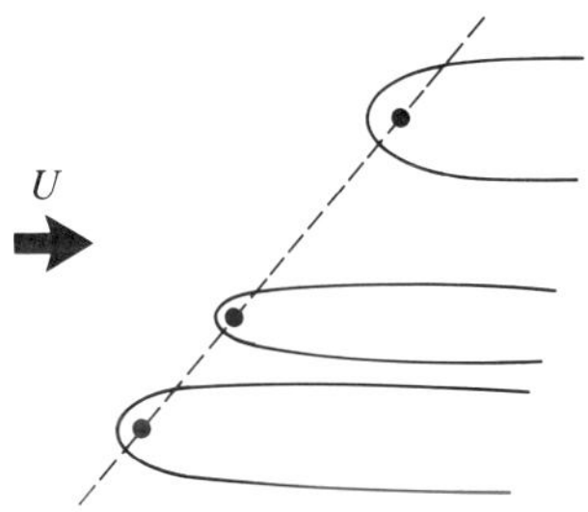

Figure 4.8. Fluid boundaries in the plane of the interface between a permeable layer and an impermeable one with flow from a series of microvents. The undisturbed interstitial fluid velocity in the permeable layer is U.

the fraction $q/(q + Uh)$ of the depth of the upper layer; the fluid is drawn from the same fraction of the top of the lower layer with the appropriate values of U and h.

When the fracture contains a linear array of microvents or separate interstitial pathways, the discharge and entry flows are three-dimensional. If Q is the volume flux from each orifice, the velocity perturbation near the orifice is described by a simple source flow with a radially outward velocity $Q/4\pi r^2$; the upstream stagnation point in a discharging orifice, separating the fracture fluid from that oncoming, is at a distance $(Q/4\pi U)^{1/2}$ from the orifice. When the volume flux per unit exposed length of the fracture nQ (where n is the number of orifices per unit length) is small compared with the oncoming volume flux in the layer Uh, the trailing regions of orifice fluid (in the absence of dispersion – we consider this shortly) are semicylindrical (Figure 4.8) with a radius that can be found by continuity considerations to be

$$r_0 = \left(\frac{2Q}{\pi(U + nQ/h)}\right)^{1/2} \approx \left(\frac{2Q}{\pi U}\right)^{1/2} \quad \text{when} \quad nQ \ll Uh.$$

This is $2\sqrt{2} \approx 2.8$ times the distance of the stagnation point upstream of the orifice.

When the permeability of the upper layer is anisotropic, the flow patterns are compressed in the direction across the layer relative to that in the plane of the layer by the factor $(k_H/k_V)^{1/2}$, but in a volume-preserving way. Thus, the scale of the efflux flow in the normal direction is reduced by the factor $(k_H/k_V)^{-1/3}$ and stretched by the factor $(k_H/k_V)^{1/6}$ in planes parallel to the bedding.

Flow in fractures can potentially disturb the local geothermal gradient, but in most cases the effect seems significant only when the flow rates are very large, as in geothermal regions. If a fracture diverts the fraction

α of the flow in a permeable layer, the Peclet number q/κ (see equation (3.5.11)), measuring the relative importance of heat convection versus conduction, is smaller by the fraction α than that of Uh/κ in the host layer; if as indicated in the preceding section Uh/κ is of order 1 or less, the Peclet number for the fissure flow is usually small and convective distortion of the isotherms negligible.

Solute concentration distributions

The dispersal of solutes *is* frequently important. Solutes contained in the fluid issuing from a fracture disperse across the streamlines such as those illustrated in Figure 4.7 and may react with constituents of the matrix or the pore fluids of the receiving layer. As shown in Section 4.2, solute can diffuse beyond the stagnation point S a distance of order $\phi D/U$, where D is the diffusivity (either molecular or the result of macroscopic dispersion) and ϕ the porosity. If this is small compared with the layer depth h, that is, if

$$\frac{Uh}{D} \gg \phi, \tag{4.6.1}$$

the solute source in the layer is essentially local. (If $Uh/D \ll \phi$, the solute rapidly diffuses across the whole layer and its concentration becomes uniform with depth within a few layer thicknesses.)

The distributions of concentration in the plume are rather different in the cases of a fracture issuing fluid continuously along its length and that of a distribution of point sources (Figure 4.8). Under steady conditions and in the absence of chemical reactions, (2.9.1) reduces to

$$\mathbf{u} \cdot \nabla c = \phi D \, \nabla^2 c. \tag{4.6.2}$$

In general, the detailed distributions of c depend on the ratio of the volume flux issuing from the fracture to that in the layer. When this ratio is small, the distribution of transport velocity u is essentially undisturbed and equal to the constant value U, as the solute diffuses into the layer. In the case of a point source in which the distribution is axisymmetrical (in the permeable layer), equation (4.6.2) simplifies further to

$$U\frac{\partial c}{\partial x} = \phi D\left\{\frac{1}{r}\frac{\partial c}{\partial r} + \frac{\partial^2 c}{\partial x^2}\right\}. \tag{4.6.3}$$

The solution to this equation has been given by Bird, Stewart, and Lightfoot (1960, p. 552) as

$$c = \frac{Q_s}{2\pi\phi D s} \exp\left\{-\frac{U(s-x)}{2\phi D}\right\}, \tag{4.6.4}$$

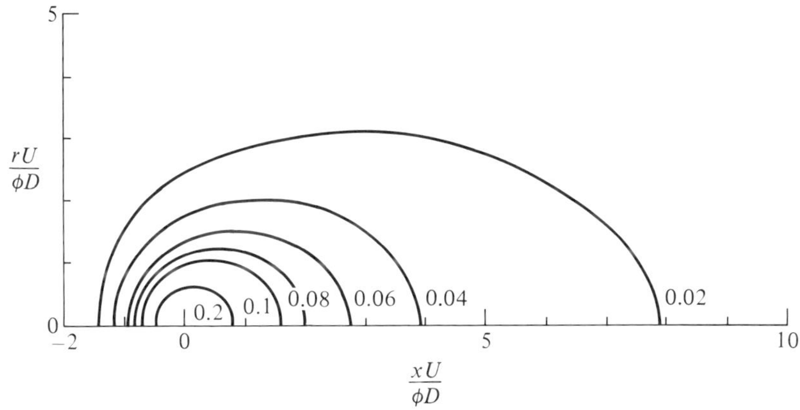

Figure 4.9. Solute concentrations from a point source at the boundary of a permeable layer. The contours are in units of $QU/(\phi D)^2$, where U is the transport velocity in the layer and Q the total solute efflux rate; the length scales are $(\phi D/U)$.

where Q_s is the flux of solute from the orifice and $s = (r^2 + x^2)^{1/2} = (x^2 + y^2 + z^2)^{1/2}$. This solution is illustrated in Figure 4.9. Solute diffuses upstream a distance of order $\phi D/U$, decreasing exponentially with distance from the source. The plume, of course, extends much farther downstream, the concentration along the axis (in the plane of the boundary) varying as x^{-1} when $xU/\phi D$ is greater than about 2. The scale of this pattern is very small; if macroscopic dispersion dominates, $D \sim l_0U/\phi$ from Sections 2.2 and 2.3, where l_0 is of the order of the grain size, so that the length scale of Figure 4.9 is l_0 itself. The concentration contours represent averages over many samples; in any individual one the pattern would be randomized considerably, though it would still display the general shape with a steep gradient upstream and a more elongated tail in the flow direction.

The solution (4.6.4) is also pertinent to the distribution of concentration around a small dissolving grain, as described by Stoessell (1987). The dissolution of plagioclase grains and the adjacent precipitation of kaolinite in a patchy asymmetrical halo commonly occurs in limestones, and Stoessell interpreted this pattern in these terms. (Note, however, that his numerical solution is for the two-dimensional equation rather than the form (4.6.3) for distributions axisymmetrical in the flow direction.)

When solute issues from a linear fracture intersecting the boundary of a permeable layer, the downstream plume decays more slowly with distance since the solute diffuses in only one lateral direction, normal to the boundary, rather than in two lateral directions as with a point source. There does not seem to be a simple analytical solution for a line source

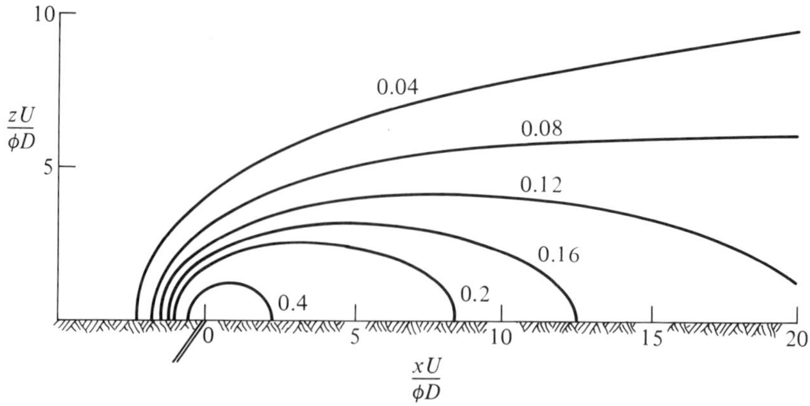

Figure 4.10. Solute concentrations from a line source at the boundary of a permeable layer in the plane normal to the boundary. Contours are in units of $q_s/\phi D$, where q_s is the flux of solute per unit length; the length scales are $(\phi D/U)$.

analogous to (4.6.4) for a point source, but since (4.6.2) is linear in the concentration c, solutions can be constructed by superposition. When the fracture is transverse to the flow direction, the distribution is two-dimensional in the x–z-plane containing the flow and normal to the line source. If q_s represents the flux of solute per unit length from the fracture, then

$$c(x, z) = \frac{q_s}{2\pi\phi D} \int_{-\infty}^{\infty} s^{-1} \exp\left\{ -\frac{U(s - x)}{2\phi D} \right\} dy, \qquad (4.6.5)$$

where $s^2 = x^2 + y^2 + z^2$. This distribution is shown in Figure 4.10. Asymptotically, when $x \gg z$, the solution (4.6.5) approaches the simpler form,

$$c(x, z) = \frac{q_s}{(\pi\phi DUx)^{1/2}} \exp\left\{ -\frac{Uz^2}{4\phi Dx} \right\}. \qquad (4.6.6)$$

The solute concentration at the base of the plume ($z = 0$) decreases with distance x from the fracture as $\{\phi DUx\}^{-1/2}$, and the height across the layer that the plume has spread by a distance x downstream of the fracture is characterized by δ such that

$$\frac{U\delta^2}{4\phi Dx} = 1$$

or

$$\frac{\delta}{h} = 2\left(\frac{\phi D}{Uh}\right)^{1/2} \left(\frac{x}{h}\right)^{1/2}. \qquad (4.6.7)$$

The solution is, of course, valid only when δ/h is appreciably less than unity or when

$$\frac{x}{h} < \frac{Uh}{4\phi D}, \tag{4.6.8}$$

but according to (4.6.1) the right-hand side is large. At distances much beyond this, the concentration has become nearly uniform across the layer and equal to q_s/Uh.

Solute concentration distributions with matrix reaction

When a solute introduced at a fracture undergoes chemical reaction with the layer matrix and a reaction product is immobile, patterns of mineral deposition are developed. For a first-order reaction, the distribution of reactant is specified by

$$U\frac{\partial c}{\partial x} = \phi D\, \nabla^2 c - \phi\gamma c \tag{4.6.9}$$

in place of (4.6.3). However, the transformation

$$c = c_1 e^{-\alpha x}, \tag{4.6.10}$$

where

$$\alpha = (2\phi D)^{-1}\{[U^2 + 4\gamma\phi^2 D]^{1/2} - U\} > 0, \tag{4.6.11}$$

reduces (4.6.9) to the form

$$U_1\frac{\partial c_1}{\partial x} = \phi D\, \nabla^2 c_1,$$

where

$$U_1 = \{U^2 + 4\gamma\phi^2 D\}^{1/2}. \tag{4.6.12}$$

This we have already solved in the preceding pages. The general patterns of the distribution of reactant are similar to those of Figures 4.9 and 4.10, but the length scales are reduced (since $U_1 > U$) and the distributions attenuate more rapidly in thc flow direction because of the $\exp(-\alpha x)$ factor in (4.6.10) representing the decay of solute in the reaction. Figure 4.11 shows concentration contours from a line source when the reaction rate $\gamma = \frac{1}{4}U^2/\phi^2 D$ and is on the same scale as Figure 4.10; both the scale of the pattern and the flow-induced asymmetry are reduced substantially. The ratio $U^2/\gamma\phi^2 D$ has a simple interpretation when the fluid disperses by meandering around grains or along fluid pathways with characteristic length scale l. Since in this case $D \sim Ul/\phi$, the ratio is of order $U/\phi\gamma l$,

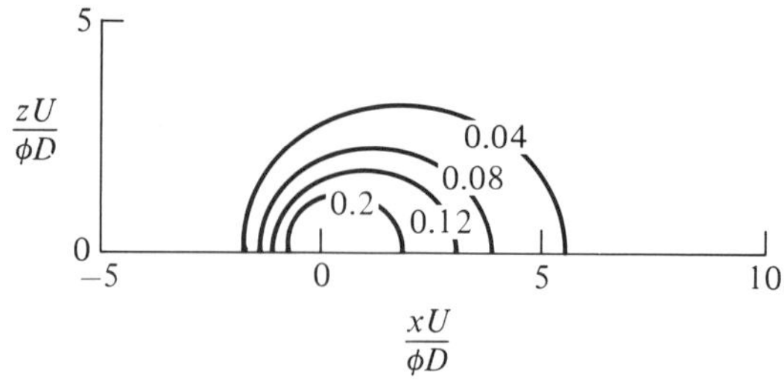

Figure 4.11. Distributions of concentration from a line source at the boundary of a permeable region of a solute reacting with the fabric, for the case $\gamma\phi^2 D/U^2 = 0.25$. The scale is the same as in Figure 4.10.

or the distance that an element of solute moves in the reaction time γ^{-1}, that is, $U/\phi\gamma$, the equilibration length (4.2.5), divided by the micro-scale l.

The rate of disappearance of solute at each point is $\phi\gamma c$, so that the rate of deposition of an immobile reaction product is proportional to this. The contours of Figure 4.9 for a point source and Figure 4.10 for a linear fracture source can thus be interpreted as distributions of alteration rates for reactions sufficiently slow that $\gamma\phi^2 D/U^2 \ll 1$, or the total amount of alteration after a fixed time; for faster reactions or lower flow rates, the patterns are compressed as in Figure 4.11. If the injected fluid is unsaturated with respect to constituents in the layer matrix, similar patterns of dissolution occur. If c_e is the saturation or equilibrium concentration in the layer, the rate of dissolution can be taken as $\phi\gamma(c_e - c)$ and the last term in (4.6.9) is replaced by $+\phi\gamma(c_e - c)$. The distributions illustrated thus also represent the distributions of $c_e - c$, or of dissolution rate.

Reaction patterns with $c_e = c_e(s)$

If the fluid issuing from a vent or fracture contains a solute with concentration s that influences the equilibrium concentration c_e of the interstitial fluid of the receiving layer, mixing zone reactions occur, as described in Section 4.5. The pattern of reaction, which is very different from those just described, can be inferred without further calculation. From (4.5.3) the rate of reaction is proportional to the *square* of the gradient of s (whether precipitation or dissolution is involved depends on the sign of the coefficients $\partial^2 c_e/\partial s^2$). From Figures 4.9 to 4.11 it is evident that the contours are closest and the gradients largest just *upstream* of the vent, so that the most intense alteration occurs there. A much less intense tail trails in the flow direction but decreases rapidly; $(\nabla s)^2$ decays rapidly as s decreases and the distribution disperses spatially.

The development of a reaction front in a thin layer

The patterns illustrated in the preceding few pages are the two-dimensional analogs of the equilibration region described in Section 4.2. As the flow and the reaction continue, the original mineral near the orifice becomes depleted and may be replaced by the reaction product in a region that gradually spreads outward as time proceeds. If the fluid vents into a thin layer, reactant diffuses laterally across it and an isothermal reaction front gradually develops downstream, with a thin diffusion front upstream.

The evolution in the pattern of solute concentration and reaction progress can be found for a first-order reaction from the two-dimensional equations analogous to (4.3.3), that is,

$$\left(\frac{n_s \phi c_e}{s_0}\right)\frac{\partial b}{\partial \tau} + \frac{\partial b}{\partial \xi} - \left(\frac{D\phi^2\gamma}{u^2}\right)\left(\frac{\partial^2 b}{\partial \xi^2} + \frac{\partial^2 b}{\partial \eta^2}\right) = -bS,$$

$$\frac{\partial S}{\partial \tau} = -bS, \qquad (4.6.13)$$

where the time scale is $s_0/(c_e n_s \phi\gamma)$ and the length scale is the advective equilibration length $u/\phi\gamma$; ξ and η are dimensionless distances along and across the permeable layer. As before, the variable $b = (c_e - c)/c_e$, where c_e is the equilibrium concentration of solute in the layer and $S = s/s_0$ is the fraction of immobile mineral in the layer that has undergone reaction. Numerical solutions to these equations are illustrated in Figures 4.12 and 4.13 with $D\phi^2\gamma/u^2 = 0.1$ and a dimensionless layer thickness of unity. To avoid the singularities associated with a point source, in this calculation solute is injected uniformly at the wall between $\xi = 0.35$ and 0.75 at a total rate of $0.5\phi D c_e$.

The distribution of solute concentration is shown in Figure 4.12 and that of degree of mineralization in Figure 4.13. After a dimensionless time of unity, the original mineral near the source is only partially depleted (Figure 4.13), and the concentration contours resemble those of Figure 4.11, though the numerical values differ because of the different normalization. By $\tau = 5$, however, the original mineral is virtually depleted near the source (inside the dashed lines of Figure 4.13), and the solute can diffuse laterally and advect downstream without reaction until it encounters significantly unaltered material with which to undergo reaction. By time $\tau = 15$, the altered mineral occupies virtually the whole layer thickness; a downstream reaction front has developed across the layer which propagates downstream and gradually becomes more one-dimensional. Upstream of the source, an inclined diffusion–advection front forms, changing very little in time, in which the flow downstream opposes the diffusion of solute upstream. As time continues, the reaction front moves farther away to the right, as described in Section 4.3.

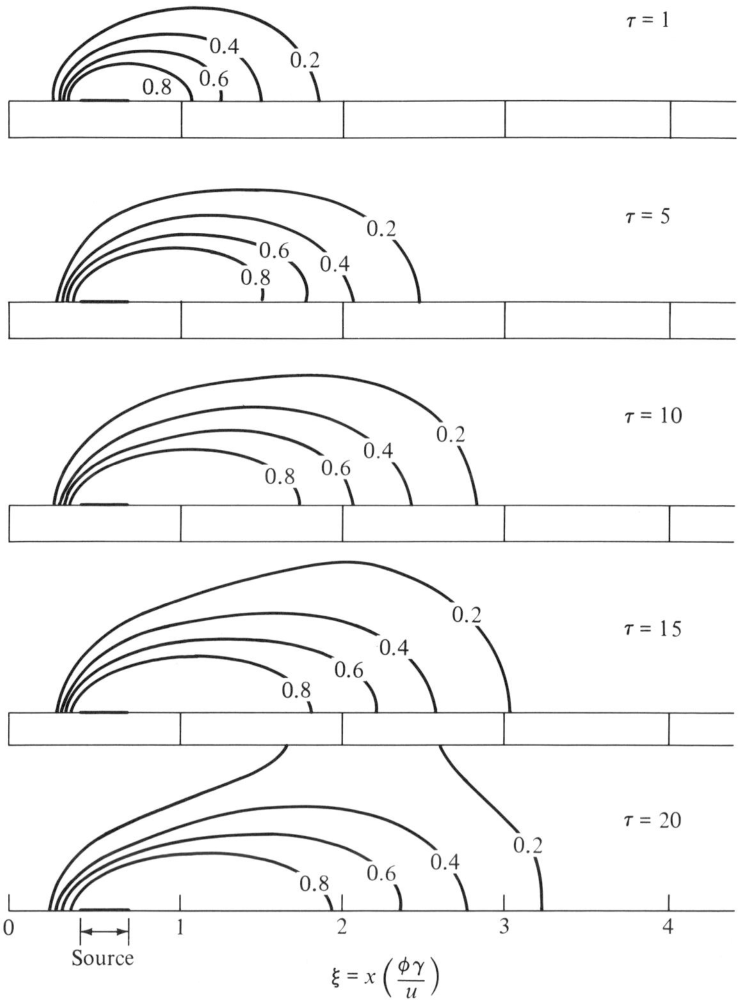

Figure 4.12. Distributions of interstitial fluid concentration, expressed as contours of $b = (c_e - c)/c_e$ produced by a line source in a permeable layer with flow from left to right. Length and time scales are specified following equation (4.6.13).

Measurements of this kind of alteration pattern could potentially be used to infer the total amount of fluid that has moved along the layer since the injection of solute commenced. The fluid–rock ratio, determined by the nature of the reaction in (4.3.17), is also related to the ratio of the fluid speed u_0 to speed of the reaction front U through (4.3.16);

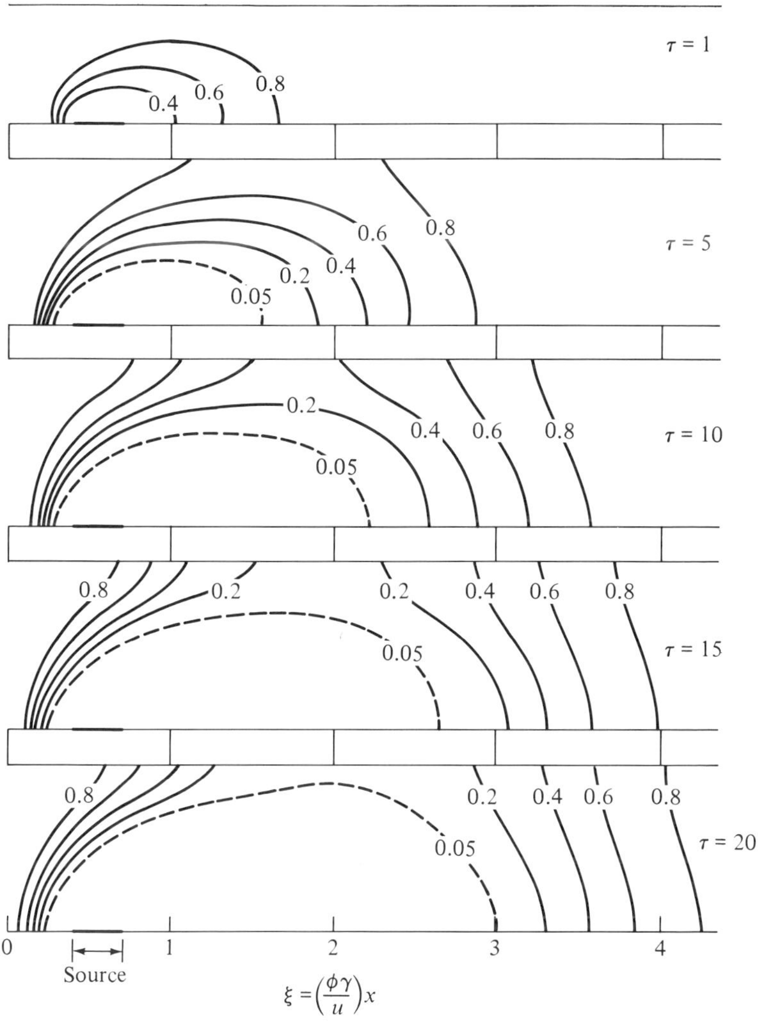

Figure 4.13. Distributions of the unaltered mineral undergoing reaction, $S = s/s_0$. The regions inside the dashed curves indicate zones of virtually complete reaction.

this is equal to the ratio of the volume of fluid that has moved past in this time interval to the volume of altered rock.

4.7 ENHANCEMENT AND DESTRUCTION OF POROSITY

Chemical reactions that occur in rocks through which fluid is percolating gradually lead to changes in porosity and so permeability. Even if the driving forces, whether hydraulic or thermal, are maintained, the flow patterns evolve as the distributions of permeability change. In simple cementation, the flow is faster in regions of high permeability, so that if the equilibration length l_e is small compared with the flow dimensions, the rate of cementation is greater there and the porosity and permeability will decrease more rapidly than in regions where they are already low and the flow slower. Cementation, therefore, tends to make the porosity more uniform throughout the region as the permeability and porosity gradually decrease with time – the deposit is disseminated. With dissolution (e.g., of carbonate rocks), the reverse occurs. Regions of initially relatively high porosity and permeability provide flow focusing, with enhanced dissolution producing enlargement of the pores, further local increase in the permeability, and ever-increasing flow, leading to tubules and ultimately perhaps to caverns. Merino, Ortoleva, and Strickholm (1983), Merino (1984), Chadham, Hoff, Merino, Ortoleva, and Sen (1986), and Ortoleva, Chadham, Merino, and Sen (1987) have discussed this kind of geochemical self-patterning or infiltrative instability in considerable detail. Ortoleva, Merino, Moore, and Chadham (1987) ascribe many instances of textural patterns such as cement bands, metamorphic layering, and oscillatory zoning to this process and give a linear instability analysis showing that a plane dissolution front is indeed unstable, small perturbations to the front developing into long fingers.

Replacement reactions, such as the dolomitization of limestone, are also generally associated with volume changes that can be of either sign. The dolomite reaction

$$2\,CaCO_3 + Mg^{2+} \rightleftharpoons CaMg(CO_3)_2 + Ca^{2+}$$

is accompanied by about a 15 percent decrease in volume of the solid phase, with consequent enhancement in porosity, increase in local flow speed, and from (4.4.5) increased dolomitization rate in a gradient reaction. It is natural to expect, then, that the distribution of dolomite deposited by this reaction will occur bimodally, with layers interbedded between unaltered limestone. This kind of structure is indeed found in atolls and in once-submerged carbonate banks such as the Latemar Massif in the Dolomites (Wilson, Hardie, and Phillips, 1990). The same general principle applies to ores or other minerals deposited by reaction in a permeable flow environment when the equilibration length is small: If the mineral formation involves a decrease in volume of the solid, the mineral will be deposited in veins or lodes; if it involves an increase in volume of the solid, the deposit is generally disseminated.

An estimate of the rates at which these alterations proceed can be found simply, as shown by Wood and Hewett (1982) and Lichtner (1985). Let Q represent the rate of addition of species to the solution (mass per unit volume of the fabric per unit time). Suppose that mineral with density ρ_1 is replaced mole for mole by another with density ρ_2. The rate of disappearance of mineral 1 into solution results in a rate of increase in the porosity

$$\rho_1 \frac{\partial \phi}{\partial t} = m_1 Q, \tag{4.7.1}$$

where m_1 is the molecular weight of this mineral. The deposition of mineral 2 leads to a rate of decrease in porosity

$$\rho_2 \frac{\partial \phi}{\partial t} = -m_2 Q \tag{4.7.2}$$

and consequently the net rate of change

$$\frac{\partial \phi}{\partial t} = \left(\frac{m_1}{\rho_1} - \frac{m_2}{\rho_2}\right) Q. \tag{4.7.3}$$

The initial rate of change in porosity in a single-mineral fabric gives a measure of the time scale in these processes. If the initial porosity is designated as ϕ_0, then from (4.7.1) with $Q > 0$ for dissolution and $Q < 0$ for deposition

$$\frac{\partial}{\partial t}\left(\frac{\phi}{\phi_0}\right) = \frac{mQ}{\rho\phi_0}, \tag{4.7.4}$$

so that the initial time scale τ for fabric alteration can be taken as $\rho\phi_0/mQ$. In the precipitation of silica, for example, Wood and Hewett, using data from Walther and Helgeson (1979), estimate that at low temperatures $\partial c_e/\partial T \sim 5 \times 10^{-3}$ kg m^{-3} °C^{-1}. With a flow velocity of 1 m yr^{-1} down a temperature gradient of 0.025°C m^{-1}, in a gradient reaction

$$\frac{\partial c_e}{\partial T} \mathbf{u} \cdot \nabla T \sim -1.2 \times 10^{-4} \text{ kg m}^{-3} \text{ yr}^{-1}.$$

If this is accepted as an order-of-magnitude estimate for mQ in (4.7.4), then if $\phi_0 = 0.25$ and $\rho = 2600$ kg m^{-3} for silica,

$$\tau = \frac{\rho\phi_0}{mQ} = 5 \times 10^6 \text{ yr}, \tag{4.7.5}$$

a fairly short geological time scale. An interesting review of quartz cementation in sandstones has been given by McBride (1989).

An example of very rapid *carbonate* cementation in the San Joaquin Basin of California has been described by Boles and Ramseyer (1987). The extraction of gas appears to have produced in the limestone such a rapid interstitial flow of isotopically light water that a dolomite cement, rich in iron, has precipitated within the past twenty years! Another study documenting calcite and iron sulfide cementation along the margins of salt domes is that of McManus and Hanor (1988).

Over time intervals of order τ or longer, however, the changing porosity modifies the flow field. The relationship between permeability k and porosity depends on the geometry of the fluid pathways. Generally, from Section 2.7, $k \sim 10^{-2}\phi\delta^2$, where δ is the characteristic diameter of the fluid pathways. If they are predominantly tubular, then from (2.2.2) $\phi \sim n\delta^2$, where n is the number of tubes per unit cross-sectional area, so that as the pore diameter changes $k/k_0 = (\phi/\phi_0)^2$, where k_0 and ϕ_0 are initial values. If, however, the fluid flows along an intersecting network of cracks or fissures, $\phi \sim \lambda\delta$ from (2.2.3), where λ is the length of crack per unit area and so $k/k_0 = (\phi/\phi_0)^3$. More generally, it might be asserted that

$$\frac{k}{k_0} = \left(\frac{\phi}{\phi_0}\right)^n,$$

where n lies between 2 and 3.

If both the pressure gradient and temperature field are maintained, the flow velocity u is proportional to k and therefore to ϕ^n; in a gradient reaction, from (4.4.4),

$$\frac{Q(t)}{Q(0)} = \left(\frac{\phi(t)}{\phi_0}\right)^n, \tag{4.7.6}$$

and (4.7.4) becomes

$$\frac{d}{dt}\left(\frac{\phi}{\phi_0}\right) = \frac{1}{\tau}\left(\frac{\phi}{\phi_0}\right)^n, \tag{4.7.7}$$

where again

$$\tau = \frac{\rho\phi_0}{mQ(0)} \tag{4.7.8}$$

is the time scale of (4.7.5), positive for dissolution and negative under cementation conditions. Since $n \neq 1$, the solution to (4.7.7) is

$$\frac{\phi(t)}{\phi_0} = \left\{1 - (n-1)\frac{t}{\tau}\right\}^{-1/(n-1)}, \tag{4.7.9}$$

which is illustrated in Figure 4.14. When τ is negative, cementation proceeds ever more slowly as the pores clog. The porosity is reduced to half its initial value in a time interval between τ and 3τ, to one-tenth between 9τ and 50τ according to the geometry of the structure. In dissolution conditions, with τ positive, however, a calamity occurs when $t = \tau/(n-1)$ (i.e., between $\frac{1}{2}\tau$ and τ) since $\phi(t) \to \infty$! *Within a finite time,* the porosity increases dramatically; in reality the pore spaces join up as the matrix dissolves and a cavity develops. The characteristic time τ then appears to be a quite accurate index of the time needed for the formation of caverns by flow through, for example, permeable limestone.

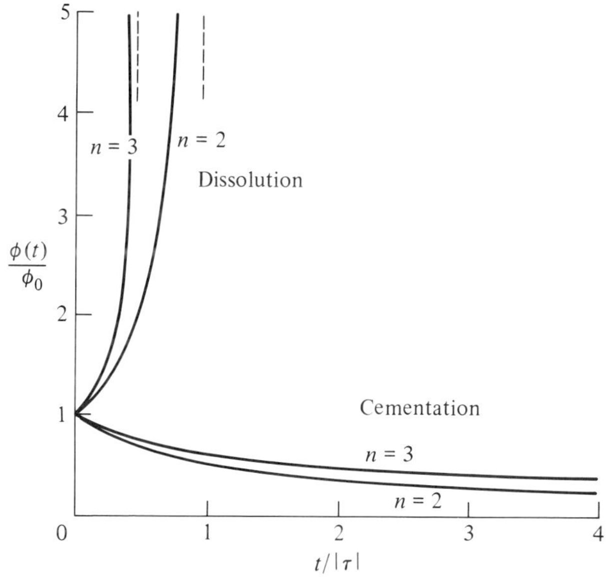

Figure 4.14. Variations in porosity under a maintained pressure gradient in conditions of segmentation and dissolution. The time scale τ is given by (4.7.8).

The time scales for replacement reactions can be found similarly. From equation (4.7.3)

$$\left(\frac{m_1\rho_2 - m_2\rho_1}{\rho_2}\right)^{-1} \frac{\partial \phi}{\partial t} = \frac{Q}{\rho_1}, \tag{4.7.10}$$

which is identical to (4.7.1) except for the additional modifying factor on the left, which simply changes the time scale but does not alter the structure of the solutions (4.7.9), illustrated in Figure 4.14. The time scale for porosity alteration is then somewhat longer:

$$\tau = \frac{\rho_1\phi_0}{m_1 Q(0)} \left\{1 - \frac{m_2\rho_1}{m_1\rho_2}\right\}^{-1} = \frac{\phi_0}{Q(0)} \left(\frac{\rho_1\rho_2}{m_1\rho_2 - m_2\rho_1}\right). \tag{4.7.11}$$

The porosity can, of course, be reduced by compaction. Wilson and McBride (1988) document the variations in porosity with depth in sandstones of the Ventura Basin, California, the most deeply buried having lost a total absolute porosity of 26 percent from a presumed initial value of 40 percent, or 0.4. In compaction, the expelled interstitial fluid moves generally upward relative to the matrix, though if the compaction is the result of continued sedimentation and the rate of sedimentation is greater than the interstitial velocity produced by compaction, the depth of indi-

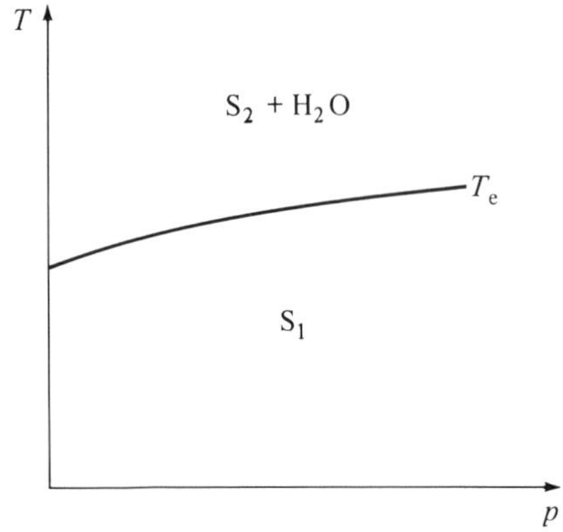

Figure 4.15. Equilibrium diagram for a dehydration reaction with a discrete $T_e(p)$.

vidual fluid elements below the depositional interface will increase. The consequences of this effect in the migration of hydrocarbons have been explored by Bonham (1980).

4.8 ISOTHERM-FOLLOWING REACTIONS

Of course, not all geochemical reactions are controlled by interstitial fluid flow. Many metamorphic reactions are the result of a history of temperature variations of a mineral assemblage, produced either by tectonics that force it to a greater depth or by volcanic intrusion. Water may be *produced* by the reaction and subsequently move through the matrix; in these dehydration reactions

$$S_1 \rightleftharpoons S_2 + H_2O.$$

An example is the dehydration of muscovite (white mica):

$$\left.\begin{array}{l} \underset{\text{Muscovite}}{KAl_3Si_3O_{10}(OH)_2} + \underset{\text{Quartz}}{SiO_2} \rightleftharpoons \underset{\text{Potassium feldspar}}{KAlSi_3O_8} \\ \qquad + \underset{\text{Sillimanite}}{Al_2SiO_5} + H_2O, \end{array}\right\} \tag{4.8.1}$$

where S_1 and S_2 of the preceding equation represent the assemblages on the left and right, respectively. The equilibrium diagram for this reaction is of the general form illustrated in Figure 4.15, the equilibrium temper-

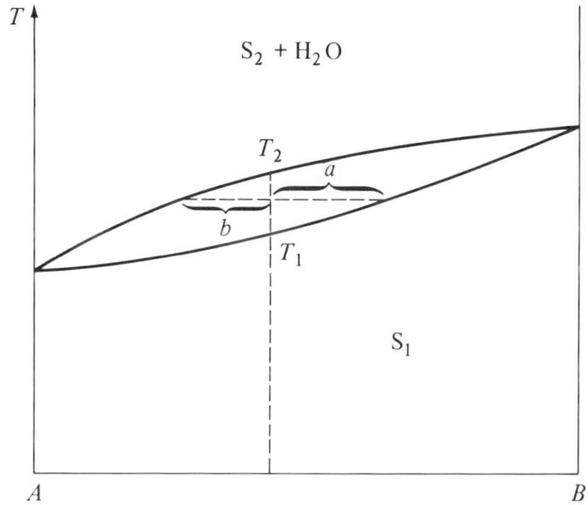

Figure 4.16. Equilibrium diagram for dehydration reactions in low-grade metamorphosis, with a range of transition temperatures.

ature being almost independent of pressure, so that as the temperature of a block of the assemblage S_1 increases, it remains stable until its temperature exceeds the equilibrium value T_e and the reaction moves to the right. In spatial terms, the isotherms move through the fabric and the reaction zone follows behind the moving equilibrium isotherm.

In low-grade metamorphism involving usually complex assemblages and a dense sequence of reactions, the equilibrium line may be replaced by a transition zone (Figure 4.16). At a given ratio A/B of major constituents in S_1, reaction commences when the temperature exceeds T_1 and is complete when it exceeds T_2. At intermediate temperatures, the assemblage containing both S_1 and S_2 is in local equilibrium, the proportion of S_2 being a/b in Figure 4.16 or approximately $(T - T_1)/(T_2 - T_1)$, so that as the temperature gradually increases – in a dehydration reaction, for example – water is gradually released. In terms of the spatial distribution of fluid generation, in the case illustrated in Figure 4.15, the source zone will be found to be limited to a restricted region following the equilibrium isotherm, whereas in that of Figure 4.16, it is more diffuse.

In either case, if water is produced by the reaction, the interstitial fluid pressure increases. If the generation rate of fluids is sufficiently rapid and the vertical permeability in the sequence above the reaction zone is sufficiently small, the interstitial fluid pressure may locally exceed the lithostatic pressure, providing a scenario for local rock fracture, chemically induced. If, however, the rock permeability above is initially relatively large, the excess pressure can be released by upward fluid migration, but this process tends to be self-sealing. The rising interstitial fluids contain solutes in approximate equilibrium with their surroundings, and if the equilibrium concentration decreases with temperature and pressure (as is the case for dissolved silica), precipitation will occur in accordance with equation (4.4.5). The porosity and permeability of the rock above the reaction zone will therefore decrease in time, increasing the vertical pressure gradient if the fluid generation rate remains even approximately constant. Under appropriate conditions and after sufficient time, local fracture may occur that temporarily relieves the excess pressure until further deposition in veins along the fractures reseals the matrix and the cycle recurs. The same cycles of events can occur when the fracturing is induced seismically, as Sibson et al. (1988) indicate – a sequence they call "seismic pumping."

It is of mineralogical interest to establish the general conditions necessary for the buildup of significant overpressure in the interstitial fluids and for the occurrence of chemically driven rock fractures. This dynamical process requires the movement of isotherms through the matrix as occurs during and following active overthrusting. How long does it take to develop pressures that approach the lithostatic value, where does this occur first, and how long do the pressures persist once the tectonic processes have become inactive?

The reaction zone

When the isotherms move through the matrix, the general pattern of the reaction is the same whether or not dehydration is involved. In the reaction, the moles s of S_1 per unit volume decreases

$$\frac{\partial s}{\partial t} = -Q_s \tag{4.8.2}$$

as it is replaced by S_2. In the situation illustrated in Figure 4.15, when $T < T_e$, the equilibrium temperature, $Q_s = 0$, but when $T > T_e$, the rate of disappearance of S_1 is proportional to the amount per unit volume present and, near equilibrium, to the temperature difference $T - T_e$, so that Q_s can be represented as $\gamma s(T - T_e)/T_e$, where the reaction rate γ has, as before, dimensions of $(\text{time})^{-1}$. Thus,

$$\begin{aligned} \frac{\partial s}{\partial t} &= 0, \qquad T < T_e, \\ &= -\gamma s\left(\frac{T - T_e}{T_e}\right), \qquad T \geqslant T_e. \end{aligned} \tag{4.8.3}$$

If the equilibrium isotherm is advancing through the mineral assemblage with speed U, say, then at any point near the equilibrium isotherm

$$T - T_e = GUt,$$

where t is the time interval since the isotherm passed that point and G (>0) is the (negative) temperature gradient in the direction of U. Thus,

$$\frac{\partial s}{\partial t} = -\frac{\gamma GUts}{T_e}$$

behind the equilibrium isotherm and zero ahead of it. In terms of the distance $\xi = Ut$ behind the equilibrium isotherm

$$U\frac{\partial s}{\partial \xi} = -\left(\frac{\gamma G}{T_e}\right)\xi s, \qquad \xi \geqslant 0, \qquad \text{behind,}$$
$$= 0, \qquad \xi < 0, \qquad \text{ahead.} \tag{4.8.4}$$

Subject to the condition that $s = s_0$ ahead of the reaction zone, these equations have the solution

$$s = s_0 \exp(-\xi^2/l^2), \qquad \xi \geqslant 0,$$
$$= s_0, \qquad \xi < 0, \tag{4.8.5}$$

where the thickness scale of the reaction zone

$$l = \left(\frac{UT_e}{2\gamma G}\right)^{1/2}. \tag{4.8.6}$$

This solution is illustrated in Figure 4.17. When the speed of advance U of the isotherm becomes very small, so does this thickness; the boundary between S_1 and S_2 approaches a discontinuity.

In lower-grade metamorphism (Figure 4.16) the thickness scale of the reaction zone is greater, since the sequence of reactions occurs over the finite temperature range (T_1, T_2). Regions of contact metamorphism surrounding magmatic intrusions frequently show gradations in mineral assemblages rather than sharp contacts, reflecting this sequence of reactions.

Dehydration

If the reaction involves dehydration, with 1 mole of the mineral or mineral assemblage S_1 releasing n_w moles of water, thc volumetric rate of fluid generation Q_w is $n_w VQ_s$, where V is the molar volume of the fluid at the ambient pressure. Thus,

$$Q_w = n_w VQ_s = -n_w V\frac{\partial s}{\partial t},$$

and in the simple case of a discrete equilibrium temperature (Figure 4.15), the rate of fluid generation per unit volume

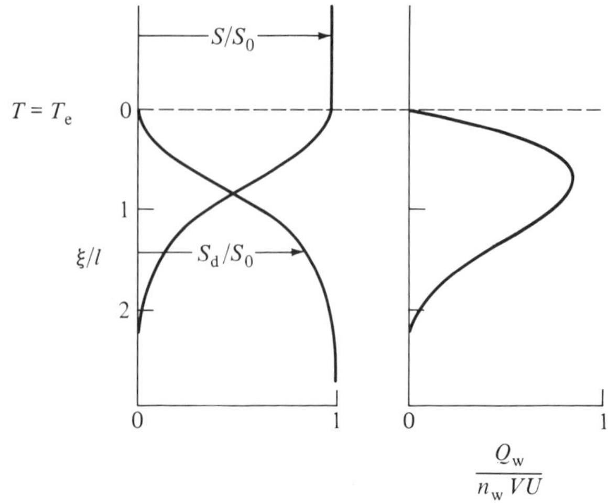

Figure 4.17. The dehydration reaction zone behind the equilibrium isotherm T_e moving vertically through a matrix with speed U. S/S_0 represents the fraction of the hydrate and S_d/S_0 that of the dehydrate; Q_w is the rate of fluid generation. The thickness scale is given by (4.8.6).

$$Q_w = -n_w VU \frac{\partial s}{\partial \xi}, \tag{4.8.7}$$

where $s = s(\xi)$ is specified by (4.8.5). This rate is a maximum at a distance $2^{-1/2}l$ behind the moving isotherm, decreasing to zero within increasing distance beyond (Figure 4.17). The total rate of generation of fluid in the zone (volume per unit time per unit area of dehydration zone) is simply

$$\mathcal{Q}_w = \int_0^\infty Q_w \quad d\xi = n_w VU s_0 \tag{4.8.8}$$

independent of the reaction rate γ and the temperature gradient G.

In lower-grade metamorphism (Figure 4.16), the distance over which the reaction occurs is greater but (4.8.8) remains valid. This can be seen either by considering the overall process as a sequence of small steps of the kind just described, or alternatively as follows. At a temperature T intermediate between T_1 and T_2, the moles of water that can still be produced by dehydration is

$$n_w s_0 \left(\frac{T_2 - T}{T_2 - T_1} \right),$$

and as the temperature increases, the volumetric rate of fluid generation is

$$n_w V s_0 (T_2 - T_1)^{-1} \frac{\partial T}{\partial t}.$$

If, again, the isotherms move with speed U, the rate of generation is therefore approximately uniform with distance between the isotherms T_1 and T_2 and equal to

$$n_w V s_0 (T_2 - T_1)^{-1} U G, \qquad (4.8.9)$$

where $G(>0)$ is the temperature gradient. The distance between these two isotherms is approximately $(T_2 - T_1)/G$, so that the total rate of fluid generation per unit area of the zone is the product of this and (4.8.9), which recovers (4.8.8).

Pressurization and fracture in deep layers

Near an isolated pluton or magmatic intrusion, fluids generated by these reactions can migrate away from the source in three dimensions, facilitated by any mechanical fracturing that may have occurred, but when they occur at the base of a deep layer, the generation of fluid may lead to an increase in pressure sufficient to produce chemically induced fracturing as described earlier.

Suppose for the sake of simplicity that the equilibrium isotherm for the dehydration reaction in a thick layer rises across an impermeable horizontal basement so that dehydration commences at the base. Fluid generated by the reaction must force its way generally upward, though it may move laterally or even down in some places to follow a higher permeability sublayer. If the porosity of the rock does not change significantly during reaction, the vertical velocity w at the top of the reaction zone is the volume of fluid produced per unit time per unit horizontal area, or $\mathcal{Q}_w$ of equation (4.8.8). The vertical *gradient* of pressure at the top of the reaction zone,

$$\frac{\partial \hat{p}}{\partial z} = -\frac{\mu}{k_V} w - \rho_w g,$$

is then constant in time once the isotherm has risen a distance equal to the thickness of the reaction zone, and the pressure perturbation induced by the flow diffuses upward, as described in Section 3.10. The reduced pressure distribution is determined by the pressure diffusion equation (3.10.6), which reduces in this case to

$$\frac{\partial p}{\partial t} = \kappa_p \frac{\partial^2 p}{\partial z^2},$$

where z is taken vertically upward from the reaction front at $z = 0$, say, at which level

$$\frac{\partial p}{\partial z} = -\frac{\mu}{k_V}\mathcal{Q}_w, \tag{4.8.10}$$

which can be taken constant in time when $t > 0$. Since the excess pressure obeys a simple diffusion equation, it diffuses upward over a region whose thickness increases as $(\kappa_p t)^{1/2}$ and the pressure in the reaction zone increases as

$$p_0 \sim \frac{\mu \mathcal{Q}_w}{k_V}(\kappa_p t)^{1/2}. \tag{4.8.11}$$

This continues until the region of pressure perturbation approaches the surface or a more permeable overlying layer at a height d, say, above the reaction zone (i.e., after a time interval of order $d^2/2\kappa_p$, as given by (3.10.8) with κ_p involving the vertical permeability). Beyond this time interval, if the local total pressure has nowhere reached the lithostatic value, the pressure distribution approaches a steady state,

$$p(z) = \frac{\mu}{k_V}\mathcal{Q}_w(d - z), \tag{4.8.12}$$

as illustrated in Figure 4.18.

The total pressure of the interstitial fluid includes the hydrostatic part:

$$\hat{p}(z) = p_0 + \left\{\frac{\mu}{k_V}\mathcal{Q}_w + \rho_w g\right\}(d - z), \tag{4.8.13}$$

where p_0 is that at the top $z = d$ of the layer – if a much more permeable overburden has depth d_1, this is $\rho_w g d_1$. The lithostatic pressure is

$$p_l = \rho_s g(d + d_1 - z)$$

when variations in the rock density ρ_s are minor, and this exceeds the interstitial fluid pressure when

$$(\rho_s - \rho_w)g(d + d_1 - z) > \frac{\mu}{k_V}\mathcal{Q}_w(d - z) = \frac{\mu n_w V s_0 U}{k_V}(d - z), \tag{4.8.14}$$

from (4.8.8). As illustrated in Figure 4.18, if k_V is imagined to decrease or if the speed U increases, this condition is closest to equality at the reaction zone $z = 0$, so that (4.8.14) requires that

$$U < \left(\frac{\rho_s}{\rho_w} - 1\right)\left(1 + \frac{d_1}{d}\right)\frac{g k_V}{\nu n_w V s_0} \tag{4.8.15}$$

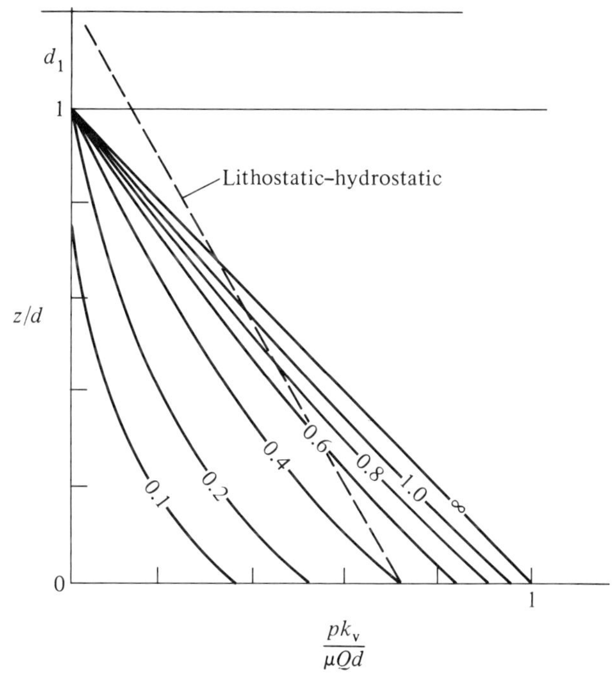

Figure 4.18. Distributions of pressure above hydrostatic when a dehydration reaction front enters the base of a moderately permeable layer of thickness d, overlain by much more permeable layer of thickness d_1. The curves are for various times in units of d^2/κ_p. If the dashed line represents the lithostatic pressure above hydrostatic, it is evident that the fluid pressure exceeds the lithostatic pressure first, just above the reaction front, where chemically induced fracturing can occur. The curves are calculated from a series solution given by Carslaw and Jaeger (1959) for an analogous thermal problem.

and under this condition, the base pressure, that at the equilibrium isotherm,

$$\hat{p}(0) = \rho_w g(d + d_1) + \frac{\mu n_w V s_0 U}{k_V}, \tag{4.8.16}$$

is sufficient, while still below lithostatic, to force the fluid that is generated by dehydration upward throughout the layer.

For a given geometry, grade of reaction (represented by s_0), and speed of rise U of the isotherms, if the initial permeability k_V is sufficiently small, the condition (4.8.15) may not be satisfied and the pressure at the reaction zone will rise above the lithostatic value before the pressure has attained a quasi-steady distribution. The mechanical strength of the rock matrix at the reaction temperature is not great, and when this occurs, chemically induced fracturing must be expected in the region just above the equilibrium isotherm. Even if the initial permeability is sufficiently great for (4.8.15) to be satisfied, the upward movement of solute-bearing fluid (e.g., silica) leads to deposition in the interstices or along the fractures by the gradient reactions discussed in Section 4.4, with a gradual reduction in porosity and permeability over the time scale τ of Section 4.7, which is generally long compared with the pressure diffusion time $(d^2/2\kappa_p)$. The detailed analysis of Section 4.7 is concerned with flows in which the pressure gradient driving them remains constant as the porosity changes, whereas in this situation, since the reaction progress is dependent only weakly on pressure, the vertical velocity $w = \mathcal{Q}_w$ remains essentially constant; the pressure gradient increases as the porosity decreases. In this case then, in equation (4.7.4), namely,

$$\frac{\partial}{\partial t}\left(\frac{\phi}{\phi_0}\right) = \frac{mQ}{\rho\phi_0} = \tau^{-1}, \tag{4.8.17}$$

the rate of dissolution Q (negative for precipitation) in gradient reactions is constant in time, according to (3.5.15). In a deposition situation it is convenient to define the time scale $\tau_s = -\tau$ as positive, so that from (4.8.17),

$$\frac{\phi}{\phi_0} = 1 - \frac{t}{\tau_s}, \tag{4.8.18}$$

in place of (4.7.9). The time τ_s is that necessary to seal the matrix completely if no fractures were to occur; in fact, however, since $k_V \to 0$ as $\phi \to 0$, the condition (4.8.15) is necessarily violated before $t = \tau_s$. The time scale τ_s is therefore an upper limit to the time needed to initiate fracture of the matrix above the dehydration zone.

Values of the in situ permeability can be inferred from the geochemistry and the distribution of interstitial pressure with depth in deep bore holes. Below the dehydration zone, there is no vertical flow, so that the vertical pressure gradient is simply hydrostatic (lithostatic if collapse has occurred). If *above* the reaction zone the pressure gradient is also close to lithostatic, maintained by a general balance between fracturing and sealing,

$$\frac{\partial \hat{p}}{\partial z} = -\left\{\frac{\mu}{k_V}\mathcal{Q}_w + \rho_w g\right\} \simeq -\rho_s g,$$

then in this zone

$$k_V = \frac{\mu\mathcal{Q}_w}{(\rho_s - \rho_w)g}. \tag{4.8.19}$$

In the unfractured region above, the pressure is less than lithostatic but the pressure *gradient* is larger, so that the permeability there is smaller than that below by the inverse ratio of the pressure gradients.

If the isotherms become stationary, dehydration and its upward flow cease and the pressure distribution relaxes on a pressure diffusion time scale determined usually by the thickness and pressure diffusivity of the region in which sealing but not fracturing has occurred.

5 Instabilities

5.1 RAYLEIGH–DARCY INSTABILITY

In considering the flow through permeable fabrics or the temperature regime in a saturated fabric, we are concerned not only with finding a solution to the governing equations but also with whether that solution is dynamically stable. The uniqueness theorem of Section 3.6 assures us that for a given distribution of permeability and given boundary conditions, when the interstitial (single-phase) fluid density is constant, the flow *is* stable, but instabilities may arise when variations in fluid density are present as a result of variations in temperature, salinity, or both. A clear distinction must be drawn between buoyancy-driven flows that, as shown in Section 3.5, necessarily occur when the interstitial fluid isopycnals are not horizontal and the self-generating and self-patterning instability flows that may develop from a state of rest or steady motion even when the isopycnals *are* horizontal. For example, a local hot spot at the basement of a horizontal permeable layer *necessarily* generates a convective circulation above it whose strength is roughly proportional to the temperature excess. In contrast, above a uniformly heated horizontal basement, interstitial fluid can remain in equilibrium at rest with the heat flux being purely conductive. If, however, the Rayleigh number (defined in Section 3.5) associated with the *uniform* heating becomes too large, an organized cellular motion develops as a result of an internal instability and convective heat flux becomes significant. The preceding chapter explored the local flow and reaction characteristics associated with preexisting geological structures that guide the flow; in this chapter we examine the flows that can arise spontaneously by the instability of a state initially at rest or in uniform motion and the self-organized patterns that can result. It is important to be able to recognize the conditions under which these instabilities can occur.

The geometries that we consider are usually simplifications of nature: Interfaces between different fabrics are never quite smooth, nor the fabrics themselves quite homogeneous – random perturbations abound. The general question of dynamical stability concerns the response of the flow patterns and temperature distributions to such perturbations: Are the temperature and flow perturbations suppressed or smoothed out, or do they amplify and grow to the extent that they seriously modify the original

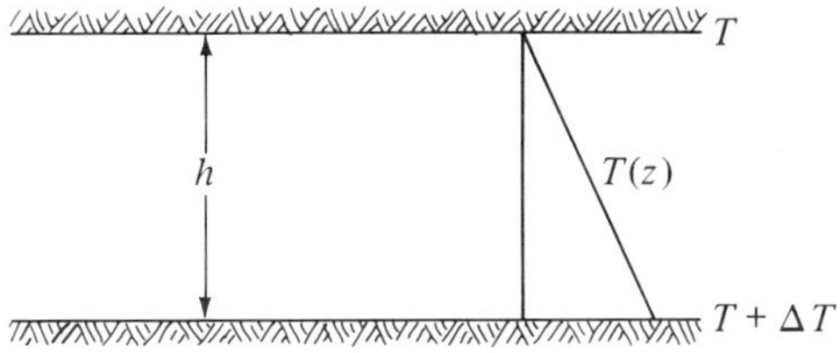

Figure 5.1. A horizontal saturated permeable layer heated from below.

state, which then moves to a different, more stable configuration? Mathematical solutions that, when disturbed by a certain class of perturbations, always return to the original solution are said to be dynamically stable to those perturbations. If, however, *any* member of the class of perturbations is found to amplify in time (even though all others may decay), the solution is unstable.

The technique for deciding whether a given solution is stable (or, more commonly, under what conditions it is stable) involves three essential steps. First, the basic state that we wish to examine, the distribution of flow, temperature, and so, must be specified as a solution to the governing equations and the appropriate boundary conditions. Second, the solution is perturbed slightly in as general a manner as possible consistent with the boundary conditions. The combined basic state and perturbations must also satisfy the governing equations, which then specify the subsequent evolution. Finally, expressions are developed for the time evolution of the perturbations to determine the conditions under which they all decay or fail to grow, that is, the conditions for stability.

The archetypical thermal stability situation is illustrated in Figure 5.1. Consider a horizontal layer of viscous, heat-conducting fluid confined between parallel plane surfaces with separation h. The temperature of the lower surface is higher by the amount ΔT than the upper surface; in a steady state at rest there is a uniform heat flux $\kappa\ \Delta T/h$ upward. Now, the fluid near the lower surface is hotter, less dense, and so more buoyant than the fluid near the top, so that it may tend to rise. However, fluid motion is resisted by viscosity, and hot spots tend to be diffused away by thermal conductivity. Under what conditions will the destabilizing buoyancy variations overcome the stabilizing viscous and conductive effects?

The case in which the gap is entirely filled with viscous fluid is the subject of Rayleigh's (1916) classical study; the somewhat more pertinent situation in which the gap is occupied by a porous medium with $k_{\mathrm{H}} = k_{\mathrm{V}} = k$ saturated with an interstitial fluid of uniform composition was solved by Horton and Rogers (1945) and by Lapwood (1948). The results are often quoted in geological contexts, but they should be interpreted with caution, as the following section shows.

Let us consider a slightly more general case than did they, supposing that k_H and k_V are uniform but not necessarily equal. The equations governing any motion in the layer are Darcy's equation, the incompressibility condition, the heat balance without internal heat sources, and the equation of state for a homogeneous fluid:

$$\begin{aligned} \frac{\partial p}{\partial z} &= -\frac{\mu}{k_V} w - \rho g, \\ \nabla_h p &= -\frac{\mu}{k_H} \mathbf{u}_h, \\ \nabla \cdot \mathbf{u} &= 0, \\ \Gamma \frac{\partial T}{\partial t} + \mathbf{u} \cdot \nabla T &= \kappa \nabla^2 T, \\ \rho &= \rho_0(1 - \alpha T), \end{aligned} \tag{5.1.1}$$

where Γ is the specific heat ratio (2.8.5) and $\mathbf{u} = (\mathbf{u}_h, w)$. The mass balance for dissolved material is satisfied trivially since $c =$ const. At the lower boundary $z = 0$, $\mathbf{u} \cdot \mathbf{n} = w = 0$ and $T = T_0 + \Delta T$, while at $z = h$, $w = 0$ and $T = T_0$. The steady state of rest with uniform temperature gradient,

$$\begin{aligned} \mathbf{u}_h = w = 0, \qquad p = p_0(z) \qquad \text{with} \quad \partial p_0/\partial z = -\rho g, \\ T = T_0 + \left(1 - \frac{z}{h}\right) \Delta T, \end{aligned} \tag{5.1.2}$$

satisfies these equations without restrictions on the parameters, but under what condition is this state of rest stable?

To examine this question, suppose that the basic state is perturbed slightly. Let

$$\begin{aligned} \mathbf{u} &= (\mathbf{u}_h, w) = \mathbf{u}'(x, y, z, t), \\ p &= p_0(z) + p'(x, y, z, t), \\ T &= T_0 + \left(1 - \frac{z}{h}\right) \Delta T + T'(x, y, z, t), \end{aligned} \tag{5.1.3}$$

where the now evolving flow, temperature, and pressure distributions must also satisfy (5.1.1). Substitution of (5.1.3) into (5.1.1) gives, after subtracting out the basic state pressure balance,

$$\begin{aligned} \frac{\partial p'}{\partial z} &= -\frac{\mu}{k_V} w' + \rho_0 g \alpha T', \\ \nabla_h p' &= -\frac{\mu}{k_H} \mathbf{u}_h', \\ \nabla \cdot \mathbf{u}' &= 0, \\ \Gamma \frac{\partial T'}{\partial t} + \mathbf{u}' \cdot \nabla T' - w' \frac{\Delta T}{h} &= \kappa \nabla^2 T', \end{aligned} \tag{5.1.4}$$

with $w' = T' = 0$ on $z = 0, h$. The equation of state has here been incorporated into the last term of the first equation.

These equations govern the evolution of the perturbations that we have imposed. An important simplification occurs when the perturbations are *small,* in particular when $T' \ll \Delta T$; as long as this remains valid, the term $\mathbf{u}' \cdot \nabla T'$ is very much less than $w' \, \Delta T/h$ and can be neglected in the last equation of (5.1.4). If that term is omitted, the set of equations has important characteristics: It is a set of coupled partial differential equations that is *linear* in the field variables primed (no squares or products), and all the coefficients are *constant* (independent of x, y, z, and t). Just as in the theory of ordinary linear differential equations with constant coefficients, the general solutions of such a set are exponential in form, so that we should be able to find solutions for each of the variables that are proportional to

$$e^{nt} e^{i(kx+ly+mz)}. \tag{5.1.5}$$

This choice is not arbitrary. In our perturbation, we are free to impose at the initial instant any spatial distribution of disturbance consistent with the boundary conditions, but by Fourier's theorem this can be represented quite generally as the superposition of spatially oscillatory terms, as in a Fourier series or Fourier integral. There is consequently no loss of generality in this spatial representation of the disturbance–if any one of these Fourier modes is unstable, the disturbance will grow. The quantity n (which is to be determined) may be complex, but if its real part is negative for all (k, l, m), all disturbances will die away. If it is positive for at least some (k, l, m), those modes will grow exponentially in time until the neglected term $\mathbf{u}' \cdot \nabla T'$ in (5.1.4) becomes comparable with the others, and some new, more complex stable state may evolve.

Two final simplifications help. First, spatial periodicity in the horizontal direction represents an oscillation in the direction of the wave-number vector $\mathbf{k} = (k, l)$. There is no preferred horizontal direction in the geometry of the problem, so that we can choose our x-axis in the direction of $\mathbf{k}$, causing the dependence on y to disappear and the problem to become two-dimensional. We can then use the stream-function representation of Section 3.1 and eliminate the pressure from the first two equations of (5.1.4):

$$0 = -\frac{\mu}{k_V}\frac{\partial^2 \psi'}{\partial x^2} - \frac{\mu}{k_H}\frac{\partial^2 \psi'}{\partial z^2} - \rho_0 g \alpha \frac{\partial T'}{\partial x}. \tag{5.1.6}$$

The linearized heat balance becomes

$$\Gamma \frac{\partial T'}{\partial t} + \frac{\partial \psi'}{\partial x}\frac{\Delta T}{h} = \kappa \nabla^2 T', \tag{5.1.7}$$

where ψ' represents the stream function of the perturbed flow and (3.1.2) is used. The boundary conditions are now

$$T' = \frac{\partial \psi'}{\partial x} = 0 \qquad \text{on} \quad z = 0, h. \tag{5.1.8}$$

Second, to satisfy these boundary conditions without further ado and to use the separation h as a natural geometrical length scale, (5.1.5) can be particularized further by letting

$$\begin{aligned} T' &= \hat{T} e^{nt} \sin \frac{m\pi z}{h} e^{ilx/h}, \\ \psi' &= \hat{\psi} e^{nt} \sin \frac{m\pi z}{h} e^{ilx/h}, \end{aligned} \tag{5.1.9}$$

where m is now a nonzero integer and the wave number l dimensionless.

The rest of the calculation is simple algebra. The subsitution of (5.1.9) into (5.1.6) and (5.1.7) gives, after cancellation of terms, a pair of simultaneous homogeneous linear algebraic equations for $\hat{\psi}$ and $\hat{T}$. The condition for existence of a nontrivial solution (i.e., one other than $\hat{\psi} = \hat{T} = 0$) is that the determinant of the coefficients vanishes, and this leads to a determination of n:

$$\frac{nh^2\Gamma}{\kappa} = \frac{l^2}{rl^2 + m^2\pi^2} \mathrm{Ra_H} - (l^2 + m^2\pi^2), \tag{5.1.10}$$

where $r = k_H/k_V$, the ratio of permeabilities parallel and normal to the bedding. The dimensionless ratio

$$\mathrm{Ra_H} = \frac{g\alpha \, \Delta T \, k_H h}{\nu\kappa} \tag{5.1.11}$$

is called the Rayleigh–Darcy number or, more simply, the Rayleigh number.

It is apparent from (5.1.10) that the growth or decay rate n is, in this instance, purely real. For stability, $n < 0$, which requires that

$$\mathrm{Ra_H} \leqslant \frac{(l^2 + m^2\pi^2)(rl^2 + m^2\pi^2)}{l^2} = f(l^2, m^2), \text{ say,} \tag{5.1.12}$$

for all values of l and integral values of m. For any given choice of m, the function on the right of (5.1.12) increases as rl^2 for large l and as $(m\pi)^4 l^{-2}$ for small l, and so has a minimum at some intermediate value. At this minimum

$$\frac{\partial}{\partial l^2} f(l^2, m^2) = 0;$$

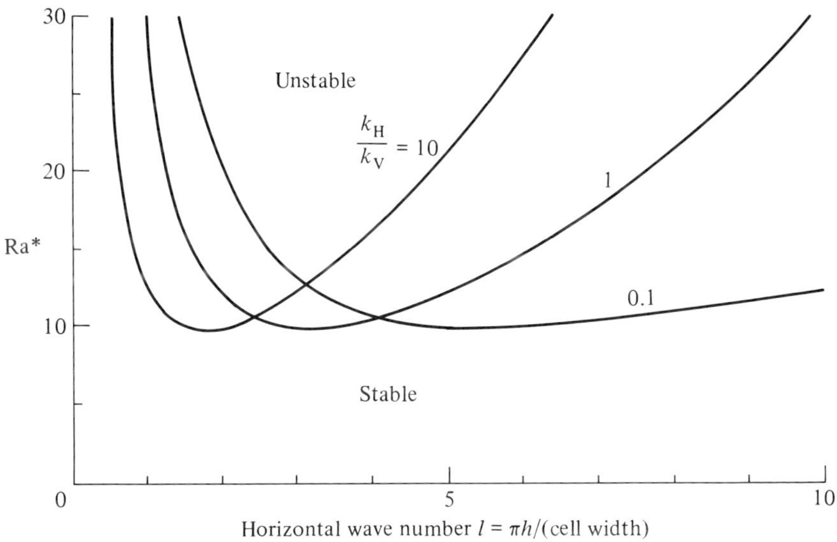

Figure 5.2. The lowest-mode stability curves for a horizontal saturated permeable layer for three values of the permeability ratio. The critical value of the modified Rayleigh–Darcy number is in each case $\pi^2 \sim 9.87$.

this occurs when $l^2 = r^{-1/2}m^2\pi^2$, at which point the function f has the value $(1 + \sqrt{r})^2m^2\pi^2$. This stability limit is most restrictive when $m = 1$, so that the critical Rayleigh number based on the horizontal permeability at which instability can first occur is

$$(\mathrm{Ra_H})_{\mathrm{crit}} = (1 + \sqrt{r})^2\pi^2. \tag{5.1.13}$$

This criterion can be expressed in terms of a modified Rayleigh–Darcy number,

$$\mathrm{Ra}^* = \frac{g\alpha\,\Delta T\,h}{\nu\kappa}\,\frac{k_H k_V}{(k_H^{1/2} + k_V^{1/2})^2}; \tag{5.1.14}$$

the condition for stability is that

$$\mathrm{Ra}^* \leqslant \pi^2. \tag{5.1.15}$$

The stability domains are illustrated in Figure 5.2. When the permeabilities are equal, $k_H = k_V = k$ and we recover Lapwood's (1948) critical Rayleigh number of $4\pi^2 = 39.48$. When either of the horizontal or vertical permeabilities is much greater than the other, the modified Rayleigh number involves essentially the *lesser* of these permeabilities. The most unstable motion when Ra* just exceeds its critical value is in the form

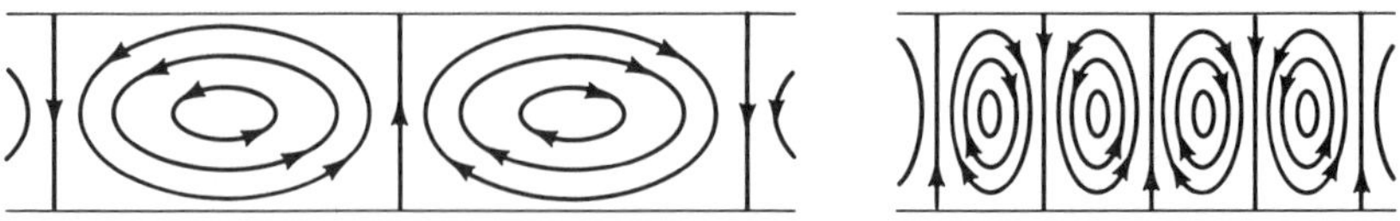

Figure 5.3. Cell shapes for the most unstable disturbances for k_H/k_υ = 10 and 0.1. When $k_H = k_\upsilon$, the cells are square.

of roll cells with horizontal width of π/l equal to $(k_H/k_V)^{+1/4}$ times the layer depth. When $k_H = k_V$, these are square; when $k_H \gg k_V$, they are elongated horizontally; and when $k_H \ll k_V$, they are compressed (Figure 5.3). The speeds of the fluid motion in these first unstable modes at slightly supercritical Rayleigh numbers gradually increase with time at the rate e^{nt}, where

$$n = \frac{\kappa}{\Gamma h^2 r^{1/2}} \{\mathrm{Ra}^* - (\mathrm{Ra}^*)_{\mathrm{crit}}\}, \qquad (5.1.16)$$

as can be seen from (5.1.10)–they ultimately stabilize at a finite speed that can be shown to be proportional to $\{\mathrm{Ra}^* - (\mathrm{Ra}^*)_{\mathrm{crit}}\}$ (see, e.g., Joseph 1976).

Experiments on the onset of Rayleigh–Darcy instability have been conducted by Elder (1967), Katto and Masuoka (1967), Combarnous and LeFur (1969), and Murray and Chen (1989), all in media for which $k_H = k_V$. Measured values of $\mathrm{Ra}_{\mathrm{crit}}$ were generally within 20 percent of the value $4\pi^2$ predicted by the theory, the major uncertainties, as Murray and Chen point out, being in the precise determination of the permeability k. The convective cells that developed were in each case two-dimensional, approximately square in cross section.

When the Rayleigh number is substantially supercritical, an increasingly wide band of wave numbers beomes unstable. Notable contributions by Palm, Weber, and Kvernvold (1972), Rudraiah and Srimani (1980), Borkowska-Pawlak and Kordylewski (1985), and others have shown that in the isotropic case at least, the patterns of convection become less regular and much more complicated as newly unstable modes grow to larger amplitudes than the roll cells and supplant them.

The numerical value $4\pi^2$ for the critical Rayleigh number in an isotropic medium is specific to the particular boundary conditions of constant temperatures and zero volume fluxes at the upper and lower surfaces. Nield (1968) found the critical values appropriate to a variety of other circumstances, showing, for example, that when the *heat flux* rather than the temperature is prescribed at upper and lower impermeable boundaries, the critical Rayleigh number is 12. Also, when the upper surface is at constant pressure (so that fluid can enter or leave across it) while the lower is impermeable, if the upper and lower temperatures are fixed, the critical value is 27.1; if the upper temperature and the lower heat flux are fixed, it is $\pi^2 = 9.87$; and if the heat fluxes across both are fixed, it is as small as 3.

When the porous layer is not horizontal, but slopes at an angle θ, the basic fluid state is not one at rest unless the isotherms happen to be horizontal. The particular case in which the temperatures at the upper and lower surfaces are uniform but with a temperature difference ΔT between them has been studied extensively, both theoretically and experimentally (see, e.g., Bories and Combarnous 1973, Combarnous and Bories 1975, and Caltagirone and Bories 1985). In this case, the isotherms also slope at an angle θ, and in a uniform isotropic medium, the undisturbed flow is one of uniform shear with vorticity given by (3.4.9):

$$\omega = -\frac{kg\alpha\,\Delta T \sin\theta}{h\nu}.$$

Various types of instability have been found to be possible, the simplest being longitudinal roll cells with their axes upslope. For these, the equations governing the transverse motions are identical to (5.1.6) and (5.1.7) (since $\partial/\partial y$, the derivative upslope, vanishes) except that $g\cos\theta$ replaces g. This type of instability then sets in when, for a uniform isotropic medium,

$$\mathrm{Ra}\cos\theta \leqslant 4\pi^2, \tag{5.1.17}$$

as confirmed experimentally by Bories and Combarnous (1973). At much higher values of Ra cos θ, the flows become oscillatory and geometrically more complex; the various states and transitions among them having been considered in detail by Caltagirone and Bories (1985). When the layer is vertical and a constant temperature is maintained across it, the flow is entirely stable, as shown by Gill (1969).

These analyses by Lapwood and later authors have been concerned with the problem of pure thermal convection, but the results can be extended immediately to the case of continuously saturated solutions in equilibrium with a major constituent of the fabric. The equation of state (2.10.4) replaces the last of the set (5.1.1); in the equations and therefore in the conclusions, the thermal density coefficient α_s of the continuously saturated solution replaces the ordinary thermal expansion coefficient α. For stability

$$\frac{g\alpha_s\,\Delta T\,h}{\nu\kappa}\,\frac{k_H k_V}{(k_H^{1/2} + k_V^{1/2})^2} \leqslant \pi^2. \tag{5.1.18}$$

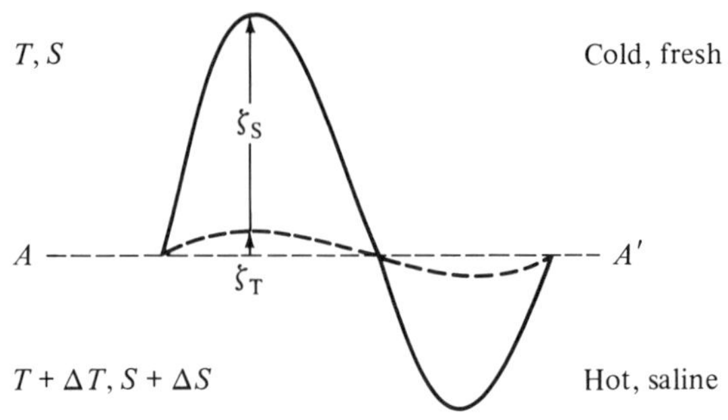

Figure 5.4. The displacements of the heat and salt interfaces in a non-diffusive saturated permeable medium.

Here, however, α_s may be negative, as it is for potassium salts and for Na_2SO_4 over temperatures less than 32° C, and if T is positive (heated at the bottom) the basic state is *unconditionally stable!* The increased density resulting from increased dissolution at the higher temperature overcomes the effect of thermal expansion. When $\alpha_s < 0$, the basic state, however, can be destabilized by heating at the *top*, so that $\Delta T < 0$, $\alpha_s \, \Delta T > 0$, and again the critical value of (5.1.18) can be exceeded if the temperature difference is sufficiently large.

5.2 SALINITY STABILIZATION

The instabilities described in the preceding section are clearly important, but the results should be applied to geological situations only with caution. When the salinity, as well as the temperature, of the interstitial fluid varies, the criteria for stability or instability can be quite different, particularly in rocks of moderate or low porosity. The two quantities that influence the interstitial fluid density, dissolved salts and heat, each advect and diffuse through the permeable fabric at different rates, and these differences give rise to a variety of phenomena of geological interest.

Consider first the consequences of the differential *advection* rates of dissolved salts and heat, the most notable of which is the stabilization of an apparently thermally unstable situation by variations in salinity. The essence of the matter can be seen most simply in a two-layer case, when, for example, a layer containing fresh interstitial water lies over a hot, brine-saturated substrate. Suppose, as in Figure 5.4, the lower layer has a temperature $T + \Delta T$ and a salinity of the interstitial fluids $S + \Delta S$, while above it the temperature and salinity are T and S. Initially, suppose the interface AA' between the two regions is horizontal. If the density of the upper fluid is ρ_0, that in the lower layer is $\rho_0(1 - \alpha\,\Delta T + \beta\,\Delta S)$ from (2.10.3). When $\beta\,\Delta S > \alpha\,\Delta T$, the upper density is smaller and one would guess that the basic state is stable, but in fact a salinity difference can stabilize the state even if this condition is not satisfied and the *upper fluid is more dense than the lower*. To demonstrate this, imagine that at some point the interstitial fluid is displaced upward by a distance ζ_S, carrying its salinity with it. The volume of fluid per unit area involved in this displacement is, however, $\phi\zeta_S$, where ϕ is the porosity; this carries upward excess heat per unit area of the amount $(\rho C)_f\,\Delta T\,\phi\zeta_S$. If the displacement of the temperature front is ζ_T, the excess heat above the initial level is $(\rho C)_m\,\Delta T\,\zeta_T$, so that

$$\zeta_T = \frac{(\rho C)_f}{(\rho C)_m}\,\phi\zeta_S = \frac{\phi}{\Gamma}\,\zeta_S < \zeta_S, \tag{5.2.1}$$

with the notation of (2.8.5). Now, in the interstitial fluid above the original interface, the excess mass of the disturbed column per unit horizontal area is

$$-(\alpha\,\Delta T)\zeta_T + (\beta\,\Delta s)\zeta_S = \left\{-\frac{\phi}{\Gamma}\,\alpha\,\Delta T + \beta\,\Delta s\right\}\zeta_S,$$

which, if positive when ζ_S is positive, tends to restore equilibrium. Accordingly, the equilibrium is stable if

$$\beta\,\Delta s > \frac{\phi}{\Gamma}\,\alpha\,\Delta T = \phi\,\frac{(\rho C)_f}{(\rho C)_m}\,\alpha\,\Delta T. \tag{5.2.2}$$

Note that the numerical value of Γ is about 1, so that when the porosity ϕ is small, even a small salinity difference can offset the destabilizing effect of the temperature difference.

The basic physics involved in this salinity stabilization is simple: Individual elements of saline fluid, carrying dissolved salts, move on average at the *interstitial* velocity $\mathbf{v}$ of Section 2.4, which is larger by the factor ϕ^{-1} than the transport velocity, adjusted by the specific heat ratio, with which the heat is advected. When the porosity ϕ is small, the salt front far outstrips the heat front, and stability is stronger. Note, again, that the undisturbed interstitial fluid density at *greater depth may be less than that above* while maintaining stability. The condition that the density of the interstitial fluid in the lower layer be greater than that above, $\beta\,\Delta s > \alpha\,\Delta T$, is much more restrictive than (5.2.2), particularly when the porosity is small. Finally, note that diffusion is not essential to this process as it is in the Rayleigh–Darcy stability criterion, though the diffusion of heat (but not salt) into each grain is implicitly at the heart of the difference in advection velocities.

A general nondiffusive stability criterion

This simple two-layer situation exhibits clearly the stabilizing effect, but a much more general nondiffusive stability criterion can be developed for arbitrary vertical distributions of temperature $T(z)$ and interstitial fluid salinity $S(z)$. As before, the analysis can, without loss of generality, be restricted to two dimensions; the governing equations are the Darcy vorticity balance (3.4.5), the equation of state, and the linearized conservation equation for small perturbations of temperature and dissolved salt (2.10.3). Molecular diffusivity is not essential to the process, so that it will be neglected for the moment – the modifications it makes will be considered in the next section. The perturbation equations are

$$\frac{1}{k_V}\frac{\partial^2\psi'}{\partial x^2} + \frac{1}{k_H}\frac{\partial^2\psi'}{\partial z^2} = \frac{g}{\nu}\left(-\alpha\frac{\partial T'}{\partial x} + \beta\frac{\partial S'}{\partial x}\right),$$
$$\Gamma\frac{\partial T'}{\partial t} - \frac{\partial\psi'}{\partial x}\frac{\partial T}{\partial z} = 0, \qquad (5.2.3)$$
$$\phi\frac{\partial S'}{\partial t} - \frac{\partial\psi'}{\partial x}\frac{\partial S}{\partial z} = 0,$$

where now the ambient temperature and salinity gradients $\partial T/\partial z$ and $\partial S/\partial z$ may be any assigned functions of z, measured upward. Since the coefficients of the perturbed quantities in this set are independent of x and t, we may seek solutions of the form

$$(\psi', T', S') = (\varphi(z), \theta(z), \sigma(z))e^{nt}e^{ilx}, \qquad (5.2.4)$$

where φ, θ, and σ are (complex) eigenfunctions whose form depends on $\partial T(z)/\partial z$ and $\partial S(z)/\partial z$. The substitution of (5.2.4) into (5.2.3) gives

$$\left(\frac{1}{k_H}\frac{d^2}{dz^2} - \frac{l^2}{k_V}\right)\varphi = \frac{ilg}{\nu}(-\alpha\theta + \beta\sigma), \qquad (5.2.5)$$

$$\Gamma n\theta - il\frac{\partial T}{\partial z}\varphi = 0, \qquad (5.2.6)$$

$$\phi n\sigma - il\frac{\partial S}{\partial z}\varphi = 0.$$

The eigenfunctions θ and σ can be eliminated from (5.2.5) by substitution from (5.2.6), whence

$$n\left(\frac{1}{k_H}\frac{d^2\varphi}{dz^2} - \frac{l^2}{k_V}\varphi\right) + \frac{gl^2}{\nu}\left(-\frac{\alpha}{\Gamma}\frac{\partial T}{\partial z} + \frac{\beta}{\phi}\frac{\partial S}{\partial z}\right)\varphi = 0,$$

or

$$n\left(\frac{1}{k_H}\frac{d^2\varphi}{dz^2} - \frac{l^2}{k_V}\varphi\right) - \frac{gl^2}{\nu}\Pi(z)\varphi = 0, \qquad (5.2.7)$$

where

$$\Pi(z) = \frac{\alpha}{\Gamma}\frac{\partial T}{\partial z} - \frac{\beta}{\phi}\frac{\partial S}{\partial z}. \qquad (5.2.8)$$

Boundary conditions are needed to complete the specification. If the porous layer is confined between horizontal impermeable strata, then $\varphi = 0$ on $z = 0, h$, say. Alternatively, if the stratified region is of finite depth with the porous medium extending well above and below, the disturbances vanish far away from the stratified region and $\varphi \to 0$ when $z \to \pm\infty$. Again, the geometry may be a combination of these–there may be a tongue of saline water near the base of a thick layer above an impermeable basement, so that $\varphi = 0$ at $z = 0$ and $\varphi \to 0$ as $z \to \infty$. Another possibility is that the upper surface of the permeable layer may be covered with water, so that the pressure there is constant and now $\partial\varphi/\partial z = 0$ on this surface. The explicit solutions to (5.2.7) depends on the distribution $\Pi(z)$ and on which of these boundary conditions are pertinent, but stability conditions can be obtained without actually finding the solution.

Since the physical variable ψ' is real, φ in (5.2.4) is necessarily complex. Multiply (5.2.7) throughout by its complex conjugate φ^*, whence

$$n\left\{\frac{1}{k_H}\frac{d^2\varphi}{dz^2}\varphi^* - \frac{l^2}{k_V}\varphi\varphi^*\right\} - \frac{gl^2}{\nu}\Pi(z)\varphi\varphi^* = 0,$$

or

$$n\left\{\frac{1}{k_H}\frac{d}{dz}\left(\frac{d\varphi}{dz}\varphi^*\right) - \frac{1}{k_H}\frac{d\varphi}{dz}\frac{d\varphi^*}{dz} - \frac{l^2}{k_V}\varphi\varphi^*\right\} - \frac{gl^2}{\nu}\Pi(z)\varphi\varphi^* = 0. \qquad (5.2.9)$$

Now integrate vertically throughout the layer. Under any type of boundary condition described above, $\varphi^*\, d\varphi/dz$ vanishes at the extremes, and after rearrangement, (5.2.9) becomes

$$n\int\left\{\frac{1}{k_H}\frac{d\varphi}{dz}\frac{d\varphi^*}{dz} + \frac{l^2}{k_V}\varphi\varphi^*\right\}dz = -\frac{gl^2}{\nu}\int\Pi(z)\varphi\varphi^*\,dz. \qquad (5.2.10)$$

The product of any function and its own complex conjugate is real and positive. Consequently, both the integrals are real and so is the growth rate n – the time history is a simple exponential growth or decay with marginal stability at $n = 0$. Moreover, the first integral is definitely positive; if n is to be positive (unstable situations), then $\Pi(z)$ must be negative over at least part of the range. To express the same thing in another way, if $\Pi(z)$ is positive everywhere (or possibly zero over some intervals), then n is necessarily negative and the situation is *stable*. Thus, a condition that guarantees stability when thermal diffusivity can be neglected is that

$$\Pi(z) = \frac{\alpha}{\Gamma}\frac{\partial T}{\partial z} - \frac{\beta}{\phi}\frac{\partial S}{\partial z} \geqslant 0 \qquad (5.2.11)$$

throughout the domain. If $\Pi(z) < 0$ throughout the domain, *instability* is guaranteed.

There are a couple of important things to note about this result. First, it is a *local* criterion, not a global one like the Rayleigh number criteria of the preceding section; it pertains to the distribution throughout the region rather than some overall parameter. Second, in mathematical terms, (5.2.11) is a sufficient condition to guarantee stability when heat conduction and salt diffusion effects can be ignored, but there may be some situations in which it is not satisfied, yet the basic state is in fact stable – in the Rayleigh–Darcy problem $\partial T/\partial z < 0$ and $\partial S/\partial z = 0$; yet the basic state is stable at small Ra*. Finally, a different class of instabilities can, in fact, be driven by the differences in thermal and salt diffusivities, as will be described in Section 5.3.

One field example in which salinity stabilization appears to occur has been described by Bjørlykke and Palm (1988). Analysis of formation waters from sedimentary basins like the North Sea shows evidence of continued stratification of the pore waters with respect to salinity and isotopic composition in spite of the increase in temperature with depth, so that the positive salinity gradient with depth appears to have inhibited vertical mixing.

5.3 DOUBLE-ADVECTIVE, DOUBLE-DIFFUSIVE INSTABILITIES

The results of the preceding section pertain to the nondiffusive limit κ, $\kappa_s \to 0$. Dissolved salts advect more rapidly through the fabric than does heat, but in an ambient state of little or no motion, they generally diffuse less rapidly. When the basic state is truly at rest, the diffusivity for dissolved salts κ_s may be taken as the molecular value; for this $\kappa_s/\kappa \sim 10^{-2} \ll 1$. In any geological example, however, there may be some feeble background drift, so that a macroscopic dispersion coefficient may be more appropriate, especially if it is enhanced by mixing in fabric inhomogeneities as described in Section 3.9. A general expectation, however, is that κ_s/κ will remain small, at least as any initial instability develops. When this is so, not only are the nondiffusive stability criteria (and those of Section 5.1) modified somewhat but also some new modes of double-diffusive instability occur that have potential geological application.

For example, when a saline layer of interstitial fluid is overlain by a fresher layer in the presence of a maintained thermal flux, the interface between the two regions may be stabilized by the strong local salinity gradient, whereas the individual layers may be thermally unstable and develop convective circulations. Across the interface between the two layers, heat is *diffused* much more rapidly than salt, so that the thermal flux continues to be supported, even though the salt layering persists. Important laboratory experiments by Griffiths (1981) have demonstrated the persistence of these layers; although the Rayleigh numbers involved (of order 10^2 to 10^3) are much higher than would be expected in most geological situations, it is possible, as he suggests, that they may be found in some geothermal systems, such as that in Wairakei, New Zealand.

In order to explore the interplay among the differences in advective velocities and diffusivities for heat and salt, it is instructive to reconsider the Rayleigh–Lapwood problem with a *uniform* temperature gradient $-\Delta T/h$ and a coexisting salinity gradient $-\Delta S/h$. This problem was first considered by Nield (1968) although in his analysis the advection rates for heat and salt were taken to be the same. His results, then, may be pertinent to highly porous media (like fiberglass) but should not be interpreted in geological contexts where ϕ is small and the differential advection speed ratio is large. The governing equations are the same as (5.2.3) but with the diffusive terms included:

$$\begin{aligned} r\frac{\partial^2\psi'}{\partial x^2}+\frac{\partial^2\psi'}{\partial z^2} &= \frac{gk_H}{\nu}\left(-\alpha\frac{\partial T'}{\partial x}+\beta\frac{\partial S'}{\partial x}\right),\\ \Gamma\frac{\partial T'}{\partial t}+\frac{\Delta T}{h}\frac{\partial\psi'}{\partial x} &= \kappa\,\nabla^2 T',\\ \phi\frac{\partial S'}{\partial t}+\frac{\Delta S}{h}\frac{\partial\psi'}{\partial x} &= \phi\kappa_s\,\nabla^2 S', \end{aligned}\tag{5.3.1}$$

where $r = k_H/k_V$. It simplifies the algebra to define dimensionless variables at this stage. Let

$$\begin{aligned} S' = (\Delta S)\sigma, \qquad T' &= (\Delta T)\theta, \qquad (x, z) = (h\xi, h\eta),\\ \psi' &= \frac{k_H g\alpha\,\Delta T\,h}{\nu}\psi = (\kappa\,\mathrm{Ra}_H)\,\psi,\\ t &= (h^2/\kappa)\tau, \end{aligned}\tag{5.3.2}$$

where Ra_H is the Rayleigh number (5.1.11) based on the horizontal permeability. In terms of these, (5.3.1) becomes

$$\begin{aligned} r\frac{\partial^2\psi}{\partial\xi^2}+\frac{\partial^2\psi}{\partial\eta^2} &= -\frac{\partial\theta}{\partial\xi}+\frac{\beta\,\Delta S}{\alpha\,\Delta T}\frac{\partial\sigma}{\partial\xi},\\ \Gamma\frac{\partial\theta}{\partial\tau}+(\mathrm{Ra})_H\frac{\partial\psi}{\partial\xi} &= \nabla^2\theta,\\ \phi\frac{\partial\sigma}{\partial\tau}+(\mathrm{Ra})_H\frac{\partial\psi}{\partial\xi} &= \frac{\phi\kappa_s}{\kappa}\nabla^2\sigma, \end{aligned}\tag{5.3.3}$$

where the Laplacian operators are now in the dimensionless space variables.

It is interesting to note now that if at marginal stability $\partial/\partial\tau = 0$, then

$$\frac{\phi\kappa_s}{\kappa}\nabla^2\sigma = \nabla^2\theta;$$

the magnitude of the temperature perturbation as a fraction of ΔT is smaller than that of the salinity fluctuation referred to ΔS by the factor $\Phi\kappa_s/\kappa \ll 1$; temperature perturbations advect more slowly by the factor ϕ and diffuse more rapidly by the factor κ/κ_s and so are smoothed out more effectively.

To examine the stability characteristics, the procedure is the same as in Section 5.1, but the algebra is somewhat more tedious. Since the coefficients in (5.3.3) are independent of ξ, η, and τ, solutions of the type $e^{n\tau} \sin m\pi\eta \exp il\xi$ can be sought, and this leads to a quadratic equation for n, the growth rate,

$$an^2 + bn + c = 0, \tag{5.3.4}$$

where

$$\begin{aligned}
a &= \Gamma \sim 1,\\
b &= (l^2 + m^2\pi^2)\left(1 + \Gamma\frac{\kappa_s}{\kappa}\right) - \frac{l^2}{rl^2 + m^2\pi^2}\mathrm{Ra_H}(1 - \Gamma\Sigma),\\
c &= (l^2 + m^2\pi^2)^2\frac{\kappa_s}{\kappa} + (l^2 + m^2\pi^2)\frac{l^2}{rl^2 + m^2\pi^2}\mathrm{Ra_H}\left(\Sigma - \frac{\kappa_s}{\kappa}\right),\\
\Sigma &= \frac{\beta\,\Delta S}{\phi\alpha\,\Delta T},\\
\mathrm{Ra_H} &= \frac{k_H g\alpha\,\Delta T\,h}{\nu\kappa}.
\end{aligned} \tag{5.3.5}$$

It is not now obvious from the algebra that the roots of (5.3.4), namely,

$$n = (2a)^{-1}\{-b \pm (b^2 - 4ac)^{1/2}\}, \tag{5.3.6}$$

are necessarily real and there are four different cases that must be considered separately. They are illustrated in Figure 5.5.

Temperature destabilizing, salinity stabilizing

In this case, cooler, fresher interstitial water lies above warmer, more saline water. When the salinity contrast is zero, it reduces to the Rayleigh–Lapwood problem of Section 5.1; one would expect intuitively that the salinity gradient would enhance the stability, and this can be shown simply. Here $\alpha\,\Delta T > 0$, $\beta\,\Delta S > 0$, so that both $\mathrm{Ra_H}$ and Σ are positive. When $\Sigma > \kappa_s/\kappa$ ($\ll 1$), $c > 0$ for all values of l and m, so that if the roots (5.3.6) are real, they have the same sign, opposite to that of

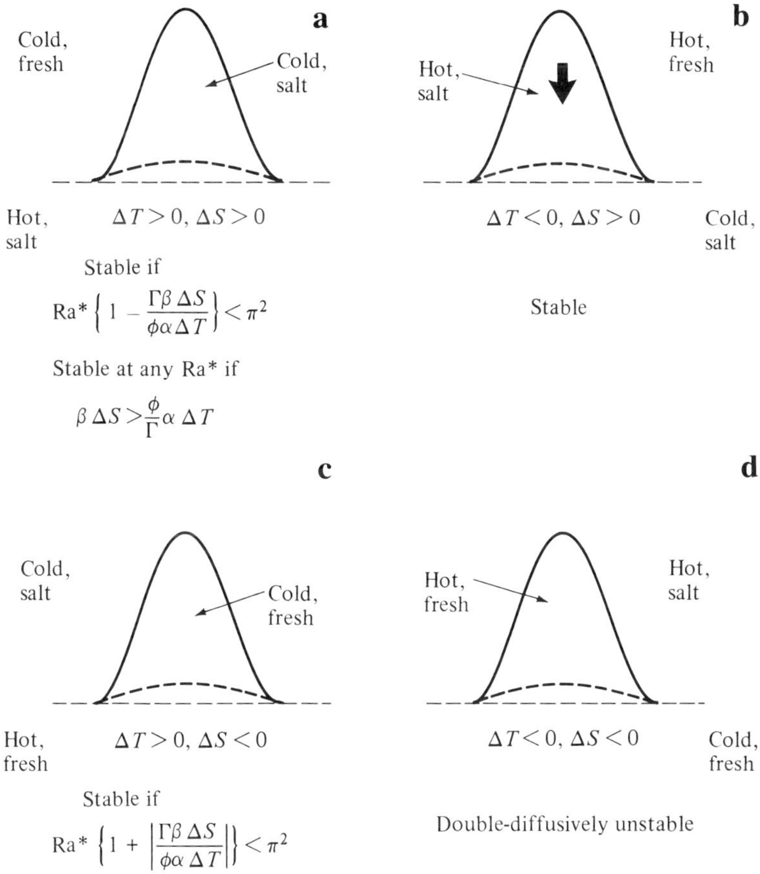

Figure 5.5. Isohaline (solid line) and isotherm (dashed line) displacements in diffusive saturated permeable media, with the associated stability characteristics when $\kappa \gg \kappa_s$. In case (c), when $\Delta T \to 0$, the stability condition given is replaced by one analogous to equation (5.1.15) but with $-\beta\Delta S/\phi\kappa_s$ replacing $\alpha\,\Delta T/\kappa$.

b. If perchance they are complex, their real parts again have sign opposite to that of b. In either event, then, the condition $b > 0$,

$$1 + \Gamma\frac{\kappa_s}{\kappa} > \frac{l^2}{(l^2 + m^2\pi^2)(rl^2 + m^2\pi^2)}\mathrm{Ra}_H(1 - \Gamma\Sigma)$$

guarantees stability to a disturbance of horizontal wave number l and mode m. The term $\kappa_s/\kappa \ll 1$ can be neglected, so that the stability criterion is

$$\mathrm{Ra_H}\left(1 - \frac{\Gamma\beta\,\Delta S}{\phi\alpha\,\Delta T}\right) < l^{-2}(l^2 + m^2\pi^2)(rl^2 + m^2\pi^2).$$

This is of precisely the same form as (5.1.12) with the additional factor on the left-hand side, so that, as before, the condition is most restrictive when $m = 1$ and the state is stable to all wave numbers l when

$$\mathrm{Ra}^*\left(1 - \frac{\Gamma\beta\,\Delta S}{\phi\alpha\,\Delta T}\right) < \pi^2. \tag{5.3.7}$$

The term in parentheses exhibits the salinity stabilization: It is less than or equal to unity, so that the critical value of Ra* increases from its value of π^2 when $\Delta S = 0$ to infinity as $\Gamma\Sigma = (\Gamma\beta\,\Delta S/\phi\alpha\,\Delta T) \to 1$. When $\Gamma\Sigma > 1$, the factor is negative, so that stability is ensured for *any* Ra*, as is consistent with the nondiffusive ($\mathrm{Ra}^* \to \infty$) criterion developed in Section 5.2. When $\Gamma\Sigma < 1$, the motion that first develops as the modified Rayleigh number becomes supercritical is again in the form of the large roll cells of Figure 5.3. In summary, the effect of salinity variation in this case is simply to delay (when $\Gamma\Sigma < 1$) or avert (when $\Gamma\Sigma > 1$) the onset of instability as Ra* increases without any change in its basic nature.

Important experiments on double-diffusive convection with a destabilizing temperature gradient and a stabilizing salinity gradient have been described by Murray and Chen (1989). A linear temperature distribution with $\partial T/\partial z < 0$ was maintained by separate baths above and below the upper and lower conducting boundaries, whose temperatures were controlled; the salinity gradient was produced by filling the box in four layers, which were allowed to diffuse. The internal salinity interfaces disappeared fairly rapidly, but since there was no salt flux across the upper and lower surfaces, $\partial S/\partial z = 0$ there, and gradually thickening, almost isohaline layers slowly developed. In these regions, the nondiffusive stability criterion (5.2.11) is not satisfied, so that instability is clearly possible. Moreover, since the region of constant salinity gradient occupied only the central region of the depth interval, the result (5.3.7) cannot be applied, but in any event, the salinity stabilization effect was evident. With a strong salinity gradient, the critical Ra increased fourfold. When Ra was large enough to induce instability, large three-dimensional cells were formed and the heat flux increased dramatically. To account quantitatively for the measurements on the onset of instability, it was necessary to take into account both the nonlinear salinity profile and the temperature dependence of the thermal expansion coefficient α – a reminder that the simple criteria such as (5.3.7) are important conceptually but probably unreliable numerically if applied to geological situations in which the conditions envisioned in the theory (both gradients constant) are not precisely satisfied.

Both temperature and salinity stabilizing

In this situation, hot, fresh interstitial water lies over cold, more saline water; in terms of the problem definition, $\alpha\ \Delta T < 0$, $\beta\ \Delta S > 0$. Consequently, $\mathrm{Ra_H} < 0$ and $\Sigma < 0$. In (5.3.5), c is then positive for all l, m, as is b. The two roots (5.3.6) then have negative real parts, and the equilibrium is, not surprisingly, always stable.

Both temperature and salinity destabilizing

Now cold, more saline water lies over fresher, hotter water; $\alpha\ \Delta T > 0$ and $\beta\ \Delta S < 0$, so that $\mathrm{Ra_H} > 0$ and $\Sigma < 0$. From (5.3.6) both roots are negative (or have negative real parts) if $c < 0$ and $b > 0$ for all l, m, and this can be shown again to require the condition (5.3.7). Since, however, $\beta\ \Delta S < 0$ in this case, the factor in parentheses is larger than unity, so that the critical Rayleigh number decreases toward zero as the term $\Gamma\beta\ \Delta s/\phi\alpha\ \Delta T$ becomes more negative – diffusive smearing is less and less able to maintain stability.

Temperature stabilizing, salinity destabilizing

This situation, when hot, saline interstitial water lies over cooler, fresher water, is possibly the most interesting, being associated with the phenomenon of double-diffusive fingering. The latter is similar in its basic dynamics to the double-diffusive instability in ordinary viscous fluids described by Turner (1973) in his fine book *Buoyancy Effects in Fluids*. If an element of interstitial fluid is displaced downward, it loses heat by diffusion but hardly any salt; it is then denser than the surrounding fluid and tends to continue downward. If displaced upward, it becomes warmer with its surroundings but remains fresher, and so is less dense and continues upward. Interleaving fingers or vertical sheets then form and continue as long as the salt flux downward is maintained. This instability was predicted theoretically by Nield (1968), but has not yet been demonstrated in the laboratory in a heat-salt system because of the experimental difficulties associated with lateral heat losses in an apparatus of limited horizontal dimensions. Imhoff and Green (1988) have shown, however, that this kind of fingering develops in the much less efficient double-diffusive system using gradients of sugar and salt in the interstitial fluid. Salt (NaCl) has a higher molecular diffusivity than does sugar, but by a factor of only 3, and the effective advection velocities are the same for the two species, whereas in the heat-salt system, the thermal diffusivity is greater than that for salt by a factor of 100 and the salt advection velocity is larger by the factor Γ/ϕ. Nevertheless, even in this less efficient case (the more diffusive species, salt, having the same dynamical influence as cold and the less diffusive one, sugar, playing the same role as salt in the heat-salt system), fingers developed that were about ten times as long as they were wide. The net vertical fluxes were about two orders of magnitude larger than the molecular diffusion in a motionless interstitial fluid. In a heat-salt system, the motion would be expected to be more vigorous and the fingers relatively longer in view of the greater diffusivity contrast together with the more rapid salt advection compared with that of heat.

Algebraically, one of the roots of (5.3.6) is real and positive (indicating instability) when $c < 0$. In the present situation, $\alpha\,\Delta T < 0$ and $\beta\,\Delta S < 0$, so that $\Sigma > 0$ and $\mathrm{Ra_H} < 0$. Let $\mathrm{R}^- = -\mathrm{Ra_H} > 0$; for small-scale, vertically moving disturbances $e^{n\tau}e^{il\xi}$, $m = 0$ and from (5.3.5)

$$c = l^2\left\{l^2\frac{\kappa_s}{\kappa} - \frac{\mathrm{R}^-}{r}\left(\Sigma - \frac{K_s}{k}\right)\right\}$$

and $c < 0$ when

$$l^2\frac{\kappa_s}{\kappa} < \frac{\mathrm{R}^-}{r}\left(\Sigma - \frac{\kappa_s}{\kappa}\right). \tag{5.3.8}$$

Accordingly, when

$$\phi\Sigma = \frac{\beta\,\Delta S}{\alpha\,\Delta T} > \phi\frac{\kappa_s}{\kappa}, \tag{5.3.9}$$

the basic state is unstable to some wave numbers at any Rayleigh number R^-. The right-hand side of (5.3.9) is numerically very small. When $|\alpha\,\Delta T| > |\beta\,\Delta S|$, the density of the interstitial fluid decreases in the vertical and the basic state is apparently statically stable; however, it is unstable in this double-diffusive manner when

$$1 > \frac{\beta\,\Delta S}{\alpha\,\Delta T} > \phi\frac{\kappa_s}{\kappa}. \tag{5.3.10}$$

In practical terms, since $\phi\kappa_s/\kappa$ is so small, unless the salinity contrast is vanishingly small, a situation of hotter, more saline interstitial fluid lying over cooler, fresher fluid is *always unstable* even if the overall fluid density decreases vertically. The resulting circulations are in the form of interleaving small vertical fingers or sheets, providing effective vertical transport of dissolved ions in the interstitial fluid. The case of cooler, more saline water overlying fresher, warmer water is also highly unstable, so that as a general rule, instability and vertical transport are almost inevitable when more saline interstitial water overlies a fresher interstitial fluid region, whatever the temperature structure. Geological scenarios of this kind are not at all uncommon, particularly in evaporative situations, one example being described in the next section.

5.4 BRINE INVASION BENEATH HYPERSALINE LAGOONS BY DOUBLE-DIFFUSIVE FINGERING

When during the course of time, a coastal embayment in an arid climate becomes landlocked, the excess of evaporation over precipitation leads to an increasing salinity in the lagoonal water, so that it becomes hypersaline relative to a seawater-saturated fabric below. Even if the (perhaps hot) lagoonal water is less dense than in the substrate, the stage is set for double-diffusive fingering of the interstitial water, leading to vertical solute transfer and geochemical alteration of the host rock. This may well have occurred in the formation of certain bedded dolomites by the post-depositional alteration of limestone shelves of the Permian Basin in the southwestern United States.

The origin of dolomites has long been enigmatic. The principal problems, as listed by Adams and Rhodes (1960), include (1) the existence of an adequate source of chemically active magnesium to allow the degree of dolomitization observed, (2) the existence of a plausible and natural internal flow system that can transport the magnesium into, and distribute it throughout, the host limestone, while at the same time removing the displaced calcium ions, and (3) the provision of a chemically favorable environment in which the reaction can occur. Commonly, dolomite is associated with beds of evaporite such as salt, gypsum, or anhydrite. Often, however, there is no such association. Some coral atolls are extensively dolomitized at depth. This suggests that there is no unique physical or geological environment associated with dolomitization – the chemistry may be the same, but the internal flow systems needed for magnesium transport may be of several kinds, the circulation beneath hypersaline lagoons being only one (for another, see Section 7.7).

The Permian Basin of western Texas and New Mexico is a huge area, rich in petroleum, natural gas, and potash, with excellent outcrops as a result of uplift and erosion. It has been extensively drilled. The economic importance of the region has stimulated many studies reconstructing the Permian geography. Some 250 millions years ago, this region was the site of a partly landlocked embayment extending more than 1,500 km into the continental interior with shell- and skeleton-forming shallow-water fauna building the basin floor upward as subsidence lowered it. Adams and Rhodes (1960) visualized extensive shallow lagoons perhaps 300 km wide and 3 m deep, evaporating in an arid climate but sporadically replenished by new seawater, so that the salinity gradually increased from that of normal seawater (35 ‰) to values five or even ten times higher when halite (rock salt) deposits began to form. The hot, saturated solution with excessively high concentrations of magnesium, potassium, sodium, and chlorine provides the chemical source of dolomitization, but in order to distribute the solutes through the carbonate substrate, Adams and Rhodes were forced to postulate a dynamically implausible density-driven flow pattern (which they called "seepage refluxion"). This hypothesis is clearly unnecessary in the light of the results of the preceding section. The substrate with hypersaline interstitial fluid at the top and less saline fluid below is unstable to double-diffusive fingering, which provides a natural (and inevitable) mechanism for the vertical transport down of magnesium and other solutes in the downward-moving fingers and the transport up of calcium ions released in the dolomite reaction in the upward-moving ones. *The brine that provides the chemical source also drives the flow.* These simultaneous counterfluxes offer a conceptual solution to the transport problem in this environment, but the question remains as to whether the magnitudes of the fluxes are sufficient to provide the magnesium needed in a reasonable time interval.

In convective instabilities, the flow pattern that develops is in essence that of the most rapidly growing unstable mode. In a permeable matrix, as the disturbance grows, the diffusivity for salt increases as macroscopic dispersion replaces molecular diffusivity, and this relaxes the driving force for double-diffusive convection; the instability evolves toward a steady circulation. In a uniform medium, the fingers are vertical, but in a natural geological environment, the flow, though on average vertical, follows lower-permeability layers and is considerably less regular. Nevertheless, an order-of-magnitude estimate of the vertical fluxes can be found simply by balancing the vertical driving force (the horizontal variation in buoyancy produced by salinity variations) by the resistance to flow in the medium:

$$\frac{\mu}{k_V} w = \rho_0 g \beta s'.$$

If Δs represents the difference in salinity between the downward- and upward-moving fingers or sheets, the magnitude of the vertical velocities up or down is

$$w \sim \frac{k_V g \beta \, \Delta s}{\nu}, \tag{5.4.1}$$

ignoring numerical factors of order unity. The total volume of fluid entering (and leaving) the matrix per unit area per unit time is then also given by (5.4.1), and in a layer of thickness h, the flushing time, or the time required for a volume of water to pass through the system that is equal to the volume of the system itself, is

$$T = \frac{h}{w} \sim \frac{h\nu}{k_V g \beta \, \Delta s}. \tag{5.4.2}$$

Numerical values are of interest. In the west Texas, New Mexico basin, the thickness $h \sim 500$ m, and k_V may be as small as 10^{-10} cm^2, or 10^{-14} m^2. Now $\beta\ \Delta s$ represents the fractional difference in densities resulting from salinity differences in the downward- and upward-moving columns. This is of the same order but somewhat less than the density difference between the hypersaline lagoon water and the interstitial water deep in the bank. As time passes, both become more saline, but the flow continues as long as the difference remains. If, for example, we take $\beta\ \Delta s \sim 0.05$ and $\nu = 10^{-2}$ c.g.s, then

$$w \sim 5 \times 10^{-7} \text{ cm s}^{-1} = 15 \text{ cm yr}^{-1},$$

and

$$T \sim 3{,}000 \text{ yr}.$$

The time required to complete the conversion of limestone to dolomite is, of course, greater than this, since the molar concentration of magnesium in even saturated brine is less than that of the dolomite produced. An accurate calculation requires a detailed chemical model of the sort outlined by Wilson et al. (1990), but the thermodynamic parameters in this case are not yet well established. A lower limit for the time interval can be obtained by assuming that all of the magnesium in the fluid is available for dolomitization. One liter of dolomite contains about 11 mol magnesium, and if the concentration in the convecting fluid is about ten times that of normal seawater, as Adams and Rhodes (1960) suggest, its concentration is about 0.5 molal. Some twenty-two flushing times are thus required, giving a lower limit of about 70,000 years. The actual time for dolomitization is certainly much greater than this, but even if this estimate is too small by a factor of 10, it is still sufficiently high to provide the needed magnesium within a geologically short time.

Another example of apparent thermohaline overturn has been described by Hanor (1987). Preferential dissolution near the top of the Iberia salt dome, Louisiana, combined with bleeding of less saline water from a geopressurized region below, provides salinity destabilization, as described in the preceding section. Hanor used the simple (restrictive) criterion of a decrease in density of pore fluids with depth to infer the existence of convective instability, but the results of Sections 5.2 and 5.3 indicate that instability and convective circulation will occur more widely than even he anticipates.

5.5 STABILITY IN THE PRESENCE OF FLOW

Potentially unstable situations may also occur when the basic state involves fluid flow, but aside from the studies on inclined uniform layers mentioned at the end of Section 5.1, little is known (or can be inferred simply) about the modifications produced to these stability criteria. If the

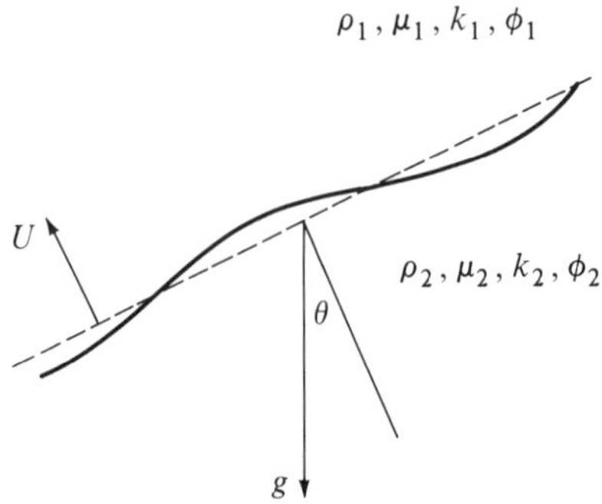

Figure 5.6. A disturbance to a moving interface between two fluids in a saturated permeable medium.

transport velocity is uniform in the flow direction but may vary across the flow in the vertical, the stability or instability of disturbances whose wave numbers are transverse (i.e., in which the disturbance is uniform in the flow direction) is unaltered by the flow. If the velocity is uniform across the layer, in the case of purely thermal instability $\Gamma\ \partial T/\partial t$ in the fourth equation (5.1.1) is replaced by $\Gamma\ \partial T/\partial t + U\ \partial T/\partial x$, but the original set of equations can be recovered by taking a frame of reference moving along the layer with speed U/Γ. The stability criteria are then again unaltered, but the modes (whether growing or decaying) propagate along the layer with this speed.

When the permeable layer contains one or a series of lenses elongated in the flow direction where the transport speed is much larger than in the surrounding matrix, modes with wave numbers in the flow direction are likely to be stabilized, though no detailed analysis of this seems to have been done. The high flow velocity in the lenses will tend to make more uniform the temperature and salinity distributions along the lenses, so that they will act rather like constant temperature and salinity boundaries, and the effective depth of the layer is the depth between lenses or between a lens and the boundary. For these modes, the effective Rayleigh number is reduced, promoting stability. The most unstable modes will then be those with transverse wave numbers in which the disturbance is elongated in the flow direction; for these, as already mentioned, the flow is irrelevant.

5.6 SAFFMAN–TAYLOR INSTABILITY

A quite different kind of instability, very important in the petroleum industry, occurs when an interstitial fluid of viscosity μ_1 and density ρ_1 is displaced by another of viscosity μ_2 and density ρ_2. Though its existence seems to have been known in the oil industry, the first account of it was published in the scientific literature by Hill in 1952, and a clear perception of the basic mechanism was provided by Saffman and Taylor (1958); the instability is generally associated with their names. Its nature can be seen most simply by considering a sharp interface across which the viscosity and density change abruptly. Capillary forces will be neglected for the moment, and the medium supposed to be locally isotropic. When the interface moves through the medium with speed U, the *transport* speed on either side is ϕU normal to the interface. Suppose now that the interface is not quite planar, but develops a small corrugation $ae^{inx+\sigma t}$, as shown in Figure 5.6, with wavelength $2\pi/n$. It grows in amplitude if σ is positive and diminishes if negative. Since the perturbation is small ($an \ll 1$), the boundary condition (3.3.2) between the two regions simplifies to $\phi\xi = w$ and

$$\xi = \frac{w_1}{\phi_1} = \frac{w_2}{\phi_2} = U + \sigma a e^{inx+\sigma t}.$$

The corresponding pressure distribution in front of and behind the interface satisfies Laplace's equation, and the disturbance pressures must vanish at infinity; they are readily found to be

$$p_1 = \rho_1 gz \cos\theta - \frac{\mu_1\phi_1}{k_1}\left(Uz - \frac{\sigma a}{n} e^{inx-nz+\sigma t}\right),$$

$$p_2 = -\rho_2 gz \cos\theta - \frac{\mu_2\phi_2}{k_2}\left(Uz + \frac{\sigma a}{n} e^{inx+nz+\sigma t}\right).$$

These must be the same on either side of the interface, and when $an \ll 1$,

$$\frac{\sigma}{n}\left(\frac{\mu_1\phi_1}{k_1} + \frac{\mu_2\phi_2}{k_2}\right) = (\rho_1 - \rho_2)g\cos\theta + \left(\frac{\mu_1\phi_1}{k_1} - \frac{\mu_2\phi_2}{k_2}\right)U. \qquad (5.6.1)$$

Now, in the Saffman–Taylor instability, the porosity and permeability are the same on either side, so that $\phi_1 = \phi_2 = \phi$, $k_1 = k_2 = k$; if $\sigma > 0$ the amplitude of the corrugation grows and the interface is unstable. The condition for instability is then that

$$(\mu_1 - \mu_2)\frac{\phi U}{k} + (\rho_1 - \rho_2)g\cos\theta > 0. \qquad (5.6.2)$$

When $\mu_1 > \mu_2$, that is, when more viscous fluid is displaced by less viscous fluid, the interface is unstable when

$$U > -\frac{(\rho_1 - \rho_2)}{(\mu_1 - \mu_2)}\frac{gk\cos\theta}{\phi}. \qquad (5.6.3)$$

For an interface moving upward, $\cos\theta > 0$. When the upper fluid is *less* dense, $\rho_1 < \rho_2$ and the gravitational (buoyancy) effect promotes stability whereas the difference in viscosities promotes instability. The latter wins when the speed of advance is greater than the critical value

$$U_c = \frac{(\rho_2 - \rho_1)gk\cos\theta}{(\mu_1 - \mu_2)\phi}.$$

The interface is then unstable. When the upper fluid is more dense as well as more viscous, both the gravitational and viscous effects promote instability; the right-hand side of (5.6.3) is negative and the interface always unstable. There are various other cases depending on the signs of $(\rho_1 - \rho_2)$, $(\mu_1 - \mu_2)$, and $\cos\theta$. A curious one is the viscous stabilization of a gravitationally unstable, downward-moving interface when heavy, more viscous fluid displaces lighter, less viscous fluid below. In this case, $\cos\theta < 0$, $\rho_1 < \rho_2$, $\mu_2 > \mu_1$, and (5.6.1) can be arranged as

$$-(\mu_2 - \mu_1)\frac{\phi U}{k} > -(\rho_2 - \rho_1)g|\cos\theta|$$

or, for *instability*,

$$U < \frac{(\rho_2 - \rho_1)gk|\cos\theta|}{(\mu_2 - \mu_1)\phi}. \tag{5.6.4}$$

When buoyancy forces are negligible, the criterion for instability is, from (5.6.2), very simply that $\mu_1 > \mu_2$, that more viscous fluid is being displaced by less viscous fluid. When $\mu_1 < \mu_2$, the interface is stable for all U; in the case exemplified by (5.6.4), the viscous stabilization overcomes the potential gravitational instability when the speed of advance is sufficiently large.

This kind of instability has been studied extensively in a Hele–Shaw cell, a device consisting of two closely spaced (transparent) parallel plates, the gap being filled with viscous liquid. When pressure differences are applied around the edges of the cell, the distribution of velocity in the interior, averaged across the gap, $\bar{\mathbf{u}}$, is specified by (see, e.g., Batchelor, 1967)

$$\nabla \cdot \bar{\mathbf{u}} = 0, \tag{5.6.5}$$
$$\nabla p = -\frac{12\mu}{b^2}\bar{\mathbf{u}},$$

where b is the gap width and p the reduced pressure of the fluid in the gap. These equations are precisely the same for *two-dimensional* flow in a *uniform isotropic* porous medium with permeability $b^2/12$, so that flow visualization in a Hele–Shaw cell, using dye to mark streamlines, provides a technique for examining flow through porous media in these circumstances. It is of limited usefulness, however – effects of inhomogeneity are difficult, and anisotropy is impossible to reproduce (by variations in b). More importantly, in two-phase flows, such as oil and water, the effects of capillarity are quite different in a Hele–Shaw cell and in a permeable medium–the forces associated with propagation of an interface through a permeable medium involve the characteristics of the individual pores whereas in a Hele–Shaw cell they involve more simply (though not trivially – see Park and Homsy, 1984, 1985) the curvatures of the interface both in the plane of the gap and across it, as well as the amount of fluid left behind and adhering to the wall. Nevertheless, experiments of this kind have provided useful concepts. Homsy (1987), in a fine review, describes the range of phenomena that occur in a Hele–Shaw cell in which the development of Saffman–Taylor instability leads to fingering of the advancing, less viscous fluid into the more viscous one, spreading of the fingers and tip splitting. Some measurements in porous media (rather than a Hele–Shaw cell) have been made, notably by Slobold and Thomas (1963) using X-rays to observe the spreading and by Habermann (1960) using a relatively thin slice of permeable medium; they also show similar fingering phenomena. Observations are made difficult by the opacity of the medium, but Chouke, Van Meurs, and van der Poel (1959) and Wooding and Morel-Seytoux (1976) describe observations with fluids having the same index of refraction as the medium. More recently, there have been many numerical, random-walk simulations, such as those of Sherwood (1986), whose development has been interpreted in terms of fractals. There continues to be intense activity in this area.

Dynamically identical but geologically very different is the infiltration instability of dissolution fronts mentioned in Section 4.7 and discussed extensively by Merino (1984), Ortoleva, Chadham, Merino, and Sen (1987), Ortoleva, Merino, Moore, and Chadham (1987), and their coworkers. The interstitial fluid densities and viscosities are now virtually identical on either side of the front, so that in (5.6.1), the condition for instability, that $\sigma > 0$, reduces to

$$\frac{\phi_1}{k_1} - \frac{\phi_2}{k_2} > 0.$$

Since the permeability $k \; \alpha \; \phi^n$, where $n > 1$, this is equivalent to $k_2 > k_1$ and $\phi_2 > \phi_1$. For infiltration instability, the permeability and porosity must be greater behind the front than ahead of it, and as the perturbations grow into fingers, flow continues to be focused into the advancing, more permeable intrusions. Hinch and Bhatt (1990) have given a detailed analysis of the stability of an acid dissolution front moving through porous rock. A precipitation front, or indeed any front in which the reaction reduces the permeability, is, in contrast, stable and perturbations tend to disappear.

6 Pressure-driven flows

6.1 FLOW IN CONFINED AQUIFERS

A confined aquifer is a saturated permeable layer lying between two saturated aquitards, or less permeable regions, and the flow in such a region is important in many hydrological as well as geological applications. As described in Chapter 3, the local permeability generally varies rapidly and substantially across the depth of the aquifer, and even if its distribution is known precisely, the task of calculating the flow patterns in a complete aquifer by numerical integration of the primitive equations of Chapter 2 is a formidable one. A dense mesh of computation points in the vertical is required if the variations of permeability are to be taken realistically into account, and unless we can be assured that the flow is two-dimensional, adequate horizontal mesh coverage is needed as well. The resulting three-dimensional problem can strain the capacity of a moderately large computer, and the results, of course, remain specific to the particular geometry considered. Nevertheless, techniques for the computation of two-dimensional flow distributions in permeable basins are quite well developed and are described in modern groundwater literature (see, e.g., Bear 1979).

Fortunately, in cases where the scale of variation along the bedding planes is sufficiently large compared with the layer thickness that $\varepsilon^2 k_H/k_V \ll 1$, the problem can be simplified considerably, and a recipe provided for a much less demanding computational program, even for three-dimensional flows beyond the reach of direct numerical simulation. In addition, and this is perhaps more important, the simplifications enable us to infer without calculation qualitative, and sometimes quantitative, associations between structural features of an aquifer and the flow patterns in them. Some of the expressions obtained in the next two sections can be derived by use of the "Dupuit approximation," in which the reduced pressure is assumed to be constant throughout the depth of the aquifer, but as explained at the end of Section 3.7, this is unnecessarily strong and frequently unrealistic. Nevertheless, it has long been known (see, e.g., Shercliff 1975 and Wooding and Chapman 1966) that the results obtained by the Dupuit approximation are frequently quite accurate when compared with precise analytical or numerical calculations, even when the approximation itself is manifestly untenable, and this has pre-

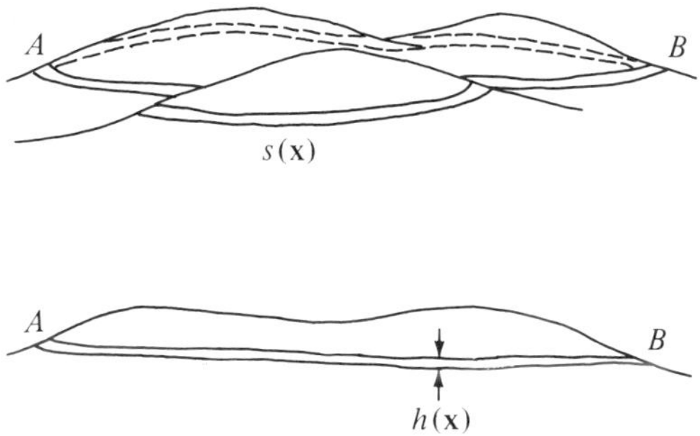

Figure 6.1. A confined aquifer.

sented something of a paradox. The fact that the assumption is not, in fact, necessary helps to resolve this curious situation.

Consider an extensive, relatively thin aquifer, such as illustrated in Figure 6.1. The actual fluid pressure can be taken to be atmospheric along the line $s(\mathbf{x})$ (on a large scale) where the aquifer intersects the surface – at points A and B in the section of Figure 6.1 – so that the reduced pressure relative to that at the lowest point of outcropping is $p = \rho g \zeta_0(\mathbf{x})$ on $s(\mathbf{x})$. In hydraulically driven flow, when the scale factor $S = \varepsilon^2 k_H / k_V \ll 1$, the velocity distribution across the aquifer at any point is given from (3.6.4) by

$$\frac{\mathbf{u}}{k_H} = \mathbf{f}(\mathbf{x}/l) = -\mu^{-1} \nabla_x p, \tag{6.1.1}$$

from Darcy's equation. Note that k_H, and so $\mathbf{u}$, may vary rapidly across the depth of the aquifer, but the pressure gradient along the bedding planes does not; it is essentially constant in z. Note again that this does not imply that the *vertical* gradient of reduced pressure is negligible; small differences in $\nabla_x p$ with respect to vertical location, maintained over large distances along the bedding planes, can lead to differences in reduced pressure in the z-direction. Though small compared with the overall pressure difference, these are, in fact, needed to drive the (small) w velocity components against the relatively small permeability for flow in this direction, as fluid moves from one lens or layer to another.

The upper and lower boundaries of the aquifer will be regarded as impermeable, so that from (3.3.3), thc appropriate boundary conditions there are

$$w = \mathbf{u} \cdot \nabla \zeta \qquad \text{at} \quad z = \zeta_u(\mathbf{x}), \zeta_l(\mathbf{x}), \tag{6.1.2}$$

where ζ_u, ζ_l represent the location of these boundaries in intrinsic coordinates. The total volume transport in the layer is

$$\mathbf{q} = \int_{\zeta_l}^{\zeta_u} \mathbf{u} \, dz. \tag{6.1.3}$$

When the fluid can be regarded as incompressible, one would anticipate that the (two-dimensional) volume transport vector would be divergence free; this can be demonstrated formally as follows.

$$\nabla_x \cdot \mathbf{q} = \nabla_x \cdot \int_{\zeta_l}^{\zeta_u} \mathbf{u}\, dz$$
$$= (\nabla_x \zeta_u) \cdot \mathbf{u}_u - (\nabla_x \zeta_l) \cdot \mathbf{u}_l + \int_{\zeta_l}^{\zeta_u} \nabla_x \cdot \mathbf{u}\, dz,$$

where u_u, u_l represent the velocity component in the bedding plane co-ordinates at the upper and lower sufaces, respectively. From (6.1.2), the preceding equation can be written

$$\nabla_x \cdot \mathbf{q} = w_u - w_l + \int_{\zeta_l}^{\zeta_u} \nabla_x \cdot \mathbf{u}\, dz,$$
$$= \int_{\zeta_l}^{\zeta_u} \left(\frac{\partial w}{\partial z} + \nabla_x \cdot \mathbf{u} \right) dz$$
$$= 0 \tag{6.1.4}$$

in virtue of the incompressibility condition. Accordingly, we can define a *transport* stream function ψ such that

$$q_x = \frac{\partial \psi}{\partial y}, \qquad q_y = -\frac{\partial \psi}{\partial x}, \tag{6.1.5}$$

and (6.1.4) is satisfied identically.

Now, since $\nabla_x p$ in (6.1.1) is independent of z,

$$\mathbf{q} = \int \mathbf{u}\, dz = -\mu^{-1} \nabla_x p \int k_H\, dz$$
$$= -\mu^{-1} (\bar{k}h) \nabla_x p, \tag{6.1.6}$$

where

$$\bar{k} = \frac{1}{h} \int_{\zeta_l}^{\zeta_u} k_H(z)\, dz$$

is the mean permeability across the aquifer for flow along it, like h, a function of (x/l). From (6.1.5) and the two components of (6.1.6),

$$\frac{\partial \psi}{\partial y} = -\frac{\bar{k}h}{\mu} \frac{\partial p}{\partial x},$$
$$\frac{\partial \psi}{\partial x} = \frac{\bar{k}h}{\mu} \frac{\partial p}{\partial y},$$

so that

$$\mu \frac{\partial^2 \psi}{\partial x\, \partial y} = -\frac{\partial}{\partial x}\left(\bar{k}h \frac{\partial p}{\partial x} \right) = \frac{\partial}{\partial y}\left(\bar{k}h \frac{\partial p}{\partial y} \right).$$

Consequently,

$$\frac{\partial}{\partial x}\left(\bar{k}h\frac{\partial p}{\partial x}\right) + \frac{\partial}{\partial y}\left(\bar{k}h\frac{\partial p}{\partial y}\right) = 0. \tag{6.1.7}$$

Similarly,

$$\frac{\partial}{\partial x}\left(\frac{1}{\bar{k}h}\frac{\partial \psi}{\partial x}\right) + \frac{\partial}{\partial y}\left(\frac{1}{\bar{k}h}\frac{\partial \psi}{\partial y}\right) = 0. \tag{6.1.8}$$

These are now equations in *two* dimensions (rather than three) to be solved subject to the boundary conditions at the basin edge. When the distribution of volume transport has been obtained, the internal distribution of velocity in the aquifer at any location x, y is given from (6.1.1):

$$\mathbf{u}(z) = -\mu^{-1}k_{\mathrm{H}}(z)\ \nabla_x p = \frac{k_{\mathrm{H}}(z)}{\bar{k}h}\mathbf{q}. \tag{6.1.9}$$

The speed at each location varies with depth in proportion to the permeability, and the direction of flow along the aquifer is independent of depth.

Equations (6.1.7) and (6.1.8) are precisely analogous to those governing two-dimensional flow in a medium of variable permeability. From (3.5.4) and (3.5.5) a flow pattern of this kind in a vertical section is governed by

$$\frac{\partial}{\partial x}\left(k_{\mathrm{H}}\frac{\partial p}{\partial x}\right) + \frac{\partial}{\partial z}\left(k_{\mathrm{V}}\frac{\partial p}{\partial z}\right) = 0,$$

or (6.1.10)

$$\frac{\partial}{\partial x}\left(\frac{1}{k_{\mathrm{H}}}\frac{\partial \psi}{\partial x}\right) + \frac{\partial}{\partial z}\left(\frac{1}{k_{\mathrm{V}}}\frac{\partial \psi}{\partial z}\right) = 0.$$

In the present situation, the product $\bar{k}h$ of the mean horizontal permeability (which can usually be considered locally iotropic in the "horizontal" plane) and the layer depth, replaces the permeability itself in (6.1.10). Patterns of transport streamlines in the plane of the aquifer produced by variations in $\bar{k}h$ correspond, for similar geometries, to the two-dimensional patterns described in Sections 3.8 and 3.9 in vertical sections produced by variations in permeability alone. The volume transport is increased in regions of higher permeability and/or layer thickness. It is decreased in thinner or lower-permeability regions. One notable difference is that the condition for validity of the "long-lens" approximation in Section 3.9 is relaxed considerably. The sands and gravel of an ancient riverbed may produce a more permeable lens whose length l is much greater than the width d of the aquifer, but because of the horizontal isotropy of the local permeability, the long-lens approximation for the horizontal transport requires only that $(d/l)^2 \ll 1$ rather than the more restrictive condition on (3.7.11).

In a region where k_H is uniform, the horizontal velocity distribution is neccessarily irrotational, as pointed out in Section 3.4; if $\bar{k}h$ is uniform, the distribution of transport $\mathbf{q}$ is likewise irrotational, since $\nabla \times \mathbf{q} = 0$ everywhere.

If the aquifer is overlain by a layer whose permeability, if not zero, is much smaller than that of the aquifer itself, some fluid may move across it. The volume transport $\mathbf{q}$ in the aquifer is then not divergence free and (6.1.4) is replaced by

$$\nabla_x \cdot \mathbf{q} = -w, \tag{6.1.11}$$

where w is the vertical velocity in the overlying aquitard. The transport stream function in the aquifer, ψ, then does not exist, but the flow can be expressed by an extension of (6.1.7) involving the reduced pressure. Let $\zeta(\mathbf{x})$ represent the elevation of the aquifer roof at an interior point above a given datum, say the lowest outcropping, and $h(\mathbf{x})$ the thickness of the overlying aquitard. At the surface, the actual pressure p is atmospheric, say zero, so that the reduced pressure $p + \rho g z$ is $\rho g(\zeta(\mathbf{x}) + h(\mathbf{x}))$. The reduced pressure in the aquifer is $p(\mathbf{x})$, so that the vertical reduced pressure gradient in the aquitard

$$\frac{\partial p}{\partial z} = -\frac{\mu}{k_V^*} w = \frac{\rho g(\zeta(\mathbf{x}) + h(\mathbf{x})) - p(\mathbf{x})}{h(\mathbf{x})},$$

where k_V^* is the small permeability in the layer above. From (6.1.6), (6.1.11), and the proceding equation, it follows that

$$\frac{\partial}{\partial x}\left(\bar{k}h \frac{\partial p}{\partial x}\right) + \frac{\partial}{\partial y}\left(\bar{k}h \frac{\partial p}{\partial y}\right) = k_V^*\left(\frac{p(\mathbf{x})}{h(\mathbf{x})} - \rho g\left\{1 + \frac{\zeta(\mathbf{x})}{h(\mathbf{x})}\right\}\right), \tag{6.1.12}$$

where $k_V^* \ll \bar{k}$. Care must be taken at the ends of the aquifer, where $h(\mathbf{x}) \to 0$, but at these points, the reduced pressure $p(\mathbf{x}) = \rho g \zeta(\mathbf{x})$, so that the right-hand side of this equation remains finite.

Lenses, basement highs, and basement lows

One set of configurations that can be described simply in analytical terms includes a circular elevation or depression of the aquifer basement, with the total thickness being uniform inside and outside, or a circular columnar intrusion of different permeability rising through the aquifer or a disk-shaped, more permeable lens at some level. In any of these cases, the product $\bar{k}h$ has different values, $(\bar{k}h)_i$ and $(\bar{k}h)_o$ inside and outside a circle of radius a, say, in the plane of the aquifer, with a transition region at the circumference that is presumed thin. In each domain, the stream func-

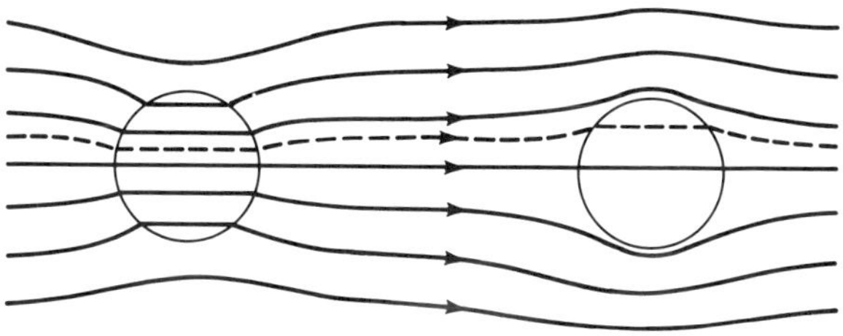

Figure 6.2. Volume transport streamlines in a confined aquifer with a local circular basement high or low or inclusion. The ratio of $(\bar{k}h)$ inside to that outside is 3 on the left and 0.3 on the right; an intermediate streamline is indicated as the dashed curve.

tion satisfies a two-dimensional Laplace's equation subject to the boundary conditions (i) that the volume transport approach a uniform value Q far from the region, (ii) that the radial component of the volume transport be continuous at $r = a$, and (iii) that the pressure p and consequently $\partial p/\partial\theta$ be continuous at $r = a$. In plane polar coordinates, then,

$$\psi \to Qr \sin\theta \qquad \text{as} \quad r \to \infty,$$

$$\left.\begin{aligned} q_r &= -\frac{1}{r}\frac{\partial\psi}{\partial\theta}, \\ \frac{1}{r}\frac{\partial p}{\partial\theta} &= \frac{\mu}{\bar{k}h}\frac{\partial\psi}{\partial r}, \end{aligned}\right\} \quad \text{continuous at } r = a.$$

The solution is the two-dimensional analog of the axisymmetrical flow near a spherical inclusion, described in Section 3.8. It is

$$\Psi_i = 2\,Qr \sin\theta\, \frac{(\bar{k}h)_i}{(\bar{k}h)_i + (\bar{k}h)_o} \qquad \text{when} \quad r < a,$$

$$\Psi_o = Qr \sin\theta \left\{ 1 + \frac{a^2}{r^2}\left(\frac{(\bar{k}h)_i - (\bar{k}h)_o}{(\bar{k}h)_i + (\bar{k}h)_o} \right) \right\} \qquad \text{when} \quad r > a. \qquad (6.1.13)$$

Transport streamline patterns for large and small values of $(\bar{k}h)_i/(\bar{k}h)_i$ are shown in Figure 6.2. They resemble, but are not identical to, those of Figure 3.11, representing axisymmetrical flow patterns. The total volume transport focusing ratio, the ratio of the transport inside the region to that far outside,

$$G = \frac{2(\bar{k}h)_i}{(\bar{k}h)_i + (\bar{k}h)_o}, \qquad (6.1.14)$$

is 2 when $\bar{k}h$ is much greater inside the circular region than outside, and of course approaches zero when it is much less. Over a circular basement high ($h_i < h_o$) in an aquifer with uniform permeability, the total transport is reduced by the factor $2h_i/(h_i + h_o)$ – the flow is partially deflected around the high – but since this transport occurs over a depth that is smaller by the ratio h_i/h_o, the interstitial fluid velocity is *increased* by the ratio $2h_o/(h_o + h_i)$, which is greater than one when $h_i < h_o$.

In a confined aquifer, a disk-shaped, more permeable lens focuses fluid both vertically and horizontally. When its horizontal extent is large compared with the aquifer thickness, the horizontal focusing expressed by equation (6.1.14) augments the vertical focusing by a factor of up to 2 beyond the two-dimensional value (3.9.4).

Another simple type of configuration involves one-dimensional basement relief, a linear basement ridge or depression or a sequence of them. If the y-axis is chosen along the ridge, then $h = h(x)$, and if the total transport is uniform outside the deformed region, then in it, $\mathbf{q} = \mathbf{q}(x)$. Since

$$\frac{\partial q_x}{\partial x} + \frac{\partial q_y}{\partial y} = 0,$$

it follows that

$$q_x = \frac{\partial \psi}{\partial y} = \text{const} = Q_x, \quad \text{say},$$

and the total transport stream function must be of the form

$$\psi = Q_x y + f(x).$$

Equation (6.1.8) reduces to

$$\frac{d}{dx}\left(\frac{1}{\bar{k}h} f'(x)\right) = 0,$$

so that

$$f'(x) = -q_y = C\bar{k}h(x), \tag{6.1.15}$$

where C is a constant. If the total transport approaching the relief has magnitude Q in the direction θ relative to the crests, it is found that

$$q_x = Q \sin \theta,$$
$$q_y = Q \cos \theta \, \frac{\bar{k}h(x)}{\bar{k}h_o}, \tag{6.1.16}$$

where h_o is the aquifer thickness beyond the relief zone. When the aquifer permeability is uniform, the $\bar{k}$ factors, of course, cancel. The total transport is thus deflected to flow *along* basement lows and *across* basement highs (Figure 6.3). The mean interstitial fluid velocity, $q/h(x)$, has components

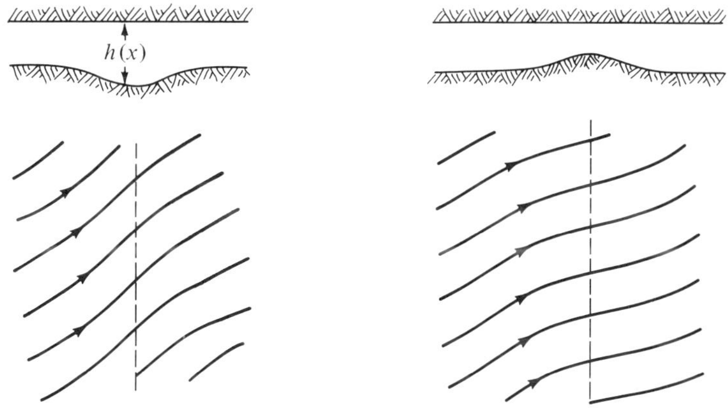

Figure 6.3. Volume transport streamlines in a confined aquifer past a basement ridge (on the right) and trough (left).

$$u = \frac{Q \sin \theta}{h(x)}, \qquad v = \frac{Q \cos \theta}{h_o}; \tag{6.1.17}$$

the component along the ridges is constant in x, whereas that across the relief is greater above the basement highs where $h(x)$ is smaller and less above the lows. If the isotherms are generally horizontal, interstitial fluid flow rises through the isotherms on the one side and sinks on the other; the rock alteration index changes sign, so that a given gradient reaction proceeds in opposite directions on the two flanks.

If the configuration is as shown in Figure 6.3 with the upper surface horizontal and if the permeability is stratified horizontally, the vertical component of fluid motion can be found readily from (6.1.16), (6.1.9), and the incompressibility condition $\nabla \cdot \mathbf{u} = 0$. For

$$u(x, z) = Q \sin \theta \frac{k_H(z)}{\bar{k}h(x)},$$

and since $\partial v/\partial y = 0$ (along the crests),

$$w(x, z) = \int_z^{\text{top}} \frac{\partial u}{\partial x} \, dz.$$

The rock alteration index

$$\begin{aligned} \text{RAI} = \mathbf{u} \cdot \nabla T &= w \frac{\partial T}{\partial z} \\ &= GQ \sin \theta \left(\frac{1}{h^2(x)} \frac{dh}{dx} \right) \int_z^{\text{top}} \frac{k_H(z)}{\bar{k}} \, dz, \end{aligned}$$

where z is measured down from the top and G represents the temperature gradient $-\partial T/\partial z$. The distribution across the basement relief is represented by the factor in parentheses, and the distribution in the vertical by the integral, which is zero at the top and greatest at the basement interface.

Another example of the association between rock alteration and overall flow patterns has been given by Wood (1987a). He considers specifically the formation of dolomite from calcium carbonate in a shallow marine carbonate platform with pore fluid moving from a recharge area above the intertidal zone, mixing with seawater as it moves across the isotherms, leading to a distribution of reaction as described earlier.

Isolated fractures in an aquifer

Total volume transport patterns in the vicinity of an isolated, highly permeable vertical fracture of finite length in a confined aquifer are precisely the same as the streamline patterns (a) to (c) in Figure 3.12. The total transport steamlines are now, of course, to be interpreted in the plane of the aquifer rather than in a vertical plane. A fracture in the direction of flow captures the total transport across a transverse direction equal to the fracture length, so that flow-controlled reactions occur more rapidly. The effect is smaller in fractures inclined to the direction of flow and disappears in transverse fractures.

Elongated lenses

Ancient riverbeds may leave elongated, more permeable lenses imbedded in a confined aquifer, either continuous or segmented, and again the total volume transport patterns correspond to those described in Section 3.9. Figure 3.16 now refers to a long aquifer, confined both above and below and laterally, with elongated lenses in the interior or at one side. Note, however, that the lateral focusing depends upon variations in *mean* permeability, averaged over the depth of the aquifer, so that the focusing influence of a thin lens of high permeability is attenuated by the ratio of its thickness to the total aquifer thickness. Likewise, Figure 3.17 now pertains to an inclined aquifer, confined above, below, and on one side; the other side rises to the surface water table. Though aquifer plan forms are usually more complex than these simple geometries, the qualitative characteristics exhibited here provide a useful guide to interpreting field observations and, possibly, three-dimensional numerical simulations when they become available.

Slowly varying conduits

A long aquifer, confined in the vertical and lateral dimensions, and varying slowly along its length, can be considered in essence to be a one-

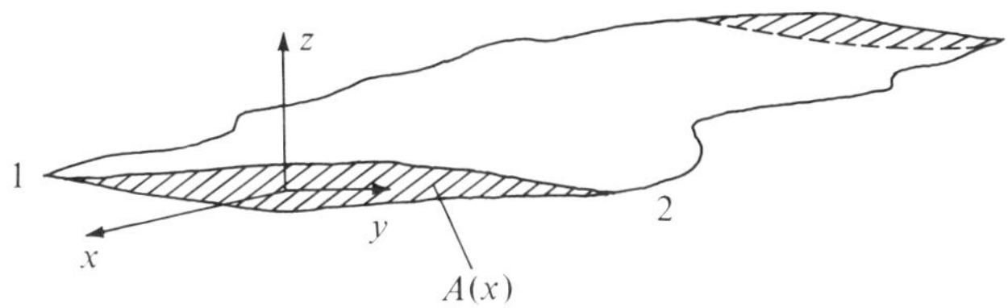

Figure 6.4. A slowly varying permeable conduit.

dimensional conduit. Since the flow is primarily axial (in the x-direction, say) the reduced pressure gradient $\partial p/\partial x$ is independent of the lateral coordinate y, and if equation (6.1.7) is integrated over y from one side to the other,

$$\frac{\partial}{\partial x}\left(\frac{\partial p}{\partial x}\int \bar{k}h\,dy\right) + \left[\bar{k}h\,\frac{\partial p}{\partial y}\right]_1^2 = 0.$$

The second term vanishes at both limits since $h_1 = h_2 = 0$ (Figure 6.4); the integral in the first term is just $\bar{k}A(x)$, where $A(x)$ is the cross-sectional area. Consequently,

$$\frac{\partial p}{\partial x}\,\bar{k}A(x) = \text{const} = B, \text{ say},$$

whence

$$p = B\int_0^x \frac{dx}{\bar{k}A(x)} + \text{const}.$$

If ζ_0 is the hydraulic head different between the ends $x = 0, l$, then

$$\rho g\zeta_0 = B\int_0^l \frac{dx}{\bar{k}A(x)}$$

and

$$p = \rho g\zeta_0 \frac{\int_0^x [\bar{k}A(x)]^{-1}\,dx}{\int_0^l [\bar{k}A(x)]^{-1}\,dx} + \text{const}. \qquad (6.1.18)$$

The total volume flux Q_T over any cross section of the aquifer is, of course, constant; from (6.1.6) and (6.1.18)

$$Q_T = -\frac{1}{\mu}\frac{\partial p}{\partial x}\int_1^2 \bar{k}h(x, y)\,dy = -\frac{1}{\mu}\frac{\partial p}{\partial x}\,\bar{k}A(x)$$
$$= \frac{g\zeta_0}{\nu}\left\{\int_0^l [\bar{k}A(x)]^{-1}\,dx\right\}^{-1}. \qquad (6.1.19)$$

If a conduit or long aquifer such as this has a "throat" somewhere along its length where the cross section $A(x)$ or permeability $\bar{k}$ is small, then in (6.1.19) the integrand $[\bar{k}A(x)]^{-1}$ is largest there and the value of the integral is dominated by this region. In physical terms, most of the pressure drop occurs across such a throat, which then limits the volume transport throughout the whole aquifer. The interstitial fluid velocity is largest at the throat, being proportional to $\{A(x)\}^{-1}$, so that if there is a temperature gradient in the flow direction, the rock alteration index is correspondingly large there.

6.2 FLOW IN UNCONFINED AQUIFERS

General characteristics

When a permeable rock fabric is not overlain by an impermeable cover but is exposed to the atmosphere (at the water table), the dynamical controls on interstitial flow are quite different, as are the resulting flow patterns in the aquifer plan form. In a *confined* aquifer, the hydraulic head differences along the boundary provide the overall driving force; the internal pressure and flow distribute themselves in accordance with the (elliptic) equations (6.1.7) and (6.1.8). The flow at any location depends on the geometry of the whole aquifer. In an *unconfined* aquifer, however, the pressure gradient and so the flow at any site are determined locally, in essence by the configuration of the water table in the vicinity. The mathematics for slowly varying aquifers is simpler. However, since interstitial water may enter or leave the matrix across the water table, the vertically integrated volume transport is no longer divergence free, and the transport stream-function representation of Section 6.1 cannot be used.

At the water table $z = \zeta(\mathbf{x})$, the (actual) pressure $\hat{p}$ can be considered to be atmospheric, so that

$$\nabla_x \hat{p} + \frac{\partial \hat{p}}{\partial z} \nabla \zeta = 0 \qquad \text{on} \quad z = \zeta, \tag{6.2.1}$$

and the lower boundary condition at the impermeable base $z = \zeta_l$, say, is

$$w + \mathbf{u} \cdot \nabla \zeta_l = 0 \qquad \text{at} \quad z = \zeta_l. \tag{6.2.2}$$

The incompressibility condition can be integrated over the depth of the aquifer as in Section 6.1, now to relate the "horizontal" divergence of the total volume transport $\mathbf{q}$ to the transport velocity w_0 in the z-direction out across the water table (leading to runoff or evaporation):

$$\nabla_x \cdot \mathbf{q} = -w_0. \tag{6.2.3}$$

The dynamical balance is simplest when the topography of the water

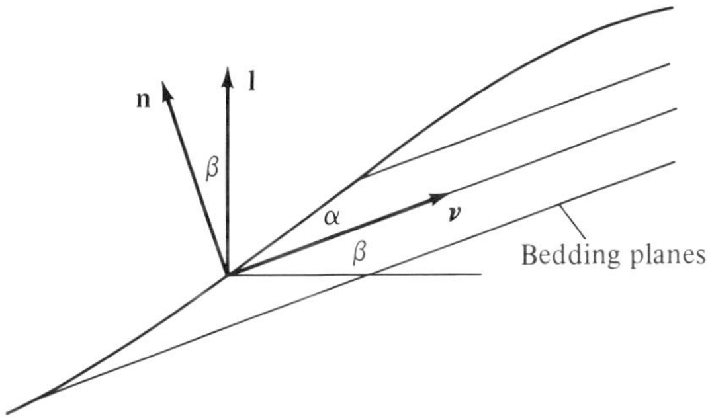

Figure 6.5. The intersection of sloping bedding planes at the surface of an unconfined aquifer or permeable basin.

table is sufficiently gentle that $\varepsilon^2 k_H/k_V \ll 1$, where ε now is the characteristic surface slope. (The effects of steep slope are discussed at the end of this section.) When the bedding planes are horizontal, the z-axis is vertical and for these geometries we have as before

$$\frac{\mathbf{u}}{\bar{k}_H} = -\mu^{-1}\,\nabla_x \hat{p} = \mathbf{f}(\mathbf{x}/l). \tag{6.2.4}$$

The horizontal pressure gradient is independent of z, and so is equal to its value (6.2.1) at $z = \zeta$,

$$\nabla_x \hat{p} = \frac{\partial \hat{p}}{\partial z}\,\nabla\zeta \approx \rho g\,\nabla\zeta,$$

since the vertical interstitial pressure gradient is in essence hydrostatic. Consequently,

$$\mathbf{u}(\mathbf{x}, z) = -\frac{g k_H(z, \mathbf{x})}{\nu}\,\nabla\zeta(\mathbf{x}), \tag{6.2.5}$$

and the total volume transport is

$$\mathbf{q} = -\frac{g\bar{k}h}{\nu}\,\nabla\zeta(\mathbf{x}). \tag{6.2.6}$$

An immediate consequence is that if $\bar{k}h$ is uniform in $\mathbf{x}$, then $\nabla \times \mathbf{q} = 0$. The transport vector $\mathbf{q}$ is irrotational, so that flow paths cannot involve closed loops or shear in the aquifer plane.

If, as in Figure 6.5, the bedding planes (normal to n) are not horizontal, the volume transport can be shown to be given by

$$\mathbf{q} = -\frac{g\bar{k}h}{\nu}\,(\mathbf{l}\cdot\boldsymbol{\nu}\boldsymbol{\nu} + \mathbf{l}\cdot\mathbf{n}\,\nabla\zeta), \tag{6.2.7}$$

where the surface slope is measured relative to the bedding planes in which the unit vector $\mathbf{n}$ lies ($\nabla\zeta = \tan\alpha$) and $\mathbf{l}$ is vertical. If the water table is locally horizontal, then in Figure 6.5 interpreted in the vertical plane normal to the bedding planes, $\alpha = -\beta$, $\nabla\zeta = -\tan\beta$, $\mathbf{l}\cdot\mathbf{n} = \cos\beta$, $\mathbf{l}\cdot\boldsymbol{\nu} = \sin\beta$, and the right-hand side of (6.2.7) vanishes. As one might expect, under a horizontal water table there is no fluid transport, whatever the slopes of the bedding planes or the configuration of the underlying bedrock. If the bedding planes are horizontal, $\mathbf{l}\cdot\boldsymbol{\nu} = 0$, $\mathbf{l}\cdot\mathbf{n} = 1$ and we recover (6.2.6).

From (6.2.6) it is evident that the subsurface transport velocity is down the gradient in any two-dimensional topography, just as is the surficial runoff. The interstitial flow velocity, but not the total volume transport, is independent of the basement configuration. Over a basement high, for example, the horizontal flow vector is undeflected (in contrast to that in a confined aquifer, as illustrated in Figure 6.2). Fluid rises across the water table above the upstream slope and is drawn down when the underlying basement slopes away. The effect of horizontal flow focusing encountered in confined aquifers has disappeared, in essence because of the localness of the flow control. High local values of k_H and $\bar{k}h$ are associated with increased local flow velocities and volume transport, but the fluid to provide it moves across the water table (as shown in Figure 3.17), not laterally. The volume flux or velocity out across the water table is, from (6.2.6),

$$w_0 = -\nabla\cdot\mathbf{q} = \frac{g}{\nu}\{\nabla(\bar{k}h)\cdot\nabla\zeta + \bar{k}h\,\nabla^2\zeta\}. \tag{6.2.8}$$

If the upper boundary slope is locally uniform so that $\nabla^2\zeta = 0$,

$$\begin{aligned} w_0 &= \frac{g}{\nu}\,\nabla(\bar{k}h)\cdot\nabla\zeta \\ &= -\frac{\nabla(\bar{k}h)}{\bar{k}h}\cdot\mathbf{q}, \end{aligned} \tag{6.2.9}$$

demonstrating explicitly that when $\bar{k}h$ increases in the flow direction (downslope), $w_0 < 0$ and water must be added at the water table; when it decreases, water rises across the upper surface. At a local maximum of the water table, $\nabla\zeta = 0$ and $\nabla^2\zeta < 0$; groundwater flow must be downward to maintain this maximum. At the bottom of a valley, $\nabla^2\zeta > 0$, and if the slope along the valley is small, w_0 is positive – the groundwater flow rises to the surface.

These expressions can also be used inversely, to determine the shape of the water table in terms of the flow. For example, in the two-dimensional groundwater flow toward a drainage basin, as illustrated in Figure 6.6, if the volume flux $-Q$ toward the basin is conserved and ζ is measured

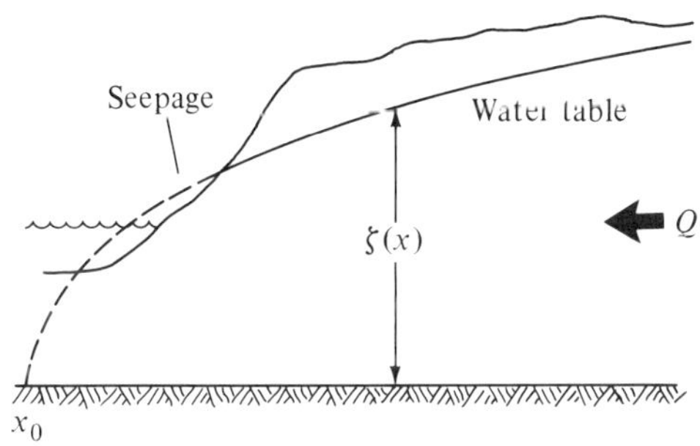

Figure 6.6. Depression of the water table by a drainage basin.

upward from impermeable basement rock, then when the slope $d\zeta/dx$ is small,

$$-Q = -\int_0^l u(x, z)\, dz = -\frac{\bar{k}g}{\nu}\zeta\frac{d\zeta}{dx},$$

from (6.2.6), so that

$$\frac{d}{dx}\zeta^2 = \frac{2\nu Q}{\bar{k}g}$$

and

$$\zeta(x) = \left\{\frac{2\nu Q}{\bar{k}g}(x - x_0)\right\}^{1/2}, \qquad (6.2.10)$$

where x_0 is a virtual origin (the point where $\zeta = 0$ if the solution is continued beyond the point of emergence of the water table at the drainage basin). This expression is sometimes called the Dupuit–Forchheimer equation.

Local topographic circulation

In an unconfined aquifer or permeable sedimentary basin as a whole, the inequality between overall horizontal and vertical scales may be sufficiently strong that the preceding analysis gives a good representation of the deep, basin-wide circulation. At the same time, local topographic features on a smaller scale can produce local circulations that can substantially modify the overall pattern or, if they extend deep enough, disrupt it.

To gain an appreciation for the flow patterns involved in local topographic circulations, consider a region in which the water table consists of a series of parallel ridges and valleys with spacing l as indicated in

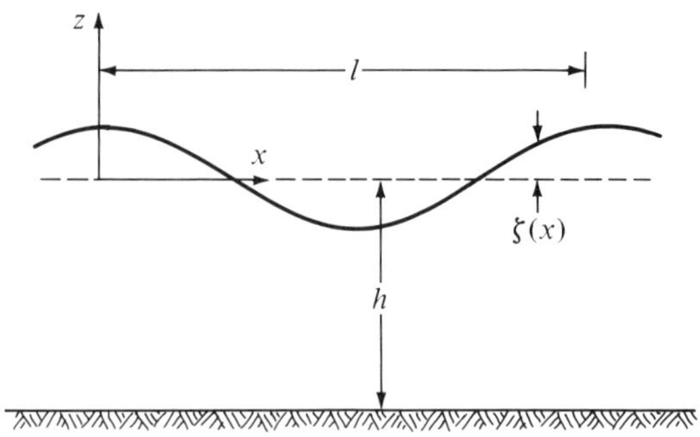

Figure 6.7. An undulating water table – definition sketch.

Figure 6.7. The average depth to bedrock is h and ζ_0 is the amplitude of the variation in water-table level. For simplicity, suppose that both the bedding planes and the impermeable basement are horizontal, so that from Section 3.4, the distribution of reduced pressure is given by

$$k_H \frac{\partial^2 p}{\partial x^2} + k_V \frac{\partial^2 p}{\partial z^2} = 0, \qquad (6.2.11)$$

where usually $k_H \gg k_V$; we do *not* now require that $S = (h/l)^2(k_H/k_V)$ be small. Across the basement $z = -h$, there is no flow, so that

$$\frac{\partial p}{\partial z} = 0 \quad \text{at} \quad z = -h, \qquad (6.2.12)$$

while at the water table $z = \zeta(x)$, the surface pressure condition (6.2.1) must be satisfied. In terms of the reduced pressure

$$\frac{\partial p}{\partial x} + \left(\frac{\partial p}{\partial z} - \rho g\right)\frac{\partial \zeta}{\partial x} = 0 \qquad (6.2.13)$$

at $z = \zeta = \zeta_0 \cos(2\pi x/l)$.

The solution can be found simply when the water-table slope $2\pi\zeta_0/l$ is small. The reduced pressure

$$p = \rho g \zeta_0 \cos\left(\frac{2\pi x}{l}\right) \frac{\cosh\{(k_H/k_V)^{1/2}\,[2\pi(z+h)/l]\}}{\cosh\{(k_H/k_V)^{1/2}(2\pi h/l)\}}, \qquad (6.2.14)$$

and the corresponding stream function

$$\psi = \frac{g\zeta_0(k_H k_V)^{1/2}}{\nu} \sin\left(\frac{2\pi x}{l}\right) \frac{\sinh\{\ \}}{\cosh\{\ \}}, \qquad (6.2.15)$$

where the arguments in braces are the same as the preceding equation.

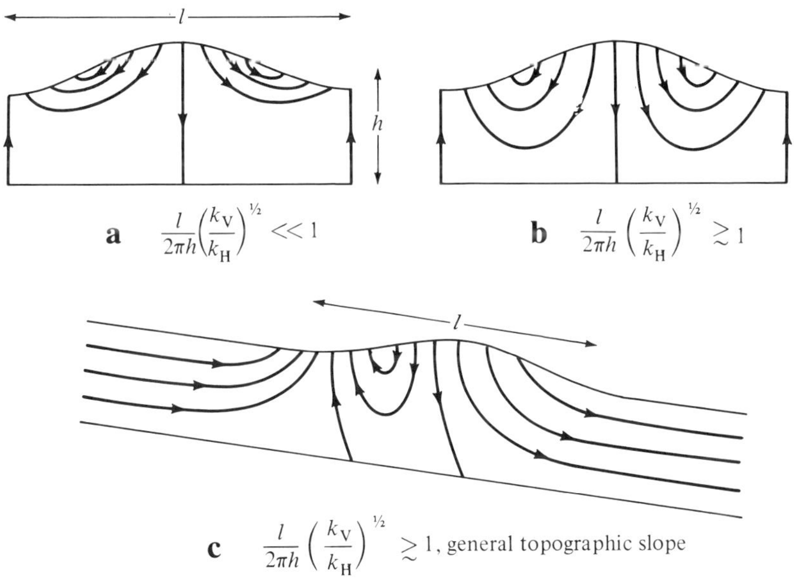

Figure 6.8. Local topographic circulations are confined to the upper part of an unconfined aquifer when $(l/2\pi h)(k_V/k_H)^{1/2} \ll 1$ but penetrate to the basement for those components of the topography whose scale l is such that this ratio is of order unity or larger. An incident, basin-wide flow is then scoured to the surface (c), producing possibly significant local values of the RAI; the flow downstream is replenished by new fluid percolating downward in the lee.

Now, when the horizontal scale l of the topographic variations is such that

$$l \ll 2\pi h(k_H/k_V)^{1/2}, \tag{6.2.16}$$

the hyperbolic functions reduce to exponentials and

$$\psi = \frac{g\zeta_0(k_H k_V)^{1/2}}{\nu} \sin\left(\frac{2\pi x}{l}\right) \exp\left\{\left(\frac{k_H}{k_V}\right)^{1/2} \frac{2\pi z}{l}\right\}. \tag{6.2.17}$$

The scale depth of the local topographic circulation is

$$\frac{z_s}{h} = \frac{l}{2\pi h}\left(\frac{k_V}{k_H}\right)^{1/2} \ll 1. \tag{6.2.18}$$

If, for example, $k_H/k_V = 10$, then for topographic scales significantly smaller than twenty layer depths, the local circulation is essentially confined to the upper part of the basin (Figure 6.8a); its influence on the deep circulation is minor. However, for topographic scales

$$l > 2\pi h(k_H/k_V)^{1/2}, \tag{6.2.19}$$

the circulation extends to the basement (Figure 6.8b) and fluid, entering the water table near the ridge crests, circulates throughout the depth of the basin and reemerges near the valley floors.

This effect has great significance in consideration of basin-wide solute transport and formation of mineral deposits. If there is a gentle topographic slope along the valleys (into the plane of Figure 6.8b), the motion is three-dimensional with a uniform component downslope parallel to the ridges and valleys superimposed on this lateral cross-flow. Fluid elements spiral down from the ridge crests, scour the whole depth of the basin, and reemerge near the valley floors, further downstream. Clearly, with this configuration, fluid elements cannot move throughout the whole basin, but circulate in and out (then being removed as runoff), dissolving minerals only along this path. A similar scouring or purging of basin-wide flows occurs when the general slope is transverse to a region of ridges and valleys (Figure 6.8c). If the slope of the local topography exceeds the general basin slope, the local topographic circulation and the basin-wide flow combine to produce the pattern illustrated, the larger scale flow, with its solute load, being forced to the surface, moving across isotherms and enhancing gradient reactions near the local upslope surface. The basin fluid is replenished with new, initially unsaturated fluid from the region of the ridge. If the basin flow is in a direction oblique to the local topographic structure, the combined flow is again three-dimensional, being a superposition of the scouring and spiraling flows just described for the incident components normal and parallel to the crests. In like manner, an isolated local water-table maximum, with horizontal scale satisfying (6.2.19) in a basin with a general surficial slope, scours and purges to the surface the flow incident upon it, leaving behind a wake of new water that has entered in the vicinity of the local maximum.

Finally, when the topographic scale l is sufficiently large that

$$l \gg 2\pi h(k_H/k_V)^{1/2},$$

the expression (6.2.15) for the stream function becomes equivalent to (6.2.5), valid for small scale factor S, since the sinh factor becomes equal to its argument and the cosh factor is unity.

6.3 COMPUTATION OF THREE-DIMENSIONAL BASIN-WIDE FLOWS

The preceding two sections have demonstrated that the dynamical constraints on interstitial flow fields are very different in the two limits of confined and unconfined aquifers, so that the most effective computational techniques for calculating basin-wide three-dimensional flows are correspondingly different. The greatest obstacles to finding a realistic and accurate solution in a given geological structure are likely to lie in the difficulty of determining accurately the structure itself and the distribution of permeabilities within it. The solution will be only as good as this specification, and no sophistication of analysis can make up for deficiencies in this basic information. The difficulties are compounded if we are interested in paleocirculations, since the structures themselves must be reconstructed as well as possible from present information – a task that calls for all the skills the structural geologist can bring to bear. No realistic, three-dimensional calculations of this kind seem to have yet been made for a real basin, but the steps involved can be outlined briefly.

Confined aquifers

In an aquifer that is internally layered, the first choice concerns the vertical scale of internal structures (and corresponding flow variations) that one wants to resolve. From the variations in permeability on scales finer than this, the horizontal and vertical permeabilities to be used in the calculation are estimated using (3.7.6) and (3.7.8).

The minimum horizontal resolution $l_{\min}$ for which the geometrical scaling remains accurate is found by requiring that the scale factor involving $l_{\min}$ be fairly small, say not more than 0.2 or 0.3. For example, in a confined aquifer of length 100 km and thickness 100 m with $k_H = 10\,k_V$, the overall scale factor is 10^{-5}, which is satisfactorily miniscule. If $(h/l_{\min})^2 k_H/k_V < 0.2$, then the minimum horizontal resolution is about $7h$, or 700 m.

The crucial step, as already mentioned, is the determination of the aquifer structure, its thickness, and the distribution of permeability with depth over a horizontal grid of points whose spacing (greater than or equal to $l_{\min}$) determines the horizontal resolution of the solution. This is the primary data base.

The elevation of the water table along the length of the outcropping or outcroppings of the aquifer is found, say $\zeta(s)$, where s is the distance along the outcropping. This provides a pressure boundary condition along the exposed edge or edges.

The next step is to calculate the transport pattern in the plane of the aquifer using *either* the reduced pressure equation (6.1.7) or the transport stream function equation (6.1.8).

(a) In terms of the pressure, we rewrite (6.1.7),

$$\frac{\partial}{\partial x}\left(\bar{k}h\,\frac{\partial p}{\partial x}\right) + \frac{\partial}{\partial y}\left(\bar{k}h\,\frac{\partial p}{\partial y}\right) = 0, \tag{6.1.7}$$

where from (6.1.6) the product

$$\bar{k}h = \int k_{\mathrm{H}}(z)\, dz,$$

the integral being across the layer, its value at every (x, y) point being found from the data base. This is to be solved numerically subject to the following boundary conditions:

(i) Along outcroppings whose altitude above a given datum level is specified by $\zeta(s)$, the actual pressure $\hat{p} = p - \rho_0 g z$ is atmospheric and can be taken as zero, so that

$$p = \rho_0 g \zeta(s) \qquad \text{along outcrops.} \tag{6.3.1}$$

(ii) Along internal edges or boundaries of the aquifer (whether it thins to zero or terminates at a fault), if there is no leakage the pressure gradient normal to the edge vanishes:

$$\partial p/\partial n = 0 \qquad \text{on impermeable edges.} \tag{6.3.2}$$

For a given configuration and distribution of $\bar{k}h(>0$, of course) and specified boundary conditions (6.3.1) and (6.3.2), a solution for the reduced pressure at the horizontal grid points (x, y) is calculated; the solution is unique and well behaved.

The distribution of volume transport $\mathbf{q}$ at the horizontal grid points is then found from equation (6.1.6), again making use of the data base (or by preserving in the calculation the quantities inside the parentheses in (6.1.7)). The internal velocity distribution with depth at each horizontal grid point is then found from (6.1.9), involving the vertical distribution of k_{H}.

(b) Alternatively, the transport stream function equation (6.1.8) can be used:

$$\frac{\partial}{\partial x}\left(\frac{1}{\bar{k}h}\frac{\partial \psi}{\partial x}\right) + \frac{\partial}{\partial y}\left(\frac{1}{\bar{k}h}\frac{\partial \psi}{\partial y}\right) = 0. \tag{6.1.8}$$

This has the same structure as (6.1.7). The appropriate boundary conditions are equivalent to (6.3.1) and (6.3.2) but appear in terms of ψ. Differentiation of (6.3.1) along the direction of the outcropping yields

$$\frac{\partial p}{\partial s} = \rho g \frac{\partial \zeta}{\partial s} = -\frac{\mu}{\bar{k}h} q_{\mathrm{s}},$$

where q_{s} is the component of the transport along the sloping outcrop, or $\partial\psi/\partial n$, where n is the inward normal. Thus,

$$\frac{\partial \psi}{\partial n} = -\frac{g\bar{k}h}{\nu}\frac{\partial \zeta(s)}{\partial s} \tag{6.3.3}$$

along the outcropping, while at impermeable edges, ψ = const. From the solution $\psi(x, y)$ at the horizontal grid points, the distribution of transport is found from (6.1.5) and the vertical velocity profile at each from (6.1.9) as before. If the upper aquitard has small but nonzero permeability, the same steps can be followed but with equation (6.1.11) replacing (6.1.7).

Unconfined aquifers

Numerical calculation of these flows is a good deal simpler than in the preceding case. The same data base must be assembled specifying the vertical distribution of permeability at each horizontal grid point together with the water-table topography $\zeta(\mathbf{x})$. The principal difficulty is that it is now essential to identify topographical features on the scale $2\pi h(k_H/k_V)^{1/2}$, which, as seen in the preceding section, purge to the surface the flow incident upon them. Yet this scale may be smaller than a reasonable horizontal grid spacing for the basin as a whole. If these small-scale irregularities are not ubiquitous throughout the basin, one solution is to smooth or filter the water-table topography on a scale large compared with $2\pi h(k_H/k_V)^{1/2}$, and calculate the volume transport point by point in the basin from (6.2.6) and the vertical velocity distribution from (6.2.7). The locations of the subgrid small-scale irregularities are then regions where this basin-wide flow is scoured to the surface and replaced by new fluid infiltrating down. If these small-scale variations *are* ubiquitous, there is, in fact, no basin-wide flow but only a dense distribution of the local topographical circulations discussed earlier.

6.4 TEMPERATURE DISTRIBUTIONS

The general scaling of Section 3.5 indicated that when the transport velocity U along an aquifer of length l and thickness h is such that

$$\frac{Uh}{\kappa} \ll \frac{l}{h}, \tag{6.4.1}$$

where κ is the thermal diffusivity, then, overall, the geothermal temperature distribution is influenced very little by the flow. Since l/h is generally large, this overall condition is usually satisfied easily – essentially horizontal flow has little influence on an essentially vertical temperature gradient ($\mathbf{u} \cdot \nabla T \sim 0$). However, in many places there do exist significant local modifications to the geothermal temperature field. In recharge or discharge regions, for example, flow moves vertically across the isotherms over a region whose scale may be of order h. If $Uh/\kappa \geq 1$, advection of heat by the flow now becomes significant and isotherms are

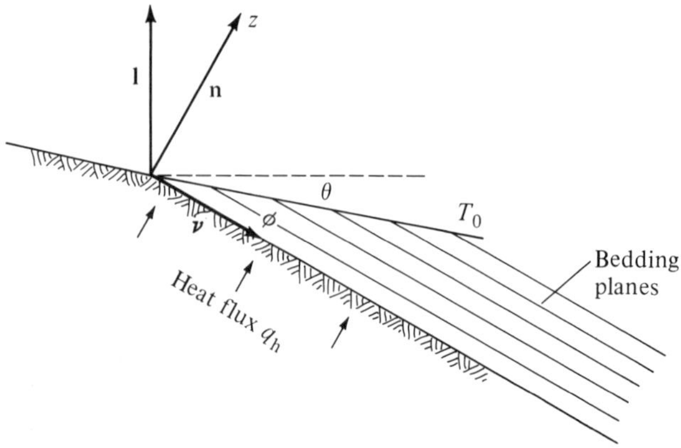

Figure 6.9. Definition sketch: flow entering a permeable layer with basement heat flux q_{h}.

displaced in the flow direction. This condition is not particularly stringent – for example, if $\kappa \sim 4 \times 10^{-3}$ cm^2 s^{-1}, a representative value for sediments, and $h \sim 1$ km $= 10^5$ cm, then a velocity of 4×10^{-8} cm s^{-1}, or about 1 cm yr^{-1}, cutting directly across the isotherms is sufficient to distort the temperature field on this scale.

This can be appreciated in a little more detail by considering a permeable basin as illustrated in Figure 6.9 with recharge along the upper surface and a relatively impermeable bed underlying it, across which the geothermal heat flux q_{h} can be supposed constant. If, near the upper corner of the basin, the permeability can be regarded as constant and the angle ϕ is small, the velocity is given from (6.2.7),

$$\mathbf{u} = -\frac{\bar{k}g}{\nu}(\mathbf{l}\cdot\boldsymbol{\nu}\boldsymbol{\nu} + \mathbf{l}\cdot\mathbf{n}\,\nabla\zeta),$$

where, remember, the slope $\nabla\zeta$ of the surface is taken relative to the bedding planes. Thus, in terms of the angles specified in Figure 6.9

$$\begin{aligned}\mathbf{l}\cdot\boldsymbol{\nu} &= -\sin(\theta + \phi).\\ \mathbf{l}\cdot\mathbf{n} &= \cos(\theta + \phi),\\ \nabla\zeta &= \tan\phi,\end{aligned}$$

and

$$u = \frac{\bar{k}g}{\nu}\frac{\sin\theta}{\cos\phi} \approx \frac{\bar{k}g}{\nu}\sin\theta, \qquad (6.4.2)$$

after a little trigonometry, the second form being approximate when ϕ is small so that $\cos\phi \approx 1$.

The temperature distribution is determined by

$$\mathbf{u}\cdot\nabla T = \kappa\,\nabla^2 T \approx \kappa\,\frac{\partial^2 T}{\partial z^2}, \tag{6.4.3}$$

when, again, ϕ is small; both the temperature T_0 at the upper surface and the heat flow q_h along the lower boundary can be taken constant. Now, as the fluid moves downward, the temperatures necessarily increase in the flow direction, so that $\mathbf{u}\cdot\nabla T > 0$ and therefore, from (6.4.3), $\partial^2 T/\partial z^2 > 0$. The temperature gradient across the sedimentary layer, negative at its base, therefore becomes less so as we move up; its magnitude decreases toward the surface. How much does the temperature gradient change across the depth of the region of flow?

Near the surface, the isotherms are essentially parallel to it, so that the temperature gradient in the direction of the bedding planes $\partial T/\partial x \approx \tan\phi\,(\partial T/\partial z)$ and consequently

$$\mathbf{u}\cdot\nabla T \approx -u\tan\phi\,\frac{\partial T}{\partial z} = \kappa\,\frac{\partial^2 T}{\partial z^2}.$$

(Note that $\partial T/\partial z < 0$.) Now if δ represents the distance scale across the layer over which the temperature gradient changes significantly as a result of flow, $\partial^2 T/\partial z^2 \sim -\delta^{-1}\,\partial T/\partial z$, and from the preceding equation,

$$\delta \approx \frac{\kappa}{u\tan\phi} = \frac{\kappa\nu}{g\bar{k}\tan\phi\sin\theta}. \tag{6.4.4}$$

If $\delta \gg h$, where h is the thickness of the flow system at the end of the recharge region (or the thickness of a confined aquifer), the temperature gradient changes little as a result of the flow; convective heat transport as the fluid moves along the basin to gradually increasing depths is unable to distort to any significant extent the geothermal temperature gradient. This condition can be stated alternatively as

$$\frac{uh\tan\phi}{\kappa} = \frac{g\bar{k}h\tan\phi\sin\theta}{\nu\kappa} \gg 1. \tag{6.4.5}$$

(Compare this with (6.4.1); in this geometry $\tan\phi \sim h/l$.)

Thermal boundary layers

If $\delta \ll h$, that is, if

$$\frac{uh\tan\phi}{\kappa} \gg 1, \tag{6.4.6}$$

the vertical temperature gradient has decreased in magnitude (as the temperature approaches T_0) well inside the thickness of the sedimentary basin

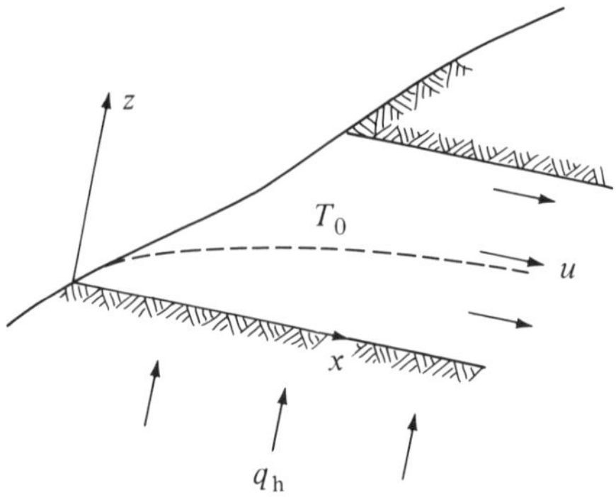

Figure 6.10. A thermal boundary layer above the base of a permeable layer; fluid at temperature T_0 enters the layer from the surface and moves to the right.

in the aquifer; this can be so at particularly large flow velocities (large $\bar{k}$ and h, large water-table slopes θ) and larger angles ϕ of dip of the floor of the basin relative to the water table. In the upper part of the permeable region, then, the temperature remains essentially the same as that at the surface, and the region of geothermal heating is confined to a relatively thin thermal boundary layer along the lower surface of the stratum of flow as in Figure 6.10. Equation (6.4.3) in the form

$$u \frac{\partial T}{\partial x} = \kappa \frac{\partial^2 T}{\partial z^2} \tag{6.4.7}$$

is therefore to be solved subject to the conditions that

$$q_h = -\kappa \frac{\partial T}{\partial z} = \text{const} \quad \text{when } z = 0 \tag{6.4.8}$$

and $T \to T_0$, the surface temperature, outside the boundary layer, which, as far as the boundary layer is concerned, is at infinity. Note that if we integrate (6.4.7) across the boundary layer at any section x, since u is independent of z,

$$\int_0^\infty u \frac{\partial T}{\partial x}\, dz = u \frac{\partial}{\partial x} \int_0^\infty (T - T_0)\, dz = -\kappa \left. \frac{\partial T}{\partial z} \right|_0 = q_h.$$

Therefore,

$$\int_0^\infty (T - T_0)\, dz = \frac{q_h x}{u}, \tag{6.4.9}$$

where x is the distance of the section from the origin of the boundary layer at the upstream corner. This equation states physically that the total heat flux across any section is equal to the total heat input upstream; together with the boundary condition (6.4.8) it constrains simply the nature of the solution. If ΔT is the temperature difference across the layer and δ the layer thickness (both functions of x), (6.4.8) requires

$$\Delta T/\delta \sim q_{\mathrm{h}}/\kappa,$$

while (6.4.9) insists that

$$\Delta T\delta \sim q_{\mathrm{h}}x/u.$$

Consequently,

$$\Delta T \sim q_{\mathrm{h}}\left(\frac{x}{\kappa u}\right)^{1/2}, \qquad \delta \sim \left(\frac{x\kappa}{u}\right)^{1/2}. \tag{6.4.10}$$

The temperature field in the thermal boundary layer is therefore of the form

$$T - T_0 = q_{\mathrm{h}}\left(\frac{x}{\kappa u}\right)^{1/2} f\left(\frac{z}{\delta}\right) = q_{\mathrm{h}}\left(\frac{x}{\kappa u}\right)^{1/2} f(\eta), \tag{6.4.11}$$

where

$$\eta = \frac{z}{\delta} = \frac{z}{x^{1/2}}\left(\frac{u}{\kappa}\right)^{1/2}. \tag{6.4.12}$$

The boundary condition (6.4.8) at $z = 0$,

$$q_{\mathrm{h}} = -\kappa\frac{\partial T}{\partial z} = -q_{\mathrm{h}}\kappa\left(\frac{x}{\kappa u}\right)^{1/2}\frac{\partial f}{\partial \eta}\frac{\partial \eta}{\partial z},$$

reduces to

$$\frac{\partial f}{\partial \eta} = -1 \qquad \text{at} \quad \eta = 0, \tag{6.4.13a}$$

while

$$f \to 0 \qquad \text{as} \quad \eta \to \infty \tag{6.4.13b}$$

since $T \to T_0$ outside the boundary layer.

Note that already we have very useful information about the temperature distribution. From (6.4.10) the temperature along the lower surface of the basin increases as $x^{1/2}$, but decreases (as $u^{-1/2}$) when the flow velocity increases. The thickness of the heated region also increases with distance along the basin as $x^{1/2}$, so that sooner or later, if the basin is long enough, it approaches the entire depth and this form of solution fails.

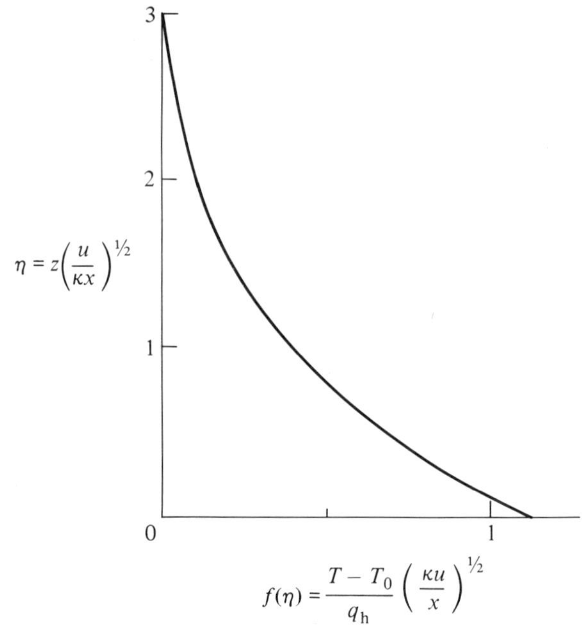

Figure 6.11. The thermal boundary layer profile, the solution to equation (6.4.15).

This occurs when

$$\delta = \left(\frac{\kappa x}{u}\right)^{1/2} \sim h$$

or when $x = x_l$ such that

$$\frac{x_l}{h} \sim \frac{uh}{\kappa} = \frac{g\bar{k}h \tan \theta}{\kappa \nu}, \tag{6.4.14}$$

which is necessarily large when (6.4.6) is satisfied. Over this "entry length" when $x/h < (uh/\kappa)$, the geothermal heat flux at the upper surface vanishes, whether we are considering a groundwater basin or a confined aquifer; in the field, the horizontal extent of this negligible heat flux region is an indication of the magnitude of the flow velocity u and consequently, from (6.4.2), the mean permeability. The temperature at the base of the basin relative to the surface temperature, at the point where $x/h = uh/\kappa$ is from (6.4.10) approximately, but fractionally less than, $q_h h/\kappa$, the temperature that would exist in the absence of flow.

The complete solution can be found by substitution of (6.4.11) into

(6.4.7), though it adds little to these conclusions except a veneer of mathematical respectability. This substitution leads to the ordinary differential equation

$$2f'' + \eta f' - f = 0, \tag{6.4.15}$$

with $f'(0) = -1$, $f \to 0$ as $\eta \to \infty$. The solution to this is somewhat untidy,

$$f(\eta) = 2\pi^{-1/2} e^{-\eta^2/4} + \eta\left\{\pi^{-1/2} \int_0^\eta e^{-y^2/4}\, dy - 1\right\},$$

where η is given by (6.4.12) and y is simply an integration variable. The solution is illustrated in Figure 6.11.

Beyond the "entry length" defined by (6.4.14), if the aquifer or sedimentary basin is sufficiently long, the temperature distribution gradually loses its curvature, and if h changes only slowly, approaches the locally linear distribution characteristic of pure thermal conduction. The detailed distributions now depend on whether the permeable layer continues to the surface, where the temperature T_0 is constant, or is overlain by another layer through which the heat must be conducted. In the former case, the isotherms follow the shape of the surface contour (on a scale large compared with h); in the latter they are modified if the relative thermal conductivities of the porous region and the overlying stratum are unequal.

When a sedimentary basin of this kind contains two or more permeable layers, the general characteristics of the temperature distribution can be found simply from these solutions. If (6.4.5) is satisfied for the layer as a whole, it is also satisfied for each sublayer separately, and the temperature distribution is conductively dominated throughout. However, if in some or all of the sublayers, (6.4.6) holds, then the upward geothermal heat flux is convected downstream with an offset equal to the total of the entry lengths in the layers below.

These thermal boundary layers, involving hotter, less dense fluid below colder, denser fluid are potentially unstable, but their stability characteristics are still largely unexplored.

Recharge areas: temperature reduction

Recharge areas associated with a local maximum in the water table may significantly modify the flow field as described in Section 5.3, and this can affect the temperature distribution. If at the maximum of the water table, its radius of curvature R is such that

$$\frac{R}{h}\left(\frac{k_V}{k_H}\right)^{1/2} \geq 1,$$

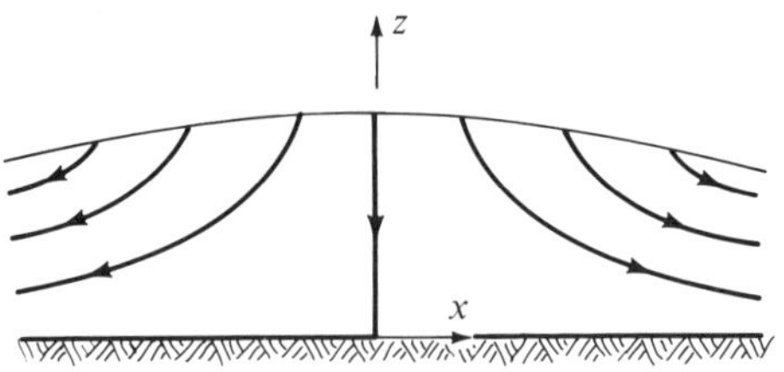

Figure 6.12. Streamline patterns in a two-dimensional recharge zone.

the topographic circulation reaches to the basement; when the inequality is strong, the velocity distribution is as specified by (6.2.5). In the vicinity of this maximum, over the horizontal extent for which the surface curvature can be considered constant, the recharge velocity is constant in x; the vertical velocity component in the basin varies linearly with z and vanishes at the lower boundary. In a two-dimensional geometry, the velocity distribution is therefore given from (6.2.5) as

$$u = \frac{g\bar{k}}{\nu}\left(\frac{\partial^2 \zeta}{\partial x^2}\right)x = \frac{g\bar{k}x}{\nu R},$$
$$w = \frac{g\bar{k}}{\nu}\left(\frac{\partial^2 \zeta}{\partial x^2}\right)z = -\frac{g\bar{k}z}{\nu R}, \tag{6.4.16}$$

where $R = -(\partial^2\zeta/\partial x^2)^{-1}$ is the radius of curvature of the water table in the vicinity of its maximum, $\bar{k}$ is the mean horizontal permeability, and z is now measured upward from the basement (Figure 6.12).

The isotherms are depressed in this area by the downward flow; since w is independent of x, the isotherms, also independent of x, are shifted downward and $T = T(z)$ is determined by

$$w\frac{\partial T}{\partial z} = \kappa\frac{\partial^2 T}{\partial z^2},$$

or

$$-\frac{g\bar{k}}{\kappa\nu R}z\frac{\partial T}{\partial z} = \frac{\partial^2 T}{\partial z^2},$$

whence

$$\frac{\partial T}{\partial z} = -\frac{q_h}{\kappa}\exp\left\{-\frac{g\bar{k}}{2\kappa\nu R}z^2\right\}, \tag{6.4.17}$$

since at the base, $\partial T/\partial z = -q_h/\kappa$ as appropriate to the geothermal flux. The scale thickness of the thermal layer is therefore

$$\delta = \left\{\frac{\kappa\nu R}{g\bar{k}}\right\}^{1/2}. \tag{6.4.18}$$

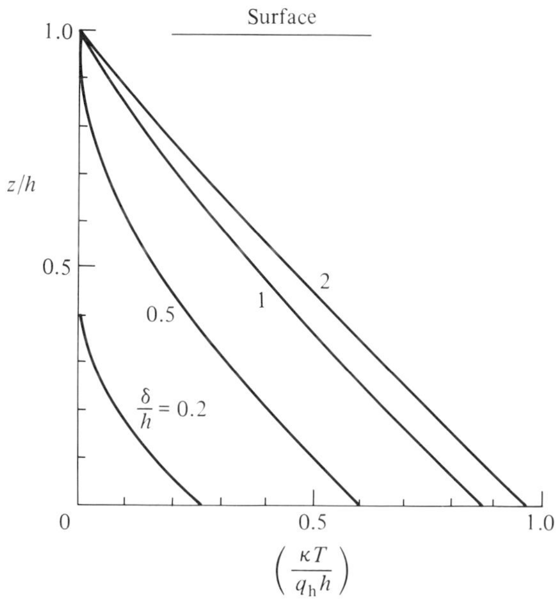

Figure 6.13. Temperature profiles in a recharge zone of depth h with constant basement heat flux; the surface is at $z/h = 1$ and δ is the thermal layer thickness (6.4.18). Temperatures are scaled by the value that the basement at $z = 0$ would have in the absence of flow.

If this is small compared with the basin depth, the heated region is confined to a relatively thin layer near the bottom of the basin, and above this, the region is essentially isothermal. The temperature at the basement is then higher than the surface temperature by the amount

$$\begin{aligned} T &= \int_0^\infty \frac{\partial T}{\partial z}\, dz \\ &= \frac{q_h}{\kappa} \int_0^\infty \exp -(z^2/2\delta^2)\, dz \\ &= \left(\frac{\pi}{2}\right)^{1/2} \frac{q_h \delta}{\kappa}, \end{aligned} \tag{6.4.19}$$

or smaller by the ratio $(\pi/2)^{1/2}\,(\delta/h)$ than it would be in the absence of flow. If, however, $\delta/h \gg 1$, the temperature gradient (6.4.17) is essentially unchanged from its value at the basement: The flow has little effect on the temperature distribution. Some representative profiles are illustrated in Figure 6.13.

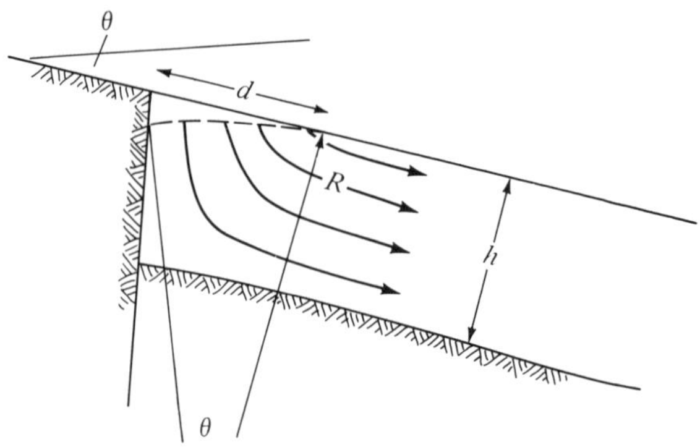

Figure 6.14. Recharge at the upper rim of a permeable basin.

For example, if $R = 40$ km, and $\kappa \sim \nu = 10^{-2}$ c.g.s. unit, in a fairly permeable sandstone with $k_H \sim 4 \times 10^{-10}$ cm^2, we find from (6.4.18) that $\delta \sim 3 \times 10^4$ cm, or 300 m, so that if the depth h of the permeable layer is substantially more than this, the heated layer is confined to a region above the basement of about this depth. The temperature at the basement is depressed to a value smaller by a factor of order δ/h than it would be above the surface value in the absence of flow. However, if the basin contains limestone with $k_H \sim 10^{-12}$ cm^2, then $\delta \sim 8$ km, and unless the basin depth is of this order or more the temperature gradient throughout approximates closely the undisturbed, purely conductive distributions.

When the water-table maximum is axially symmetrical about the vertical direction rather than two-dimensional, $\nabla^2\zeta = 2/R$, where R is the radius of curvature in any section. The thickness δ is in this case reduced by a factor $1/\sqrt{2} \approx 0.7$.

When the upper rim of a permeable basin where recharge occurs is terminated by a steeply dipping fault (Figure 6.14), the flow conditions in this vicinity resemble those near a topographic maximum in the water table. If there is no flow across the fault, the water table must be horizontal there (whatever the actual surface topography) but rising to meet the surface slope farther down the basin if the infiltration rate is sufficient. The curved shape of the water table in this region depends on the recharge fluids available at the upper corner; if this is supplied by the local infiltration rather than by runoff from further upslope, the curvature of the water table is given from (6.3.15) by

$$w = \text{average surface infiltration rate} = \frac{g\bar{k}}{\nu}\left(-\frac{\partial^2\zeta}{\partial x^2}\right)h = \frac{g\bar{k}h}{\nu R},$$

and the water table rises to be parallel to the surface of the basin a distance d from the fault such that

$$wd = uh = \frac{g\bar{k}h}{\nu} \sin\theta,$$

or

$$d \approx R \sin\theta = \frac{g\bar{k}h \sin\theta}{\nu w}. \tag{6.4.20}$$

When $x < d$, with a vertical termination of the basin, the fluid flow is identical to (one-half of) that described earlier and illustrated in Figure 6.12.

Beyond the region of water-table curvature, if the slope settles down to a constant, u becomes constant and the temperature distributions are as described earlier. When $\delta/h \ll 1$, the thermal boundary layer develops as specified by (6.4.10) and (6.4.15) with x measured from a virtual origin that is determined by the beginning thickness of the thermal boundary layer. When $\delta/h \gg 1$, the distribution of temperature is conductively dominated throughout.

Discharge areas: the temperature excess

In a discharge region when δ/h is not large, the opposite effect occurs. Even if the geothermal heat flux across the basement is no larger than it is in surrounding regions, the temperature near the basement is *greater* in a discharge region than it would be in the absence of flow, providing the opportunity for zones of mineral formation at depths shallower than would normally be expected. The reason for this is that warm fluid, near the basement and approaching the discharge area, begins to rise toward the surface; the isotherms move upward. At the basement, however, the normal geothermal *gradient* is maintained by the upward heat flux; since the temperature farther up is elevated, so is the basement temperature. The magnitude of the temperature excess over the normal geothermal value is seen later to depend primarily on the magnitude of the thermal layer thickness δ, given by (6.4.18), relative to the permeable layer depth h in the discharge region and secondarily on the horizontal extent of the discharge area relative to h.

The detailed analysis of the temperature distribution is a little more complicated than in the case of a recharge zone since the temperature of the fluid *entering* the zone varies with depth and the isotherms in the region are not horizontal. The flow field in a two-dimensional situation is as given by (6.4.16) but with the radius of curvature positive, so that $R = (\partial^2\zeta/\partial x^2)^{-1}$ and

$$u = -\frac{g\bar{k}}{\nu R}x,$$
$$w = \frac{g\bar{k}}{\nu R}z. \tag{6.4.21}$$

The temperature field is determined by

$$u\frac{\partial T}{\partial x} + w\frac{\partial T}{\partial z} = \kappa\left(\frac{\partial^2 T}{\partial x^2} + \frac{\partial^2 T}{\partial z^2}\right),$$

or in view of (6.4.21) by

$$-x\frac{\partial T}{\partial x} + z\frac{\partial T}{\partial z} = \left(\frac{\kappa\nu R}{g\bar{k}}\right)\left(\frac{\partial^2 T}{\partial x^2} + \frac{\partial^2 T}{\partial z^2}\right). \tag{6.4.22}$$

In terms of dimensionless coordinates $\xi = x/l_0$, $\eta = z/h_0$, and with the temperature scaled by the value it would have (above the surface temperature) at the basement axis in the absence of flow, that is qh_0/κ, equation (6.4.22) becomes

$$-\xi\frac{\partial \tau}{\partial \xi} + \eta\frac{\partial \tau}{\partial \eta} = P^{-1}\left(A^2\frac{\partial^2 \tau}{\partial \xi^2} + \frac{\partial^2 \tau}{\partial \eta^2}\right), \tag{6.4.23}$$

where $\tau = T(\kappa/qh_0)$, A is the aspect ratio h_0/l_0, and

$$P = \frac{g\bar{k}h_0^2}{\kappa\nu R}. \tag{6.4.24}$$

The flow parameter P can, in view of (6.4.21), be interpreted more conveniently in terms of the total discharge per unit length of the region,

$$Q = 2h_0(-u)_{x\approx l_0} = \frac{2g\bar{k}h_0 l_0}{\nu R},$$

whence

$$P = \tfrac{1}{2}\left(\frac{h_0}{l_0}\right)\frac{Q}{\kappa}. \tag{6.4.25}$$

Therefore, P can be considered to be a Peclet number based on the average surface discharge velocity $w = Q/2l_0$ and the layer depth h_0. Alternatively, from (6.4.24), in terms of the thermal scale δ of (6.4.18),

$$P = h_0^2/\delta^2, \tag{6.4.26}$$

which can be large compared with unity in a deep, permeable layer, or small if the layer is shallow and less permeable.

The boundary conditions for the region shown in Figure 6.15 are that the basement heat flux is constant and equal to q, so that

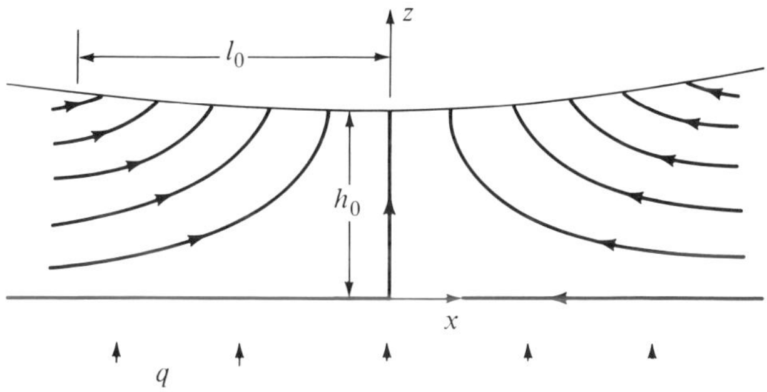

Figure 6.15. Streamline pattern in a two-dimensional discharge region, with a uniform basement heat flux q.

$$\frac{\partial \tau}{\partial \eta} = -1 \qquad \text{at} \quad \eta = 0; \tag{6.4.27}$$

the temperature τ at the surface is zero

$$\tau = 0 \qquad \text{at} \quad \eta = 1 + \left(\frac{l_0^2}{2Rh_0}\right)\xi^2, \tag{6.4.28}$$

and the temperature gradient in the nearly horizontal flow entering laterally is the normal geothermal value:

$$\tau = 1 - \eta + \left(\frac{l_0^2}{2Rh_0}\right) \qquad \text{at} \quad \xi = \pm 1. \tag{6.4.29}$$

If the fractional change in layer depth over the discharge region is small, $l_0^2/2Rh_0 \ll 1$ and the last two boundary conditions can be taken as

$$\tau = 0 \qquad \text{at} \quad \eta = 1; \qquad \tau = (1 - \eta) \qquad \text{at} \quad \xi = \pm 1. \tag{6.4.30}$$

Numerical solutions to (6.4.23) subject to the boundary conditions (6.4.27) and (6.4.30) are shown in Figure 6.16. The maximum temperature occurs at the basement along the axis and increases with the value of P (Figure 6.17). An analytical solution can be found in the case where $A^2 = h_0^2/l_0^2$ is vanishingly small. Along the vertical axis of symmetry, (6.4.23) can be approximated by

$$\eta \frac{\partial \tau}{\partial \eta} = P^{-1} \frac{\partial^2 \tau}{\partial \eta^2} \tag{6.4.31}$$

since the first term in (6.4.23) vanishes quadratically at $\xi = 0$; lateral heat conduction is then negligible. A first integral of (6.4.31) with $\partial\tau/\partial\eta = -1$ at $\eta = 0$ is

$$\frac{\partial \tau}{\partial \eta} = -\exp\left\{\tfrac{1}{2}P\eta^2\right\}.$$

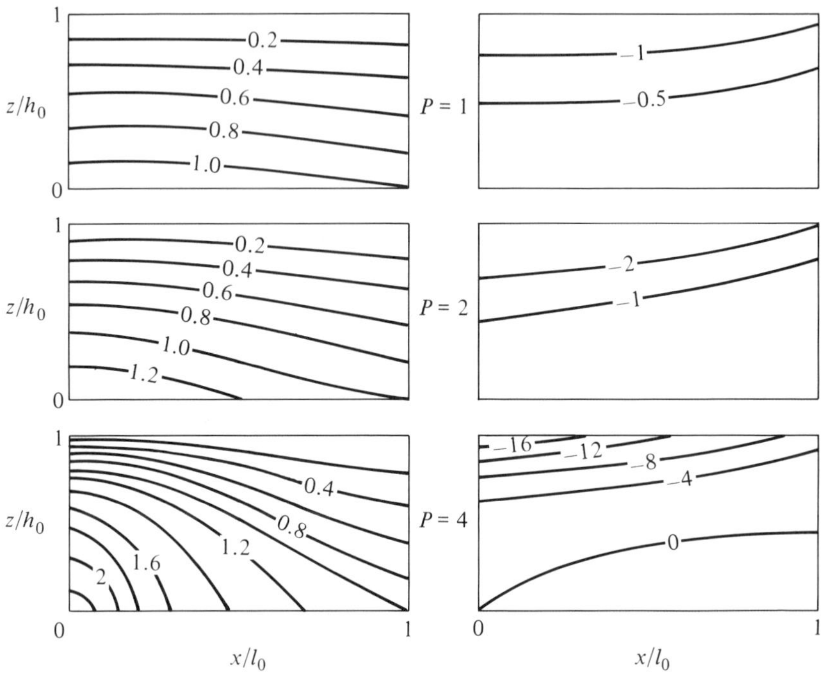

Figure 6.16. Distributions of temperature (on left) and $\mathbf{u} \cdot \nabla T$ (on right) in a two-dimensional discharge region with uniform basement heat flux $\kappa\ \partial T/\partial x = q$. The axis of symmetry is at $x/l_0 = 0$ and the aspect ratio $h_0/l_0 = 0.1$; $P = \frac{1}{2}(h_0/l_0)(Q/k)$. Temperature $T = 1$ is that which would exist at the basement in the absence of fluid flow; the units of $\mathbf{u} \cdot \nabla T$ are q/h_0.

When P is very small, the vertical temperature gradient in $0 < \eta < 1$ remains nearly constant, as appropriate to pure thermal conduction, but when P is of order unity or greater, the nondimensional gradient begins to increase (quite rapidly) in magnitude as we approach the surface and η approaches 1. The vertical temperature distributions on the symmetry plane,

$$\tau(0, \eta) = \int_\eta^1 \exp\left\{\tfrac{1}{2}P\eta^2\right\} dz, \tag{6.4.32}$$

for various values of P are illustrated in Figure 6.17. The basement temperature increases rapidly as P increases; the thermal-convective layer below the upper surface becomes thinner, and the temperature difference across it increases. The surface heat flux in this region also increases, of course, as the result of the advection of heat from both sides.

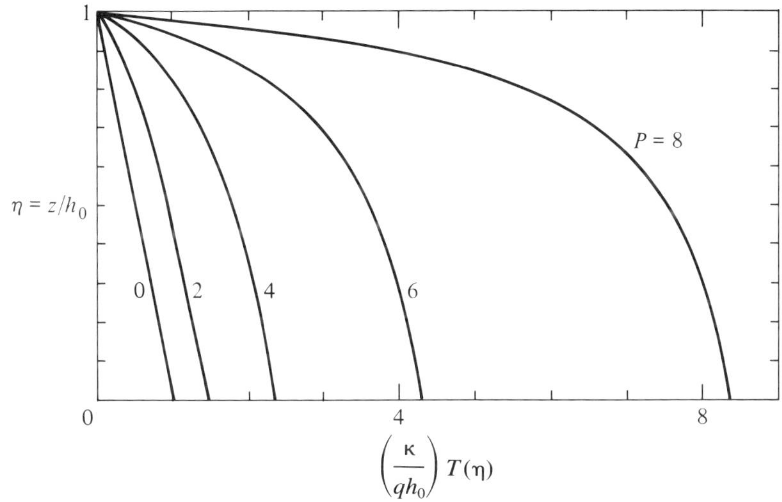

Figure 6.17. Axial temperature profiles in a two-dimensional discharge region with $l_0 \gg h_0$ above an impermeable basement at depth h_0 with constant and uniform heat flux. $P = \frac{1}{2}(h_0/l_0)(Q/\kappa)$; see equations (6.4.24) and (6.4.25). The scale of temperature is that which would exist in the absence of flow. As P increases, a thermal boundary layer develops below the upper surface, and the interior temperature rises rapidly, as does the *surface* heat flux, even though the basement heat flux is constant; heat is advected laterally.

The elevation of basement temperature is most pronounced in distributed discharge regions from deep, highly permeable layers. Because of the chronic uncertainty in estimating the effective permeability of buried rocks, possibly fractured, the Peclet number P can be found more conveniently by estimation or measurement of the total discharge Q per unit length or the mean discharge velocity $w = Q/2l_0$ and the use of (6.4.25) rather than (6.4.24). For example, in a permeable layer of thickness 1 km and thermal diffusivity 4×10^{-3} cm^2 s^{-1}, substitution into (6.4.25) shows that $P \sim 1$ when the mean discharge velocity, rising directly across the isotherms, is as small as 4×10^{-8} cm s^{-1} or 1 cm yr^{-1}. From Figure 6.16 the basement temperature is then only about 20 percent higher than it would be in the absence of flow, that is, about 36° C above the surface temperature compared with the purely conductive increase of some 30° C. If, however, $w \sim 5$ cm yr^{-1} (the horizontal inflow being greater by the ratio $l_0/h_0 \gg 1$) then $P \sim 5$ and the basement temperature along the axis is some 3.5 times when it would be in the absence of flow, that is, about

100° C above the surface temperature. The surface heat flux is increased approximately by the factor $\exp(\frac{1}{2}P) \sim 12$, even though the basement heat flux is uniform and no larger than average. The extra heat is advected laterally by the flow.

If P were to increase beyond these values, the basement temperature would increase further, but the analysis presented here rapidly becomes unrealistic. The increasing temperature at the interface between the basement and the permeable layer *reduces* the temperature gradient in the underlying basement and correspondingly reduces the upward heat flux, here taken as constant. The thermal problem then involves the whole crust, and a more complete analysis is required for each specific situation.

In any event, it is evident that converging, upwelling flow over a moderate horizontal area can result in considerably enhanced surface heat fluxes such as those found in geothermal areas. Many geothermal regions, like Wairakei in New Zealand, are tectonically active, and the additional heat flux is probably supplied by hot, igneous intrusion from below, a pluton. However, the characteristic cooling time for a pluton of reasonable size is quite short, of order l^2/κ, where l is its characteristic dimension. If $l = 1$ km and $\kappa = 4 \times 10^{-3}$ cm^2 s^{-1}, this cooling time scale is about 100,000 years – on a geological time scale, the mere blink of a cosmic eyelid. Cathels (1977) has constructed detailed numerical models of the cooling of an igneous intrusive by both natural liquid convection and vaporization; for intrusions of scale 1 km or so, he calculates rather shorter cooling times, approximately 20,000 or 30,000 years. As he points out, the existence of active geothermal regions over time scales significantly longer than this requires continued intrusion from below as in actively rifting areas. Lower-grade geothermal areas in tectonically inactive regions can, however, continue as a result of flow convergence and discharge, even if the basement heat flux is not unusually large.

When P is not small, the temperature gradient is largest at the surface, and since the fluid flow is directly down this gradient, the rock alteration index RAI $= \mathbf{u} \cdot \nabla T$ is large within a distance δ of the surface, leading to mineral deposition or dissolution according to the sign of the geochemical coefficients of equations (4.4.4) and (4.4.6). Distributions of RAI, $\mathbf{u} \cdot \nabla T$, are illustrated in the right-hand boxes of Figure 6.16 for various values of P, large negative values appearing within a distance $\delta = h_0 P^{-1/2}$ of the surface as P increases.

When the discharge region is axisymmetrical rather than two-dimensional, the axial temperature profiles can be shown to be again as illustrated in Figure 6.17 with the flow parameter or Peclet number P replaced by

$$P_{\mathrm{T}} = \frac{Q_{\mathrm{T}} h_0}{\pi R^2 \kappa}, \tag{6.4.33}$$

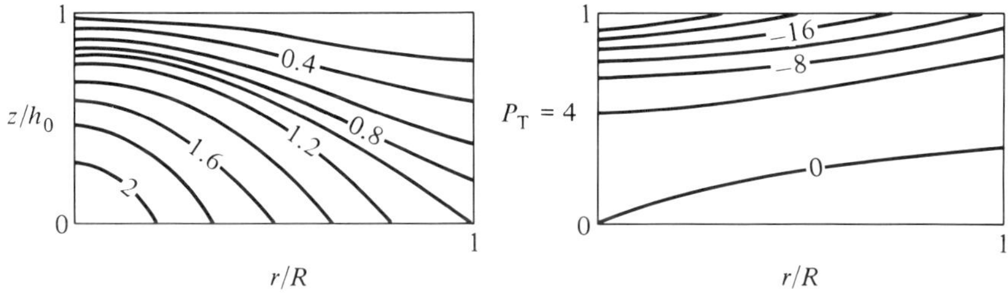

Figure 6.18. Distributions of isotherms (on the left) and rock alteration index $\mathbf{u} \cdot \nabla T$ (on the right) in an axisymmetrical discharge zone of radius R with uniform basement heat flux q. $P_T = Q_T h_0/\pi R^2 \kappa = 4$. The scales are as in Figure 6.16.

where Q_T is the total flow discharge over the circle of radius R. Thus, if the average discharge rate $Q_T/\pi R^2$ is 5 cm yr^{-1} $\sim 4 \times 10^{-7}$ cm s^{-1}, $\kappa = 4 \times 10^{-3}$ cm^2 s^{-1}, and $h_0 = 1$ km, then again $P_T = 5$. The internal distributions of temperature and of the RAI for the case $P_T = 4$ are as shown in Figure 6.18; they are generally similar to the two-dimensional case but differ in detail.

It is important to recognize that the solutions for these discharge region flows are hydraulically, not thermally, driven. The hydraulic forcing pressure gradients are $\rho g\ \nabla\zeta$, where $\nabla\zeta$ is the surface slope at the edges of the discharge region, whereas the thermal forcing term is $\rho g \alpha\ \Delta T$, where ΔT is the temperature increase at the axis. Hydraulic forcing dominates when $\nabla\zeta > \alpha\ \Delta T$ or, if $\Delta T = 100°$ C and $\alpha \sim 5 \times 10^{-4}$ °C^{-1}, when the surface slope at the edges is greater than 5×10^{-2}. Elevated temperatures at the axis produced by convergence will to some extent augment the discharge flow beyond what is produced either by "forced convection" alone as described here or by free convection as considered in the next chapter.

6.5 MIXING ZONES

In the discharge of a coastal aquifer, the groundwater hydrologist is interested primarily in the configuration of the "interface" between fresh water and the underlying salt wedge, but the geochemist, interested in patterns of reaction or dissolution associated with mixing as described in Section 4.4, requires some insight into the characteristics of the mixing zone itself. The shape of the interface is usually described to a first approximation by what is known as the Ghyben–Herzberg approximation (see, e.g., Bear 1979), in which the interstitial fluid pressure is assumed

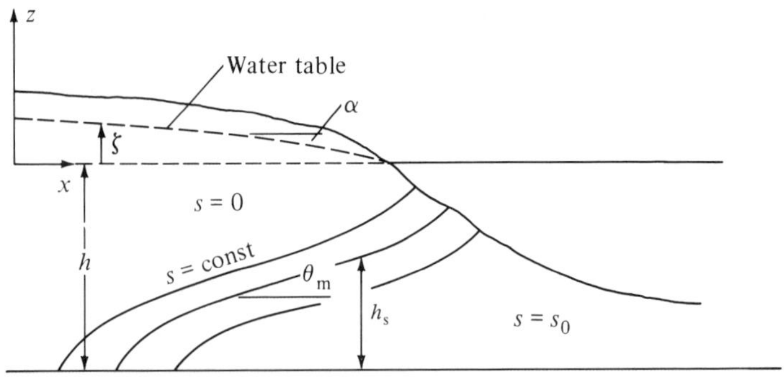

Figure 6.19. A coastal mixing layer in permeable fabric above an impermeable basement. The approximately parallel curves represent isohalines separating the fresh interstitial water on the left, $s = 0$, from the saline water on the right, $s = s_0$.

to be, in essence, hydrostatic. The solution can, however, be recovered more generally even when the interface is diffuse, provided that the slopes of the isopycnals are small.

When $(h/l)^2(k_H/k_V) \ll 1$, the second equation of (3.4.4) reduces to

$$\frac{\partial}{\partial z}\left(\frac{u}{k_H}\right) = \frac{g}{\mu}\frac{\partial \rho}{\partial x}$$

$$= \frac{g\beta}{\nu}\frac{\partial s}{\partial x} \tag{6.5.1}$$

when the variations in density are associated with salinity variations and $\rho = \rho_0(1 + \beta s)$. If, in the mixing layer illustrated in Figure 6.19, the isopycnals slope upward at the angle θ_m,

$$\frac{\partial s}{\partial x} = -\tan\theta_m(x)\frac{\partial s}{\partial z},$$

and

$$\frac{\partial}{\partial z}\left(\frac{u}{k_H}\right) = -\frac{g\beta\tan\theta_m(x)}{\nu}\frac{\partial s}{\partial z},$$

which can be integrated upward from depth $-z$ to the phreatic surface (the water table) at $z = \zeta$;

$$\frac{u(\zeta)}{k_H(\zeta)} - \frac{u(z)}{k_H(z)} = \frac{g\beta\tan\theta_m(x)}{\nu}s(z), \tag{6.5.2}$$

since $s(\zeta) = 0$. At the water table, when $(h/l)^2k_H/k_V \ll 1$, again,

$$\frac{u(\zeta)}{k_{\mathrm{H}}(\zeta)} = -\frac{g}{\nu}\frac{\partial\zeta}{\partial x} = \frac{g}{\nu}\tan\alpha(x),$$

where $\alpha(x)$ is the downward slope of the water table. Thus, from (6.5.2)

$$\frac{u(z)}{k_{\mathrm{H}}(z)} = \frac{g}{\nu}\{\tan\alpha(x) - \beta s(z)\tan\theta_{\mathrm{m}}(x)\}. \quad (6.5.3)$$

If the saltwater wedge is *assumed* to be at rest and z is taken to be just below the mixing zone where the salinity is the seawater value s_0, then $u(z) = 0$ there and (6.5.3) gives

$$\tan\theta_{\mathrm{m}}(x) = (\beta s_0)^{-1}\tan\alpha(x) = \left(\frac{\rho}{\Delta\rho}\right)\tan\alpha(x), \quad (6.5.4)$$

where $\Delta\rho$ is the density difference across the layer. The slope of the mixing layer then mirrors that of the water table but is larger by the factor $(\rho/\Delta\rho)$. This, in the case of an abrupt interface, is the classical Ghyben–Herzberg result. In a real mixing zone, however, salt water is entrained into the discharging fresh water and so is drawn into the mixing layer from the seaward direction; $u(z) < 0$ beneath the mixing layer, the right-hand side of (6.5.3) is negative, and the slope of the layer is necessarily greater than given by (6.5.4).

The fluid discharge Q per unit length of coastline is related to the water-table slope, as described at the end of Section 6.2, but here the velocity varies with depth because of the density variation. Equation (6.5.3) can be integrated from the underlying, assumed impermeable bed up to the water table, and if k_{H} is taken constant,

$$\begin{aligned} Q &= \int_{-h}^{\zeta} u(z)\,dz \\ &= \frac{gk_{\mathrm{H}}}{\nu}\left\{(h+\zeta)\tan\alpha(x) - \beta\tan\theta_{\mathrm{m}}(x)\int_{-h}^{\zeta} s(z)dz\right\} \\ &= \frac{gk_{\mathrm{H}}}{\nu}\{(h+\zeta)\tan\alpha(x) - \beta s_0 h_s\tan\theta_{\mathrm{m}}(x)\}, \end{aligned} \quad (6.5.5)$$

where h_s is the height of the middle of the mixing layer above the basement. The actual relationship between $\theta_{\mathrm{m}}(x)$ and $\alpha(x)$ depends, as already mentioned, on the entrainment and mixing characteristics of the layer; if entrainment is negligible, (6.5.4) can be used and the thickness of the freshwater wedge can be shown to decrease as the shoreline is approached, approximately as

$$h_{\mathrm{f}} \simeq \left\{\frac{2Q\nu}{g\beta s_0 k_{\mathrm{H}}}(x_0 - x)\right\}^{1/2}, \quad (6.5.6)$$

where x_0 is a virtual origin.

In the field, of course, any small or mesoscale layering will modify the general shape of the layer around the macroscopically averaged configuration (6.5.6), but the effect of these structures on mixing is overwhelming. If the bedding is horizontal, the *overall* macroscopic coefficient for horizontal dispersion may be of the order $\phi^{-1}ul$, as described in Section 3.9, where l is the horizontal length scale of the lenses or individual layers. Even though the mean flow is primarily almost horizontal, vertical dispersion arises when fluid moves from one lens or layer to another; if h_l represents the characteristic vertical interval between lenses or more permeable layers, the fluid performs a "random walk" in the vertical direction with a step size proportional to h_l and an asymptotic dispersion coefficient proportional to $\phi^{-1}uh_l$. Generally $l \gg h_l$, so that this is small compared with the horizontal dispersion. However, the mixing zone crosses the bedding planes at the angle θ_m, so that the dispersion coefficients across and along the mixing zone, D_1 and D_2, respectively, are

$$D_1 = D_h \cos^2 \theta_m + D_v \sin^2 \theta_m, \qquad (6.5.7)$$
$$D_2 = D_h \sin^2 \theta_m + D_v \cos^2 \theta_m,$$

where D_h and D_v are the principal components of the dispersion tensor in the horizontal and vertical directions, respectively. When θ_m is not large, the longitudinal dispersion coefficient may be approximated by D_h, but the lateral dispersion across the layer involves the two terms proportional to $\phi^{-1}ul \sin^2 \theta_m$ and $\phi^{-1}uh_l \cos^2 \theta_m$, whose relative influence depends on the internal geometry of the structure and the layer slope. Moreover, the geometrical scales of the mixing region cannot generally be presumed large compared with the scales of the lenses or layers (they are more likely to be of the same order), so that the very concept of a megadispersion coefficient at this scale is on somewhat shaky ground.

If, in spite of these reservations, we represent the salinity balance as

$$\mathbf{u} \cdot \nabla S \simeq \phi D_2 \frac{\partial^2 S}{\partial n^2}, \qquad (6.5.8)$$

where n is taken normal to the mixing zone, an estimate of its thickness δ can be found. The term on the left represents the transport speed times the salinity gradient along the lines of flow and so is of the order us_0/L, where L is the length of the zone; the term on the right is of order $\phi D_2 s_0/\delta^2$. Consequently,

$$\delta \sim (\phi D_2 L/u)^{1/2}$$
$$\sim (lL)^{1/2} \sin \phi_m \qquad \text{or} \qquad (h_l L)^{1/2}, \qquad (6.5.9)$$

according to whether the first or second term for D_2 in (6.5.7) dominates.

As described in Section 4.5 mixing zones between seawater and freshwater aquifer in a limestone matrix lead to dissolution of the limestone at the rate specified by (4.5.3). In this zone, the salinity gradient is of order s_0/δ, so that the rate of dissolution per unit volume of fabric can be represented as

$$Q_s \simeq -\phi D_2 \left(\frac{\partial^2 c_e}{\partial s^2}\right)\left(\frac{s_0}{s}\right)^2$$
$$= -\frac{s_0^2 u}{L}\left(\frac{\partial^2 c_e}{\partial s^2}\right), \qquad (6.5.10)$$

which, fortunately, does not involve D_2 explicitly. The dissolution rate is greatest below the freshwater outflow at the upper, right-hand end in Figure 6.19 where u is largest and the location of the mixing region relatively stable in time. The location of the toe is much more variable, advancing to the left in dry periods and receding in wet ones, so that deeper dissolution is more diffuse. The time scale τ for cavern formation at the coastline is, from Section 4.6, $\rho\phi_0/m_1 Q_s$, where ρ and ϕ_0 represent the density and initial porosity of the fabric, respectively, m_1 is the molecular weight of $CaCO_3$, and Q_s is given by (6.5.10). A prediction of numerical values requires information on $(\partial^2 c_e/\partial s^2)$ from the kinetics of the reaction, but a very rough estimate that could be an error by an order of magnitude can be made for Q_s by taking $\partial^2 c_e/\partial s^2$ as $\Delta c_e/s_0^2$, where Δc_e is the difference in saturation concentrations of $CaCO_3$ in the seawater and the fresh water of the aquifer; Q_s is then of order $u\Delta c_0/L$ and τ of order $\rho\phi_0 L/um_1\ \Delta c_e$.

6.6 TRAPPING OF HYDROCARBONS

In the absence of any overall flow along a permeable layer, denser interstitial fluids such as concentrated brines tend to collect at low points of the layer and less dense components such as hydrocarbons tend to migrate upward to the surface in an unconfined aquifer, or to high points in a confined one. The migration of brines can be treated relatively simply, but that of hydrocarbons may involve multiphase flows, which, as mentioned in Section 2.6, cannot generally be described in terms of the Darcy equation because of capillarity and possibly other interfacial effects. Once a pool of hydrocarbons has collected, however, and there is no further motion, the equilibrium configuration of the pool does not involve the pressure gradient–velocity relationship, whatever it may be, since the pool is at rest. If the underlying interstitial fluid also is not moving, the interface separating the water-saturated and the hydrocarbon-saturated regions is, on average, horizontal, though in detail it may be diffuse. However, as Hubbert pointed out in his monumental paper of

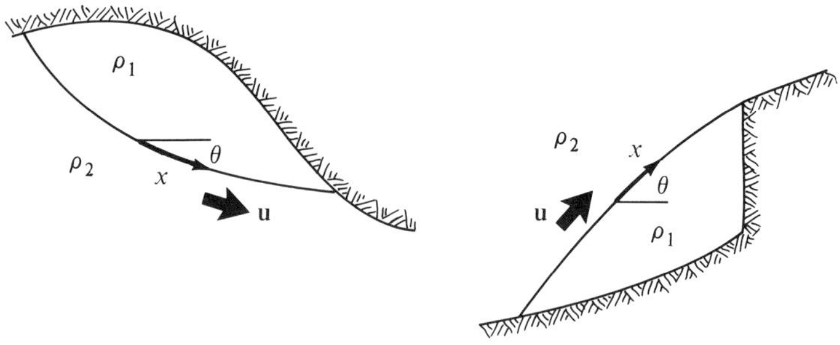

Figure 6.20. Trapped pools of less dense fluid (a) and more dense fluid (b) in a permeable layer with flow velocity **u**.

1940, aqueous fluid flow along the stratum tends to wash out pockets of dense fluids from low points and lighter fluids from high points. It is of interest to determine the conditions under which these pools can remain in equilibrium in the presence of flow, and how the location and configuration of the pools are influenced by the flow and the geometry of the geological structures.

Consider, then, the situations illustrated in Figure 6.20 with an immobile pool of less dense fluid trapped below an impermeable roof as in (a) or of more dense fluid trapped above the basement as in (b). In the immobile fluid, where the interstitial fluid density is ρ_1, the pressure gradient is hydrostatic, so that along the interface

$$\frac{\partial \hat{p}}{\partial x} = -\rho_1 g \sin \phi. \tag{6.6.1}$$

Outside the pool, the matrix is water saturated and the interstitial fluid, with density ρ_2 and viscosity μ, moving with the transport velocity u satisfies Darcy's equation:

$$u = -\frac{k}{\mu}\left(\frac{\partial \hat{p}}{\partial x} + \rho_2 g \sin \phi\right). \tag{6.6.2}$$

Since the pressure gradients along the interface must be the same on either side,

$$\begin{aligned} u &= -\frac{k}{\mu}(-\rho_1 + \rho_2) g \sin \phi \\ &= \left(\frac{\rho_1}{\rho_2} - 1\right)\frac{kg}{\nu} \sin \phi && (6.6.3) \\ &= \hat{U}_c \sin \phi, \qquad \text{say}, && (6.6.4) \end{aligned}$$

where $\nu = \mu/\rho_2$ is the kinematic viscosity of the (moving) aqueous fluid in region 2 and

$$\hat{U}_c = \frac{kg}{\nu}\left(\frac{\rho_1}{\rho_2} - 1\right). \tag{6.6.5}$$

In the case of Figure 6.20b with a denser pool, $\rho_1 > \rho_2$, $\sin\theta > 0$, and the equilibrium interface always slopes *upward* in the flow direction; in that of a less dense pool trapped at the top of a permeable layer, where $\rho_1, < \rho_2$ and $\sin\theta < 0$, it always slopes *downward* in the flow direction, in each case as a result of the pressure in the moving fluid being higher upstream than downstream. The magnitude of $\hat{U}_c$, considered without regard to sign,

$$U_c = |\hat{U}_c| = \frac{kg}{\nu}\frac{\Delta\rho}{\rho}, \tag{6.6.6}$$

where $\Delta\rho$ is the density difference between the lower and upper fluid, can be regarded as a critical speed for trapping. When $u > U_c$, no equilibrium solution is possible, and any pool of denser or lighter interstitial fluid is necessarily flushed out whatever the layer geometry.

Unconfined aquifers

When the aquifer is such that $(h/l)^2 k_H/k_V \ll 1$, the interfacial configuration of a denser pool at its base is given by the Ghyben–Herzberg expression of the preceding section, provided that the interface remains sharp and no entrainment occurs. In nature, however, a denser, miscible brine is entrained into the fluid above, and unless there is a source to replenish the pool, it will be flushed out as the brine disperses. If brine *is* emanating from a local fracture, say, it will collect as a pool (as distinct from the downstream plume described in Section 4.6) only when the local basement slope exceeds θ_m, as specified by equation (6.5.4).

Trapped pools in confined aquifers

The interfacial configuration of a trapped pool in a confined aquifer involves somewhat different constraints. Consider for the sake of definiteness a pool of lighter fluid such as hydrocarbon trapped above a water-saturated region of thickness H measured normal to the basement in which the transport velocity is u. In this case, capillarity inhibits entrainment, and the interface slopes downward in the flow direction at the angle θ (which is therefore negative) and

$$u = -U_c \sin\theta. \tag{6.6.7}$$

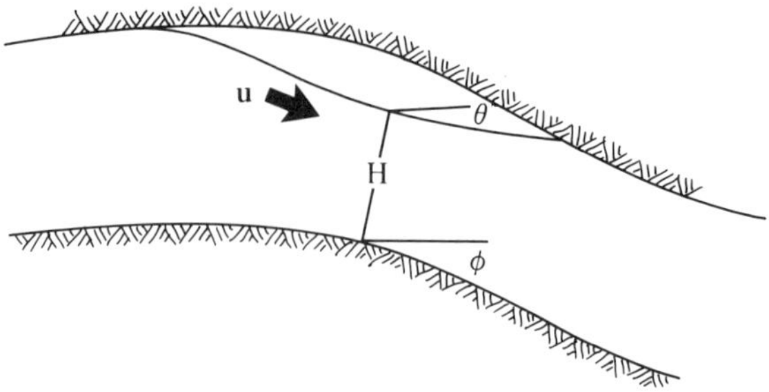

Figure 6.21. A trapped pool in a confined aquifer.

If the flow is locally two-dimensional, the total volume flux Q is independent of x, and in the simplest case of a permeability uniform in depth,

$$Q = uH. \tag{6.6.8}$$

Furthermore, from the geometry of Figure 6.21,

$$\frac{dH}{dx} = \tan\theta - \tan\phi, \tag{6.6.9}$$

where x is taken in the flow direction and the basement slope angle ϕ is negative when it dips in the flow direction. The configuration of the lower surface of a trapped pool in equilibrium is then determined by the simultaneous solution of equations (6.6.7) to (6.6.9). Note that, in general, the shape of the interface depends on the *basement* configuration, since the aqueous flow is confined between the basement and the pool's lower interface; the *extent* of a possible pool depends on the configuration of the upper boundary of the permeable region and the necessity for a downward dip or step in the flow direction to intercept the interface.

When the transport speed is strongly subcritical, $u/U_c \ll 1$, and the mathematical problem is simplified somewhat. The angle θ is then small and

$$\tan\theta \simeq \sin\theta = -u/U_c = Q/U_cH,$$

from (6.6.7) and (6.6.8). Thus,

$$\frac{dH}{dx} + \frac{Q}{U_cH} = -\tan\phi(x), \tag{6.6.10}$$

which is a simple differential equation for $H(x)$ given the basement configuration $\phi(x)$, but an awkward one to solve analytically in general. Numerical solutions can be found simply for specified geometries, though their qualitative behavior can be inferred simply from the equation itself, as follows.

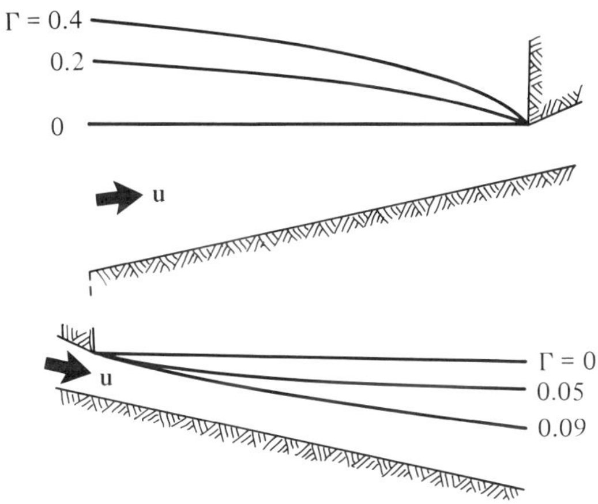

Figure 6.22. Numerical solutions to equation (6.6.10) for the interfacial profiles of trapped pools upstream of a downward displacement in a rising stream (a) and downstream of an upward displacement in a stratum with the denser fluid descending (b). To retain the pools, the profiles shown must intersect the roof at some point away from the constriction. $\Gamma = Q/U_c H_0$. The basement slope is ± 0.1, and the vertical scale in this figure is exaggerated by a factor of 2.

Physically significant solutions must intercept the impermeable roof of the permeable layer at two points: the upstream and downstream ends of the pool. When the basement slopes *upward* in the flow direction as in Figure 6.22a, $\phi > 0$, so that

$$\frac{\partial H}{\partial x} = -\frac{Q}{U_c H} - \tan\theta < 0. \qquad (6.6.11)$$

With no flow, the hydrocarbon–water interface is, of course, horizontal; $Q = 0$ and $dH/dx = -\tan\phi$, H being measured upward from the basement. With flow to the right as in Figure 6.22a, $Q > 0$ and from (6.6.11) the slope of the interface intensifies as H decreases toward the constriction. A pool can be retained only when the roof slope downward exceeds dH/dx given by (6.6.11) – for example, at an impermeable displacement fault. When the basement slopes *downward* in the flow direction, a hydrocarbon pool may be trapped downstream of a constriction but only if $\Gamma = Q/U_c H_0 < |\tan\phi|$. The basement slope $\tan\phi$ is now negative, and if the right-hand side of (6.6.11) as a whole is negative, H decreases ever more rapidly, and the solution intersects the basement rather than the roof, an unacceptable solution. In physical terms, if the thickness H of the flow region constricts below the pool, the increased

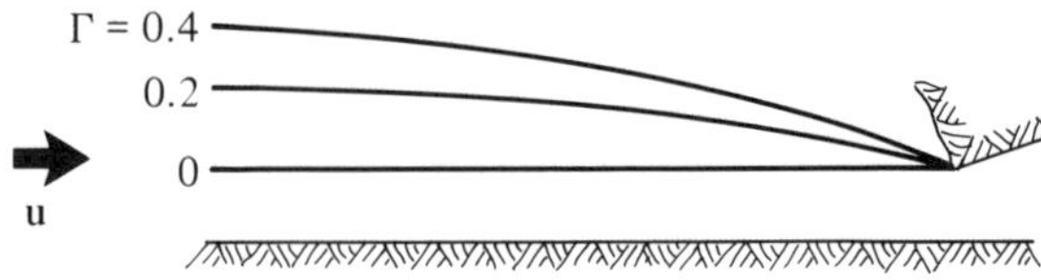

Figure 6.23. Interfacial profiles of a light stagnant pool over a horizontal basement; vertical exaggeration ×2.

pressure gradient needed to drive the flow forces the interface upward on the lee of the geological constriction, and the pool globs out.

Analytical solutions can be given for cases in which the basement slope ϕ is constant. If $H = H_0$ at some location $x = 0$ and $\xi = x/H_0$, $\eta = H(x)/H_0$, equation (6.6.10) becomes

$$\frac{d\eta}{d\xi} + \frac{\Gamma}{\eta} = -\tan\phi, \qquad (6.6.12)$$

where $\Gamma = Q/U_c H_0 (\ll 1)$. When, in particular, $\phi = 0$, (6.6.12) integrates to

$$\xi = (2\Gamma)^{-1}(1 - \eta^2),$$

so that

$$\frac{H}{H_0} = \left\{1 - \frac{2Qx}{U_c H_0^2}\right\}^{1/2}, \qquad (6.6.13)$$

which increases with distance upstream ($x < 0$), as illustrated in Figure 6.23. When $\phi \neq 0$, the solution is a little more complicated:

$$\xi = \frac{1-\eta}{\tan\phi} + \frac{\Gamma}{\tan^2\phi}\ln\left\{\frac{\Gamma + \eta\tan\phi}{\Gamma + \tan\phi}\right\}. \qquad (6.6.14)$$

giving $\eta = H/H_0$ as an implicit function of $\xi = x/H_0$. These configurations are as illustrated in Figure 6.22.

Note that for the trapping of pools of gas, the critical speed is larger than it is for the trapping of pools of liquid hydrocarbons, since in equation (6.6.6) the density difference across the interface is greater. For a given geometry, then, there is a range of aqueous fluid velocities that will allow gas pools to be trapped while flushing out pools of liquid hydrocarbons – a hydraulic separation mechanism!

6.7 SPATIAL DISTRIBUTIONS OF MINERALIZATION ASSOCIATED WITH FLOW

Generalities

The ability to predict the distributions on a basin-wide scale of dissolution and porosity enhancement, diagenesis, and ore formation are of obvious economic importance. Frequently, the available structural and geochemical information on a particular basin is far from complete, and one may be forced to rely too greatly on conjecture—an inadequate base on which to build a detailed numerical simulation. The analyses of the preceding sections do, however, provide a series of general guides in terms of which the overall characteristics of the flow fields can be inferred simply, and regions of particular interest identified.

For example, in an unconfined aquifer or sedimentary basin, the subsurface fluid velocity (on a large scale, subject to the caveats of Section 6.3) is down the gradient of the present water table (or the one that existed at the time of interest). According to equation (6.2.5) it varies with depth in proportion to the permeability, and so is preferentially focused in lenses or layers of high permeability. Note that the flow magnitude and direction follow the surface gradient, not the basement, so that with a uniform surface gradient over a local basement high, the flow is up and over the basement high, not around it.

When the stratigraphy and the flow are primarily horizontal, the local temperature gradient is usually not influenced much by the flow, since the Peclet number Uh/κ is usually sufficiently small compared with the aspect ratio l/h that the condition (6.4.1) is satisfied. Important local exceptions can occur in recharge and discharge regions of limited horizontal extent where (since, in effect, l, the horizontal scale of this region, is now small) significant subsurface temperature anomalies develop even when the geothermal heat flux is uniform. Discharge regions may be of particular geochemical interest since if the total discharge Q per unit length is large compared with $\kappa l_0/h_0$ (where κ is the thermal diffusivity, l_0 the horizontal scale of the region, and h_0 the layer depth), then according to Section 6.4 anomalously high temperature gradients occur at shallow depths and consequently, in general, reactions are more rapid.

It is important to recognize that the kinetic reaction rate does not usually influence directly the rate of mineralization. It does, in combination with the flow velocity, determine the equilibration length, as discussed in Section 4.2, and only if the reaction is so slow that this equilibration length is *large* compared with the longitudinal dimensions of the flow is the rate of mineral deposition proportional to the kinetic reaction rate and independent of the flow speed, being given by equation (4.4.7). In other cases with more vigorous reactions, the equilibration length is relatively small, the interstitial fluid is generally close to local chemical equilibrium with its geological environment, and the rate of mineralization (or dissolution, etc.) is governed by the capacity of the fluid motion to transport solutes into and out of the region of interest and is independent of the reaction rate.

The two primary types of flow-coupled reactions, the isothermal and gradient reactions, have characteristics that leave specific recognizable signatures in the geological record. Isothermal reactions produce what are in effect moving mineralogical boundaries (sometimes described as waves) that propagate in the flow direction at a rate that is proportional to the flow speed and dependent on the fluid–rock ratio, as shown by equation (4.3.16). These reactions essentially go to completion as fluid, in equilibrium with its surroundings in one assemblage, moves across the boundary, where the solute reacts to reach equilibrium with its new surroundings. The compositional change is crucial, not any temperature change. In a region that is initially chemically homogeneous but with striations in permeability, isothermal reaction fronts move through the matrix at rates proportional to the local permeability, so that after sufficient time, the matrix may consist of layers of altered mineral composition in which the reaction front has gone by at an earlier time, interspersed among completely unaltered regions of lower permeability in which the reaction fronts, moving more slowly, have not yet arrived. The bimodal distribution may be smeared somewhat on the scale of the local equilibration length by diffusion or local dispersion. If a sufficiently large region can be mapped, the flow direction is indicated unambiguously by the direction in which the altered fingers point.

In contrast, gradient reactions involve the variation in local equilibrium with temperature, pressure, or some other chemical constituent and are more pervasive. They occur *simultaneously* throughout the whole matrix, though at rates proportional to the local fluid velocity (cf. equations (4.4.4) and (4.4.6)), so that in a structure with layering of permeability, the resulting mineral alteration, though ubiquitous, is greater in layers of high permeability and less in less permeable layers. The process is reversible – if the flow direction reverses relative to the temperature and pressure gradients, the sign of Q_c changes. In discharge zones, the isotherms can be crowded toward the surface and the temperature gradient in the flow direction becomes large, a particularly favorable scenario for intense mineralization by gradient reactions. If the paleotemperature gradient can be inferred, a measurement of the net mineralization provides what is in effect a net vector flow meter, integrated over time (equation (4.4.6)).

The two kinds of reaction may, of course, have been occurring simultaneously throughout the geological history of a region. Isothermal reaction fronts produce a pattern of completely altered and unaltered lay-

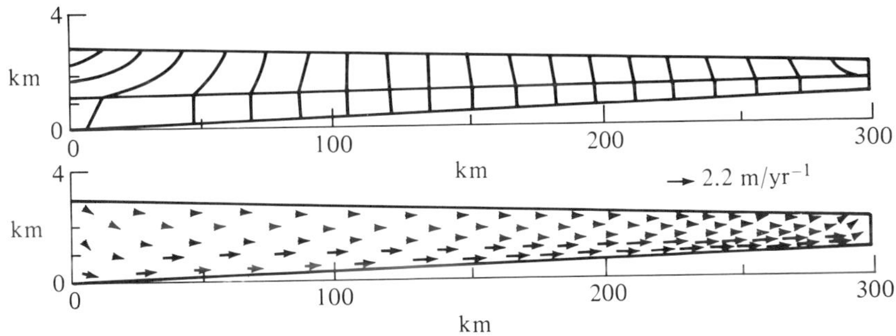

Figure 6.24. Calculations of hydraulic head contours proportional to the reduced pressure (upper figure) and flow vectors (lower figure) modeled by Garven and Freeze (1984b) for a two-layer unconfined aquifer.

ers; gradient reactions, a pattern of more altered or less altered layers; and the combination, one of completely altered layers (where isothermal reaction fronts have passed) among regions of pervasive but less intensive mineralization by gradient reactions.

Basin-wide flow and temperature characteristics

With these generalities in mind, let us turn first to the pioneering numerical experiments of Garven and Freeze (1984a,b) on the flow, temperature, and transport distributions in a simple two-layer, two-dimensional, truncated, wedge-shaped basin under various combinations of parameters. In these calculations, the overall aspect ratio l/h was generally held constant at 150 and the permeability anisotropy $k_H/k_V = 100$, so that

$$\frac{k_H}{k_V}\left(\frac{h}{l}\right)^2 \approx 4 \times 10^{-3},$$

which is certainly sufficiently small that the overall geometrical scaling can be applied. The region considered was bounded by vertical end walls, however, and near the ends, the effective horizontal scale is the distance from the wall. The results of overall scaling will then be accurate only when this distance is large compared with $(k_H/k_V)^{1/2}h$, or about 30 km from the left-hand end and 10 km from the right-hand end of Figures 6.24 to 6.26, but in the central region the flow calculated numerically agrees well with that given by (6.2.5). For example, in Figure 6.24 (fig. 3B of Garven and Freeze 1984b) the uniform surface slope is 1/300, in the base layer the hydraulic conductivity is taken as $K_{xx} = k_H g/\nu = 100$ m yr^{-1} and the porosity ϕ as 0.15; in the upper layer the values chosen are 10 and 0.1, respectively. From (6.2.5), the transport velocity $\mathbf{u} = -K_{xx}\,\nabla\xi$ in the most permeable, lower region is 0.3 m yr^{-1} and in the

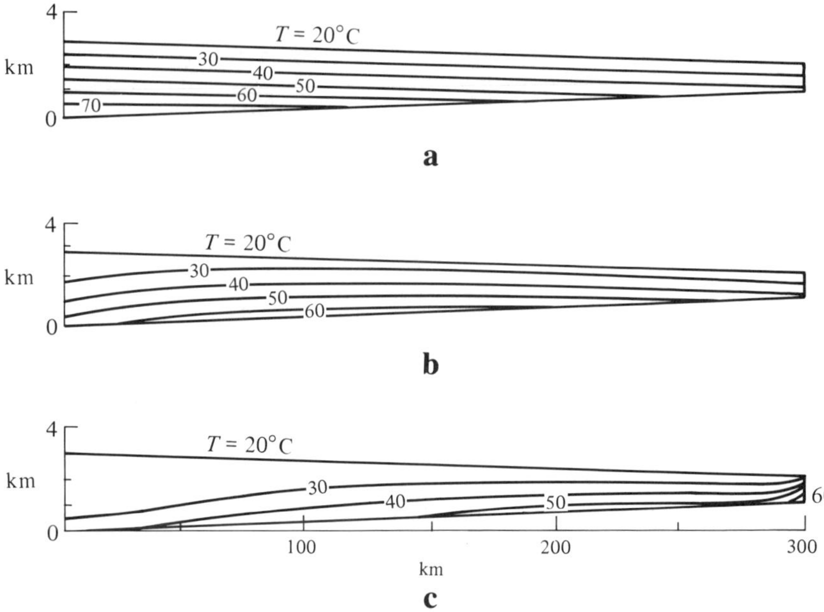

Figure 6.25. Temperature distributions calculated by Garven and Freeze (1984b) for a two-layer unconfined aquifer with an aspect ratio $l/h = 150$, where h is the mean thickness, for various values of the hydraulic conductivity $K = k_H g/\nu$ in the lower layer. In (a), $K = 10$ m yr^{-1} and the Peclet number is 30, much smaller than l/h, so that the isotherms are hardly disturbed by the flow. In (b) $K = 100$ m yr^{-1} and the Peclet number is 300, comparable to l/h. In (c) $K = 500$ m yr^{-1} and Pe = 1,500 $\gg l/h$, and a thermal boundary layer develops above the impermeable basement.

upper region, 0.03 m yr^{-1}, so that the interstitial fluid velocities are 2 and 0.2 m yr^{-1}, which are consistent with the numerical results illustrated in Figure 6.24. Other simulations show the flow up and over a basement high and the convergence at a discharge region, and are again in accordance with the patterns predicted analytically.

Garven and Freeze also calculate temperature distributions by using a uniform geothermal heat flux of 60 mW m^{-2} (1.4×10^{-6} cal s^{-1} cm^{-2}) and thermal conductivity of 3.0 W m^{-1} °C^{-1} (thermal diffusivity κ about 7×10^{-3} cm s^{-1}), and Figure 6.25 shows the changes in isotherm patterns as the hydraulic conductivities in the layers (and so the flow velocities) are increased. Figure 6.14 corresponds to the left-hand end of the flow of Figure 6.25; the Peclet number Uh/κ (taking the flow velocity U in the lower layer of 0.3 m yr^{-1} $\sim 10^{-6}$ cm s^{-1} and h as the mean depth, 2 km = 2×10^5 cm) is then about 300, which is of the same order as

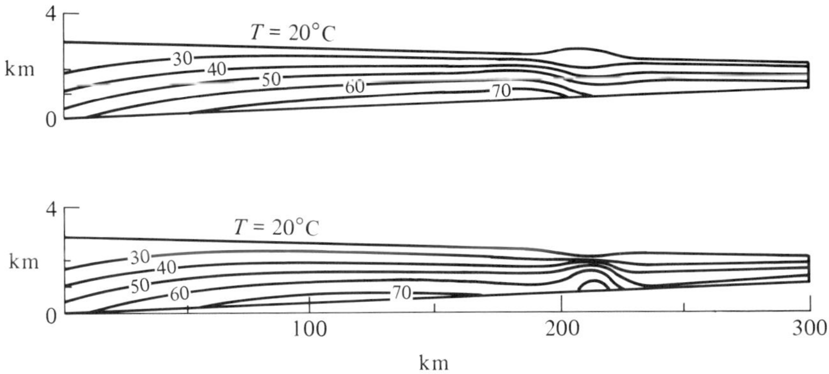

Figure 6.26. The influence on temperature distributions of a local recharge region (upper) and discharge region (lower) as modeled by Garven and Freeze (1984b). The Peclet number is 300 and the local aspect ratio is about 10.

the aspect ratio. In the condition (6.4.1), the two sides are comparable, so that the flow should modify but not dominate the temperature distribution, as, indeed, Figure 6.25b illustrates, with the isotherms depressed somewhat near the left-hand end of the basin. When the permeability and hydraulic conductivity are decreased by a factor of 10, the Peclet number is then only 30 whereas still $l/h = 150$, so that (6.4.1) *is* satisfied and the temperature distribution is determined essentially conductively, as in Figure 6.25a. In contrast, when the permeabilities are *increased* by a factor of 5, the Peclet number is 1,500, much greater than l/h. A large, nearly isothermal region appears in the recharge zone, with a well-developed thermal layer appearing along the lower boundary, as described in Section 6.4. These comparisons confirm that the simple condition (6.4.1) can usefully distinguish between conditions in which the temperature distribution is little affected by the flow from those in which it is – in the latter case the considerations of Section 6.4 must be invoked.

Another illustration of this is given by calculations of the temperature distributions produced in recharge or discharge regions, reproduced as Figure 6.26. In the upper portion, a water-table maximum of horizontal extent of about 20 km produces a *local* aspect ratio of about 10, which is small compared with the Peclet number Uh/κ of 300. In this recharge region, then, the isotherms should be displaced downward, as, indeed, the numerical calculations show. Correspondingly, in the discharge region associated with a water-table minimum, the isotherms crowd toward the surface and a temperature excess develops, as also described in Section 6.4. The analysis of that section shows that if the horizontal extent of the water-table depression were to be reduced or the layer depth or hydraulic conductivity increased, this excess temperature would increase quite rapidly (the parameter P in equation (6.4.24) increases, the effect being shown in Figure 6.17).

These simple examples demonstrate how a more extensive understanding can be developed of a detailed and perhaps complex numerical simulation and how its general correctness can be confirmed. They also illustrate how, without detailed calculation, one can infer the changes in the qualitative nature of the solutions as the numerical parameters are varied.

Distributions of mineral alteration

Consider now several general examples that illustrate the relationships between mineralization and flow, the first being the distribution of lead–zinc deposits in the Mississippi Valley. The Paleozoic sediments of the greater Mississippi Valley provide the locale for extensive major and minor deposits of galena, sphalerite, barite, and fluorite, all of which possess such striking geochemical and geological similarities among themselves and to those in similar regions on other continents that they are all generally known as Mississippi Valley–type deposits. They have been studied extensively and are the subject of a fine review by Sverjensky (1986). It appears to be well established that they are formed from hot, saline aqueous solutions some time after the lithification of the host rock, but the detailed geochemical processes involved in their formation have been a matter of some dispute (i.e., whether the precipitation reactions involved sulfate reduction or pH changes or dilution, etc.); different possibilities are reflected in differences in the chemically determined coefficients in equations such as (4.4.4).

Nevertheless, the outstanding characteristics of Mississippi Valley–type deposits, as listed by Sverjensky, include the following:

1. They occur principally in the limestone or dolostone that forms a relatively thin cover over an igneous or highly metamorphosed basement.
2. They consist of bedded replacements, vuggy ores, and veins, the ore being strongly controlled by individual strata; they are generally not associated with igneous rocks.
3. They always occur in areas of mild deformation, expressed in brittle fractures, broad domes and basins, and gentle folds.
4. The ore is never in the basement rocks, but its distribution is often spatially related to basement highs, with the ore located within sandbanks, ridges, and reef structures that surround the basement highs.
5. The ore is generally at shallow depths, generally less than 600 m

relative to the present surface, and was probably never at depths greater than 1,200 m.

6. Fluid inclusions remaining in sphalerite, fluorite, barite, and calcite always contain dense, saline aqueous fluids and often oil and/or methane.

All these characteristics seem to be consistent with the dissolution of dispersed minerals, fluid transport, and local deposition and concentration, probably by gradient reactions. An important clue is provided by the absence of ore deposits in undeformed regions. Here, any fluid flow is essentially along the isotherms, so that the solute transport is constant; in flow through regions of deformation or over basement highs, the flow crosses the isotherms, so that on the flanks of these highs, both $\mathbf{u} \cdot \nabla T$ and w in (4.4.4) are nonzero. Local deposition occurs in regions of rising fluid when the chemical coefficients in (4.4.6) are positive or in regions of sinking fluid when they are negative. The fact that they are associated with more permeable regions near basement highs is again consistent with flow focusing there and more rapid deposition in gradient reactions. It does not appear to be known whether the formation of these shallow deposits was associated with fluid discharge regions, where the near-surface temperature gradient would be anomalously large, although Cathels and Smith (1983) argue persuasively that episodic discharges would appear to account for the colorbanding observed in sphalerite (ZnS) in these deposits resulting from variations in iron content.

Nevertheless, the qualitative consistency between the characteristics listed by Sverjensky and those that would be expected from the theory does suggest strongly that gradient reactions associated with basin-wide flow were responsible for the ore deposition. More highly quantitative studies are needed, however, before the association can be regarded as tight. The rate at which minerals were deposited depends on the chemical coefficients involved in equations such as (4.4.4) and (4.4.6), and the estimation of these for particular reactions is beyond the scope of this book. Some systems have been considered recently. The well-known work of Garrels and Christ (1965) considers the carbonate system, and Wood (1987b) presents detailed calculations for the calcite–dolomite system saturated either with initially fresh water or with brine of the composition of seawater. Wood gives calculations for both isothermal and gradient reactions (which he terms irreversible and reversible, respectively), taking care to distinguish between them. In either case, however, he finds that the capacity of a seawater brine to dissolve or produce calcite and dolomite is several hundred times greater than that of initially pure water, so that dolomitization occurs in less time and involves less volume flux when brine is involved.

Finally, the close association between the spatial distribution of me-

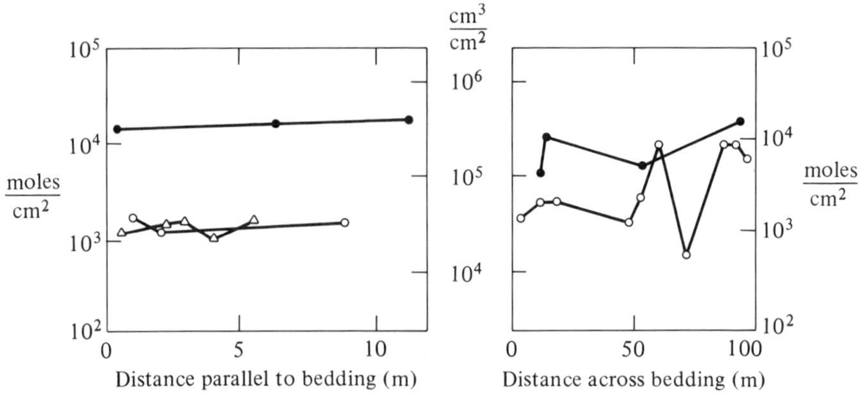

Figure 6.27. Time-integrated fluxes during metamorphism from the limestone member of the Waterville Formation, calculated by Baumgartner and Ferry (in press) from Ferry's (1987) field samples. The fluxes are constant within a single bed but vary by factors of 10 from one bed to another.

tamorphism and paleoflow has been demonstrated in a fine, detailed field study by Ferry (1987) on regionally metamorphosed biotite- and garnet-grade rocks of the Vassalboro Formation and the limestone member of the Waterville Formation in south central Maine. The biotite-forming reaction involves decarbonation:

$$\underset{\text{Muscovite}}{KAl_3Si_3O_{10}(OH)_2} + 3\ \underset{\text{Ankerite}}{CaMg(CO_3)_2} + 2\ \underset{\text{Quartz}}{SiO_2} \rightleftharpoons$$

$$\underset{\text{Biotite}}{KMgAlSi_3O_{10}(OH)_2} + \underset{\text{Plagioclase}}{CaAl_2Si_2O_8} + 2\ \underset{\text{Calcite}}{CaCO_3} + 4CO_2.$$

The geometrical distribution of reaction products, with gradual variations through the fabric rather than abrupt transitions, suggests that gradient reactions, rather than the movement of isothermal reaction fronts, were involved. Baumgartner and Ferry (1989), interpreting the metamorphism in terms of gradient reactions and using the appropriate stoichiometric coefficients, showed that the total time-integrated volume transports over the period of formation were in the range 10^4 to 10^6 $cm^3\ cm^{-2}$. Although the variations among different lithological layers were considerable (reflecting variations in permeability), the calculated fluxes from samples at different points in the same layer are very similar (Figure 6.27), indicating the channelization of fluid flow along it. By taking 10^7 years as an upper limit for the time interval involved, Baumgartner and Ferry calculated volume fluxes or transport velocities of order 10^{-2} cm yr^{-1} at least.

Another important example of metasomatism (change in the composition of a metamorphic rock) involving the alteration of peridotite from the Ivrea zone, northern Italy, has been discussed by Ferry and Dipple (in press). The alteration involved the loss of calcium and magnesium from the rock and the gain of potassium relative to silicon. Thermodynamic calculations indicated that the influence of pressure on c_e in equation (4.4.6) could be neglected, and the changes in element ratios gave three independent constraints on the two unknown variables, time-integrated volume flux and dissolved chlorine molality (influencing $\partial c_e/\partial T$). They found it possible to choose these two variables so that the three ratios predicted by (4.4.6) were consistent with those measured, which strongly suggests that gradient reactions were indeed responsible for the alteration.

These individual studies, particularly the last, give tantalizing glimpses of the insights that are beginning to be developed through the combination of flow and geochemical analyses. There remains, however, a continuing need for many more carefully documented and thoughtfully interpreted field studies if the association is to flower.

7 Thermal convection

7.1 THE OCCURRENCE OF THERMALLY DRIVEN FLOWS

In terrestrial, near-surface aquifers, an apparently small hydraulic head gradient generally dominates buoyancy forces insofar as the flow pattern is concerned. Even in regions of intense geothermal activity where buoyancy forces must certainly be taken into account, the hydraulically induced pressure gradient may modify the flow field substantially. With a horizontal temperature gradient of 100°C km^{-1} in a region 1 km thick, the hydraulic and thermal forcing of the flow, discussed in Section 3.5, are comparable when the water-table slope is of order 3×10^{-2}. Nevertheless, in large geothermal areas the buoyancy forces are certainly dominant, with the augmented heat flux provided from below rather than by lateral advection, as described in Section 6.4. A good review of the physical processes in geothermal areas has been given by Elder (1965) with specific reference to the Tuscan steam zone in Italy and the Taupo hydrothermal area in New Zealand. The detailed structure of these regions is to some extent conjectural, but hydraulic forcing associated with the water-table configuration probably only modifies an essentially convective flow.

Pure thermally driven flows are generally to be found only when the hydraulic forcing is in effect absent, in submerged banks, beds, or continental shelf regions or in isolated, totally confined permeable strata. In submerged sediments, the surface pressure is equal to the hydrostatic head of the water above, so that the reduced pressure at the sediment–water interface vanishes. In regions of intense current variation, there may be some dynamically induced variations in reduced pressure, but these are usually negligible. The fluid pressure in an isolated, confined permeable layer may be greater than the hydrostatic pressure corresponding to its depth (it may approach the lithostatic pressure or even exceed it somewhat), but as far as the flow is concerned, that is immaterial. It is the imposed pressure *gradient* that forces hydraulically driven flow, and if the region is isolated by impermeable rock on all sides with no fluid entering or leaving it, any internal motion must be driven thermally. In any of these regions, if for any reason the isotherms are not horizontal, interstitial fluid will move through a permeable stratum, as dictated by the generalized vorticity relation (3.4.4).

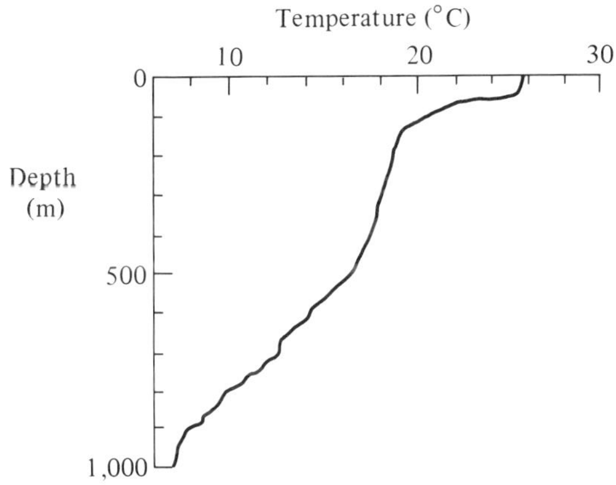

Figure 7.1. An oceanic temperature profile, measured by Hogg, Katz, and Sanford (1978) off Bermuda in mid-October 1975. The temperature in the upper 50 m is approximately constant, and it then decreases rapidly in the next 100 m. Below this, the decrease with depth is more gradual, reaching about 6°C at 1,000 m. At abyssal depths, 4,000 m or so, the temperature is generally in the range 1–2°C.

In marine sediments, the internal isotherms will slope when the sediment–water interface slopes or when the geothermal heat flux varies or when both occur, and the consequences can be discerned in ancient marine sediments, now uplifted and long dormant. At the upper surface of a submerged sedimentary bank or coral reef, the rapidity of the heat transfer into the ocean fixes the surface temperature of the sediment as essentially that of the surrounding water. Near the ocean surface, in tropical or temperate latitudes, the near-surface water temperature is approximately 20 to 25°C, varying somewhat with the season, but it decreases with depth (Figure 7.1), until below the thermocline, which may extend to 1,000 m or so, it is only a few degrees Celsius and remains so at all latitudes throughout the year. In high latitudes, the water column is cold throughout its depth; in fact, the cold, deep water in all the world's oceans is maintained by the subsidence of Arctic and Antarctic Ocean water from the surface to great depths, the gradual meander south and north, and the eventual upwelling and incorporation into the oceanic surface layers. The geothermal heat flux through the ocean floor has a generally imperceptible influence on this oceanic temperature distribution except locally in the vicinity of thermally active vents. In most areas, hot plumes from deep oceanic vents rise and mix quite rapidly with the surrounding water, though in a few regions devoid of deep ocean circulation (the Red Sea being a notable example) hot, dense brine pools, fed by vents, fill the greatest depths.

The characteristics of the overall flow patterns in marine banks depend on two of the parameters encountered earlier–the Rayleigh number Ra and the scale ratio $S = (h/l)^2(k_H/k_V)$, where h is the stratum thickness and l the length scale along it. It was shown in Section 3.5 that when Ra is sufficiently small, advection of heat by the fluid motion is negligible compared with conduction through the matrix, so that the temperature distribution is in essence the geothermal one, uninfluenced by the flow. In these cases, the overall flow patterns are generally diffuse, although with, of course, local flow concentration in any smaller-scale regions with high permeability. The overall distribution of flow relative to the isotherms depends on the second parameter $(h/l)^2(k_H/k_V)$. In the hydraulically driven flows in a sedimentary basin discussed in Chapter 6, the thickness h of the permeable stratum is often so small compared with its extent that the product $(h/l)^2(k_H/k_V)$ remains small despite a large ratio of the permeability k_H/k_V, but in marine sedimentary banks this may not be so. Coral reefs or islands are frequently composed of high-permeability reef limestone interlayered with lithified lagoonal muds and silts of low permeability, so that the ratio k_H/k_V is large while the total thickness h of the reef may not be particularly small compared with its width l; the product $(h/l)^2(k_H/k_V)$ may be large compared with unity. In physical terms, interstitial fluid can move relatively freely over the short horizontal distances involved, and the convective circulation is dependent on the capacity of the thermally induced buoyancy variations to drive vertical motion through the more resistive, less permeable layers. This calls for a different kind of geometrical scaling, as described in detail in Section 7.4.

In high Rayleigh number situations, characterized by thick, relatively permeable platforms, particularly when boosted by locally elevated basement temperatures associated with volcanism, the convective streams are more concentrated than at low Ra, with thermal boundary layers and plumes forming as described in Section 3.5. Where cold water is drawn into the platform from the surrounding ocean, the flow velocities are sufficiently large that convection overwhelms conduction – the fluid (and the matrix) remains cold until incorporated into a plume or a thermal boundary layer where temperature variations are concentrated and where the most intense fabric alterations can be expected. Fluid systems of this kind, which involve local rather than overall scaling in their analysis, are described in Sections 7.5 to 7.7.

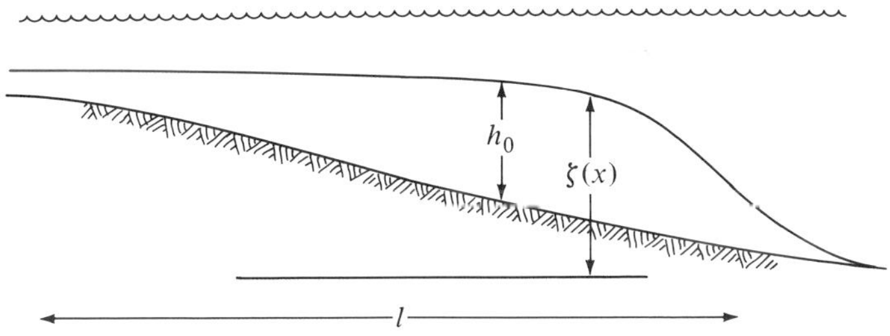

Figure 7.2. A submerged permeable bank.

7.2 SUBMERGED BANKS OF SLOWLY VARYING THICKNESS WITH SMALL RAYLEIGH NUMBER AND SCALE RATIO

Temperature distributions

Consider a submerged permeable bank such as shown in Figure 7.2, whose thickness varies slowly in the horizontal direction and is small compared with the horizontal extent l of the layer. Since buoyancy forces act in the vertical direction, it is convenient now to take the z-axis as vertical; the thickness scale of the bank will be represented by h_0, and $\zeta(\mathbf{x})$ is the height of the sediment–water interface above a convenient reference level. When the Rayleigh number is small (the precise form of this condition varies from one situation to another and will be established a posteriori), the temperature distribution in the bank is determined by $\nabla^2 T = 0$, and when $(h_0/l)^2 \ll 1$, inside the thin permeable layer this reduces to

$$\frac{\partial^2 T}{\partial z^2} = 0,$$

as shown in Section 3.6. The vertical temperature gradient is then constant over the depth of the layer but may vary slowly with horizontal position $\mathbf{x}$:

$$\frac{\partial T}{\partial z} = -G(\mathbf{x}), \tag{7.2.1}$$

where $G(\mathbf{x})$ (>0) is the geothermal gradient. The geothermal heat flux across the basement,

$$q_\mathrm{h} = -\kappa_\mathrm{m}\frac{\partial T}{\partial z} = -(\rho C)_\mathrm{f}\,\kappa\frac{\partial T}{\partial z}, \tag{7.2.2}$$

from (2.8.2) and (2.8.5), where κ_m is the thermal conductivity of the matrix, $(\rho C)_\mathrm{f}$ the product of density and specific heat for the interstitial water and κ the thermal diffusivity. If q_h is uniform across the basement, then $\partial T/\partial z$ is constant throughout the bank.

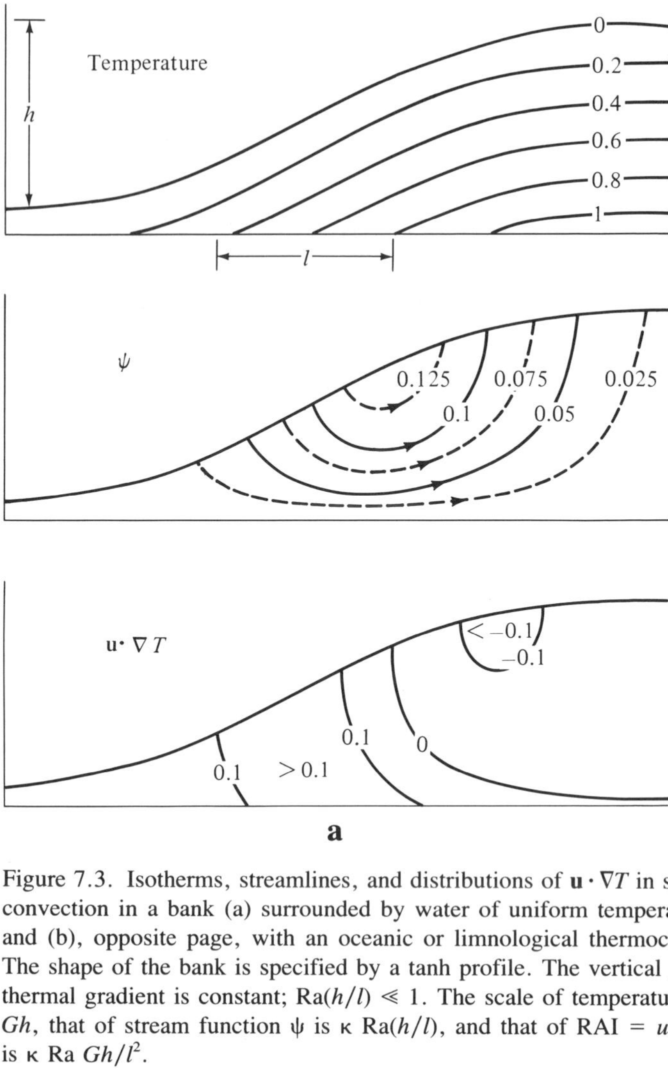

Figure 7.3. Isotherms, streamlines, and distributions of $\mathbf{u} \cdot \nabla T$ in slope convection in a bank (a) surrounded by water of uniform temperature and (b), opposite page, with an oceanic or limnological thermocline. The shape of the bank is specified by a tanh profile. The vertical geothermal gradient is constant; $\text{Ra}(h/l) \ll 1$. The scale of temperature is Gh, that of stream function ψ is $\kappa\ \text{Ra}(h/l)$, and that of RAI $= u \cdot \nabla T$ is $\kappa\ \text{Ra}\ Gh/l^2$.

At the sediment–water interface $z = \zeta(\mathbf{x})$, the temperature is that of the ambient seawater. When the interface is entirely below the thermocline, the ambient temperature is in effect constant and can be taken as the zero reference temperature. Equation (7.2.1) then integrates to give

$$T(\mathbf{x}, z) = G(\mathbf{x})(\zeta(\mathbf{x}) - z). \tag{7.2.3}$$

In contrast, if the bank lies in or reaches into the ocean thermocline, the ambient temperature is a function of the depth of the upper surface of the bank, or $\zeta(\mathbf{x})$, so that when $z = \zeta(\mathbf{x})$, $T = T(\zeta(\mathbf{x}))$. In this instance,

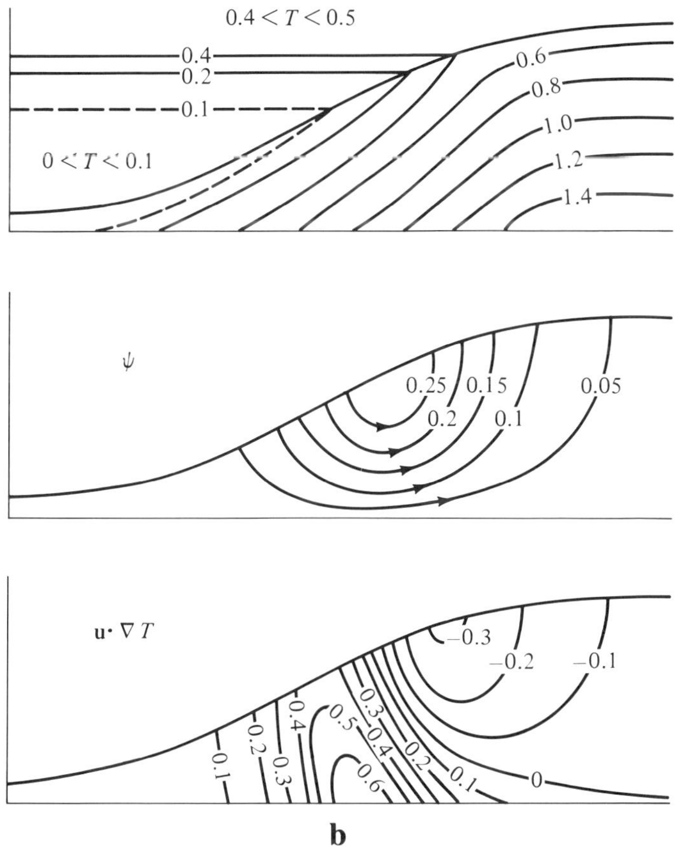

Figure 7.3, continued

$$T(\mathbf{x}, z) = T(\zeta(\mathbf{x})) + G(\mathbf{x})(\zeta(\mathbf{x}) - z). \tag{7.2.4}$$

When the geothermal heat flux, proportional to G, is constant and the bank is entirely beneath the thermocline (or surrounded by water of constant temperature), the isotherms inside the bank are parallel to the surface (Figure 7.3a). When, however, the sediment–water interface rises through the thermocline, the surface temperature near the top of the bank is warmer by perhaps 20°C than it is at greater depths, so that the internal isotherms slope more steeply as in Figure 7.3b, enhancing, as we shall see, the local convective forcing in this region. When the upper surface of the bank is horizontal, the isotherms will slope if G varies with horizontal position (Figure 7.4) surrounding a local maximum in G. Analytically, the horizontal temperature gradient in the bank is found from equation (7.2.4) to be

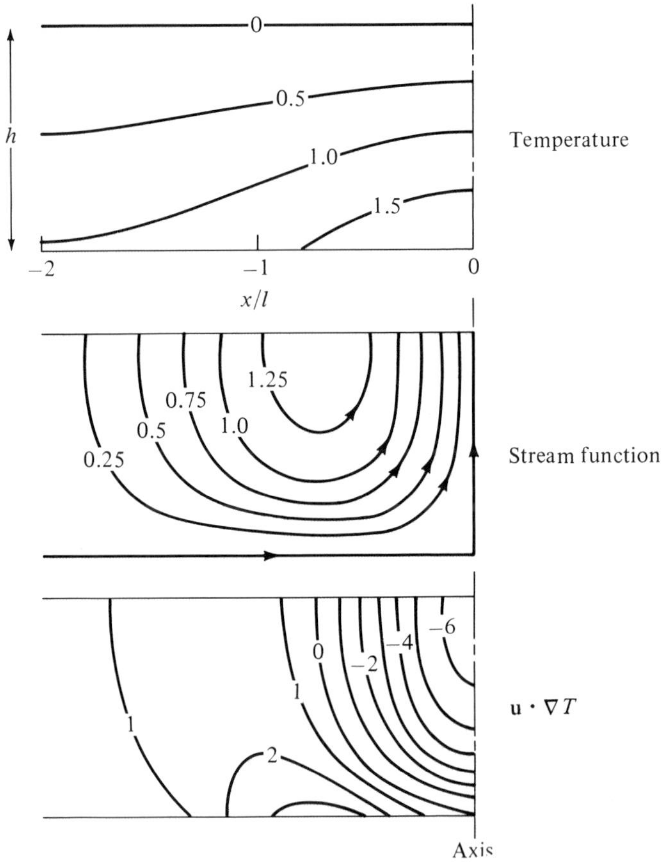

Figure 7.4. Patterns produced by a strip of locally elevated basement temperature in a submerged bank when both Ra and S are small. The temperature scale is Gh, where G is the undisturbed geothermal gradient; the upper figure is for an increment $\Delta G = G$. The scale of stream function ψ is κ Ra(h/l) and that of $\mathbf{u}\cdot\nabla T$ is κ Ra Δ Gh/l^2. Only the left-hand halves of the distribution are shown; the patterns are symmetrical about the central axis $x = 0$.

$$\nabla_h T = \left\{ \left.\frac{\partial T}{\partial z}\right|_x + G(\mathbf{x}) \right\} \nabla\zeta + \{\zeta(\mathbf{x}) - z\}\,\nabla G(\mathbf{x}), \qquad (7.2.5)$$

where $\partial T/\partial z|_x$ represents the vertical temperature gradient in the ambient water at the depth of the surface of the bank at the location $\mathbf{x}$.

Slope convection

When the geothermal heat flux across the basement is uniform, $\nabla G(\mathbf{x}) = 0$ and the internal horizontal temperature gradient

$$\nabla_h T = \left\{ \frac{\partial T}{\partial z}\bigg|_x + G \right\} \nabla \zeta \tag{7.2.6}$$

is independent of z. From (3.4.4), with $\rho = \rho_0(1 - \alpha T)$,

$$\frac{\partial}{\partial z}\left(\frac{\mathbf{u}}{k_H}\right) - \nabla_h\left(\frac{w}{k_V}\right) = \frac{-g\alpha}{\nu} \nabla_h T, \tag{7.2.7}$$

where, for simplicity, the bedding is taken to be horizontal. In a gently sloping bank, when $S = (h/l)^2 k_H/k_V \ll 1$, this can as before be approximated as

$$\frac{\partial}{\partial z}\left(\frac{\mathbf{u}}{k_H}\right) = -\frac{g\alpha}{\nu} \nabla_h T, \tag{7.2.8}$$

and since $\nabla_h T$ is independent of z in this case, an integration gives

$$\mathbf{u} = \frac{g\alpha k_H}{\nu} \nabla_h T \,(A(\mathbf{x}) - z), \tag{7.2.9}$$

where the function $A(\mathbf{x})$ is to be determined from the boundary conditions. At the upper surface of a submerged bank or slope, the reduced pressure is constant, so that from (3.3.5) and the Darcy equation,

$$\mathbf{u} + \frac{k_H}{k_V} \mathbf{w}\, \nabla \zeta = 0 \qquad \text{on} \quad z = \zeta, \tag{7.2.10}$$

and when $(k_H/k_V)\nabla\zeta \ll 1$, fluid enters the sediments almost vertically, so that $\mathbf{u} \simeq 0$ on $z = \zeta$. Consequently, from (7.2.9), $A(\mathbf{x}) = \zeta(\mathbf{x})$ and

$$\mathbf{u} = \frac{g\alpha k_H}{\nu} \nabla_h T \,(\zeta(\mathbf{x}) - z). \tag{7.2.11}$$

Note that in this solution, no assumption is made about the shape of the bank in plan form. Note also that if the averaging volume used to describe the flow is taken to be smaller than the vertical scale of any laminations in permeability, k_H in these equations is a function of z, and on this scale flow focusing in layers of high permeability will occur as described in Section 3.8. If we are content to average over the vertical scale of the bedding, the mean k_H over this scale is to be used and the distribution of $\mathbf{u}$ is smoothed in the vertical. In either event, when $S = (h/l)^2 k_H/k_V \ll 1$, *the magnitude of the horizontal transport velocity is proportional to the horizontal temperature gradient and is in the same direction, regardless of the plan form of the bank or slope*. An important characteristic of this type of convection with the geothermal flux constant is that it is confined to regions below sloping surfaces of the permeable bank; if the sediment—water interface is horizontal, $\nabla\zeta = 0$, and from (7.2.6) $\nabla_h T = 0$ and so $\mathbf{u} = 0$.

When the bank is two-dimensional, the flow can be represented in terms of the stream function ψ of (3.1.2),

$$\frac{\partial\psi}{\partial z} = \frac{g\alpha k_H}{\nu} \nabla_h T\,(\zeta(\mathbf{x}) - z),$$

and the smoothed streamline pattern, with k_H constant, is found by integration,

$$\psi = \frac{g\alpha k_H}{\nu} \nabla_h T(z\zeta(\mathbf{x}) - \tfrac{1}{2} z^2), \qquad (7.2.12)$$

when the basement is flat so that $\psi = 0$ at $z = 0$. Some representative patterns of slope convection are shown in Figure 7.3. When the surrounding fluid is at constant temperature, as in Figure 7.3a, the circulation is relatively feeble. The value of the stream function at any streamline specifies the total volume transport per unit length along the bank between that streamline and the basement where $\psi = 0$. At the surface $z = \zeta(x)$ of the bank, the value of the stream function is

$$\psi = \frac{g\alpha k_H}{2\nu} \nabla_H T\, \zeta^2(x), \qquad (7.2.13)$$

and the total rate of volume transport through the slope of the bank, or the total volume flux per unit length, is the maximum value of this expression. In the case of Figure 7.3a, the isotherms slope parallel to the surface and $\nabla_h T = G\, \nabla\zeta$, so that

$$\text{Total volume transport} = \frac{g\alpha k_H G}{2\nu} (\zeta^2\, \nabla\zeta)_{\max},$$

and the maximum of $\zeta^2\, \nabla\zeta$ occurs rather above the point of maximum slope, say at $\zeta \sim 0.7h$. The slope itself is about $0.5h/l$, so that this transport can be written as

$$\frac{g\alpha k_H G}{2\nu} (0.5h^2 \cdot 0.5h/l) \approx \frac{1}{8}\, \kappa\, \mathrm{Ra}(h/l), \qquad (7.2.14)$$

in agreement with the numerical calculation. The magnitude of the thermal convection terms in the temperature equation relative to the diffusion terms, the Peclet number, is of order $\psi_{\max}/\kappa$, which for this form of solution must be small, so that the low Rayleigh number condition in this context is equivalent to

$$\mathrm{Ra}(\nabla\zeta)_{\max} \ll 1.$$

In two-dimensional slope convection, then, the maximum volume flux per unit length of bank (with units of velocity times length) is necessarily a small fraction of the thermal diffusivity, κ, measured in the same units.

When the slope of the bank intersects a thermocline, however, the deep water is colder than the surface water, the isotherms in the bank slope much more steeply, and the circulation is more vigorous (Figure 7.3b). In this calculation, the temperature difference across the thermocline is taken as half the geothermal temperature difference over the depth interval h, and the volume transport through the bank is more than doubled.

The solution (7.2.11) for slope convection in gentle axisymmetrical banks, circular in plan form, can be represented in terms of the Stokes stream function ψ_s of equation (3.1.5):

$$\psi_s = \frac{g\alpha k_H}{\nu} r \frac{\partial T}{\partial r}\left(z\zeta(r) - \frac{1}{2}z^2\right). \qquad (7.2.15)$$

The flow patterns are very similar to those illustrated in Figure 7.3 for the two-dimensional case. The total rate at which fluid moves through the bank is 2π times the maximum value of ψ_s at its upper surface, and by steps similar to those leading to (7.2.14),

$$\text{Total volume flux} = \pi\kappa \,\text{Ra}\, r_m\left(\frac{\partial \zeta}{\partial r}\right)_{max}, \qquad (7.2.16)$$

where r_m is the radius of the bank at its point of maximum slope.

When solutes dissolved in the circulating water can undergo reaction with the matrix, or if dissolution or precipitation occurs, the patterns of rock alteration reflect the patterns of circulation and temperature. The isothermal reaction fronts described in Section 4.3 may move through the matrix, following the streamlines at speed $u(1 + r)^{-1}$ according to (4.3.11), where u is the local transport speed and r the fluid–rock ratio. As described in Section 6.8, these reactions go to completion; the fronts move relatively rapidly along layers of higher permeability and lead to mineralogical striations on a bedding-layer scale. At the same time, gradient reactions occur pervasively at rates, specified by (4.4.4), proportional to the flow velocity component across the isotherms and isobars. Patterns of the rock alteration index, $\mathbf{u} \cdot \nabla T$, are also shown in Figure 7.3. In the absence of a thermocline, the pattern is diffuse, positive below the lower part of the slope where inward- and downward-moving fluid moves to higher temperatures, and negative below the top of the slope where rising fluid moves to lower temperatures. The presence of a thermocline has the strongest effect below the lower slope. The increased slope of the isotherms there increases both the flow velocities and the temperature gradient in the direction of flow, so that $\mathbf{u} \cdot \nabla T$ is very much larger. If we are concerned, for example, with the dissolution and cementation of calcite for which $\partial c_e/\partial T < 0$ by the circulation of chemically homogeneous ambient water that is unsaturated with respect to calcite, then on the lower slope dissolution occurs in a thin surface layer whose thickness is of the order of the equilibration length, as described in Section 4.2. The then-saturated solution moves across the isotherms to higher temperatures in the zone where $\mathbf{u}\cdot\nabla T > 0$ and cementation occurs in accordance with (4.4.4). Farther upslope, as the fluid rises toward the sediment–water interface, it cools, $\mathbf{u}\cdot\nabla T < 0$, and dissolution occurs again. If the bank is laminated on a small scale, the flow is channeled on this scale and the reaction patterns are similarly striated.

Variable geothermal flux

When horizontal temperature gradients are produced by variations in the geothermal heat flux across the basement, low Rayleigh number circulations result even if the upper surface is horizontal. The simplest illustrative case involves a horizontal basement also; in equation (7.2.5) $\zeta = h =$ const, and

$$\nabla_{\mathrm{h}} T = (h - z)\,\nabla G(\mathbf{x}),$$

which, in contrast to the case of slope convection, now varies significantly with z. The distribution of horizontal velocity $\mathbf{u}$ is found by integration of (7.2.8), here giving

$$\mathbf{u} = -\frac{g\alpha k_{\mathrm{H}}}{\nu}\nabla G(\mathbf{x})\{hz - \tfrac{1}{2}z^2 + A(\mathbf{x})\}, \qquad (7.2.17)$$

where, again, $A(\mathbf{x})$ is an arbitrary function of integration to be determined by the boundary conditions. If the upper surface is submerged, the pressure is constant there and $\mathbf{u} = 0$ when $z = h$, so that $A(\mathbf{x}) = -\tfrac{1}{2}h^2$ and

$$\mathbf{u} = \frac{g\alpha k_{\mathrm{H}}}{2\nu}\nabla G(\mathbf{x})(h - z)^2. \qquad (7.2.18)$$

This expression applies whatever the horizontal distribution of geothermal flux $G(\mathbf{x})$; in all cases the direction of the horizontal flow is up the horizontal temperature gradient, toward the temperature maximum. In a layered bank, the speed of horizontal flow is proportional to the local horizontal permeability. With a two-dimensional strip of elevated temperature, the stream function is specified by

$$\frac{\partial\psi}{\partial z} = \frac{g\alpha k_{\mathrm{H}}}{2\nu}\frac{\partial G(\mathbf{x})}{\partial x}(h - z)^2,$$

and the pattern of streamlines, when $k_{\mathrm{H}} =$ const and with no flow across the basement so that $\psi = 0$ at $z = 0$, is given by

$$\psi = \frac{g\alpha k_{\mathrm{H}}}{6\nu}\frac{\partial G}{\partial x}\{h^3 - (h - z)^3\}, \qquad (7.2.19)$$

and illustrated in Figure 7.4. Water is drawn into the region on the flanks of the strip of elevated temperatures and moves toward the center, where it rises in a more concentrated stream. The maximum value of the stream function at the sediment–water interface $z = h$ occurs above the point of maximum horizontal temperature gradient at the basement; the total volume flux per unit length of the bed is just twice this (half on each side), or

$$\frac{g\alpha k_{\mathrm{H}} h^2}{3\nu}\left(\frac{\partial G}{\partial x}\right)_{\max} = \frac{1}{3}\kappa\,\mathrm{Ra}\left(\frac{h}{l}\right), \tag{7.2.20}$$

where this Rayleigh number $(g\alpha k_{\mathrm{H}} h^2\,\Delta G)/\kappa\nu$ is based on the increase ΔG in geothermal gradient at the center.

With an axisymmetrical basement hot spot, the flow can be represented, as before, in terms of the Stokes stream function (3.1.5), and in a submerged layer it is found that

$$\psi_s = \frac{g\alpha k_{\mathrm{H}}}{6\nu} r \frac{\partial G}{\partial r}\{h^3 - (h-z)^3\}, \tag{7.2.21}$$

with the total volume flux

$$\frac{\pi}{3}\frac{g\alpha k_{\mathrm{H}} h^3}{\nu}\left(r\frac{\partial G}{\partial r}\right)_{\max}. \tag{7.2.22}$$

The distribution of the Stokes stream function indicating the lines of flow is similar to that shown in Figure 7.4 but slightly more diffuse in the central region. The distribution of the rock alteration index $\mathbf{u}\cdot\nabla T$ is also shown in Figure 7.4. Positive values are found in a diffuse region on the flanks; negative values of larger magnitude are concentrated close to the axis, where more intense vertical flow moves down the enhanced vertical temperature gradient.

When the upper surface of the bank is exposed, a light, freshwater lens may be expected to collect there, preventing the circulating fluids from entering or leaving the upper surface. To obtain the solution in this case, we return to equation (7.2.17), which, in the two-dimensional case, is

$$\frac{\partial\psi}{\partial z} = -\frac{g\alpha k_{\mathrm{H}}}{\nu}\frac{\partial G}{\partial x}\{hz - \tfrac{1}{2}z^2 + A(x)\},$$

which integrates to

$$\psi = -\frac{g\alpha k_{\mathrm{H}}}{\nu}\frac{\partial G}{\partial x}\{\tfrac{1}{2}hz^2 - \tfrac{1}{6}z^3 + A(x)z + B(x)\},$$

where $B(x)$ is a second integration function. But $\psi = 0$ at $z = 0$ (no flow across the basement), so that $B = 0$. Also $\psi = 0$ at $z = h$ (no flow across the top), so that $A = -h^2/3$ and

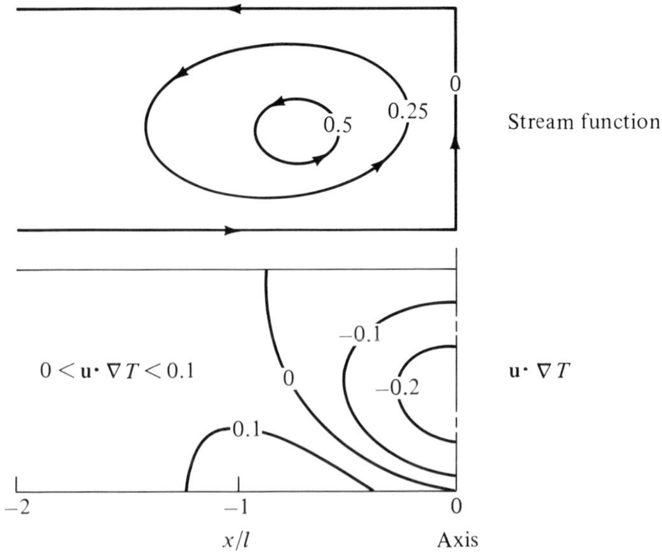

Figure 7.5. Patterns of flow and $\mathbf{u} \cdot \nabla T$ above a strip of elevated basement temperature–with no flow across the upper surface. The temperature field and scales of ψ and $\mathbf{u} \cdot \nabla T$ are as in Figure 7.4.

$$\psi = \frac{g\alpha k_H}{6\nu} \frac{\partial G}{\partial x} z(z - h)(z - 2h), \qquad (7.2.23)$$

as illustrated in Figure 7.5. In this case, interstitial fluids continue to circulate internally without introduction of outside water, and the circulation is much more feeble than when the upper surface is submerged. As a result of gradient reactions, however, solutes may be dissolved in the central region and precipitated on the flanks (or the reverse). The flow pattern is not quite symmetrical in the vertical because of the asymmetry of the driving temperature field. Circulation patterns for the axisymmetrical case can be found similarly.

7.3 CONVECTION AND DIAGENESIS IN SLOPING ISOLATED PERMEABLE STRATA

Convective circulations driven by the geothermal temperature gradient may also occur in isolated permeable strata, as shown qualitatively by Wood and Hewett (1982) and quantitatively by Davis, Rosenblat, Wood, and Hewett (1985). Consider a gently sloping layer of saturated permeable sandstone (say) lying between impermeable layers of shale (Figure

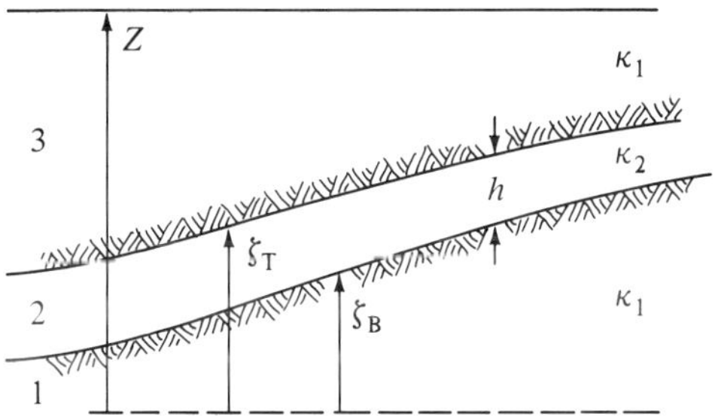

Figure 7.6. Definition sketch.

7.6). Flow in the permeable layer is produced by sloping isotherms, which, as shown later in this section are associated with variations in the thermal conductivity of the matrix when the ambient geothermal gradient is vertical. The temperature distributions in the shale and, at low Rayleigh numbers, in the saturated layer also, are determined as in Section 7.2 by

$$\partial^2 T/\partial z^2 = 0 \tag{7.3.1}$$

in each region, with a fractional error of order $(h/l)^2$. Note, however, that in this situation l represents the length of a horizontal traverse of the layer, so that h/l is the characteristic dip of the layer, which we assume to be small. At the upper and lower surfaces of the permeable region, the temperatures must be continuous, as must the normal heat fluxes,

$$T_1 = T_2,$$
$$\kappa_1\left(-\frac{\partial T_1}{\partial x}\sin\theta + \frac{\partial T_1}{\partial z}\cos\theta\right) = \kappa_2\left(-\frac{\partial T_2}{\partial x}\sin\theta + \frac{\partial T_2}{\partial z}\cos\theta\right), \tag{7.3.2}$$

at $z = \zeta_B$, with a similar set for the upper surface $z = \zeta_T$ involving T_2 and T_3, where the subscripts 1, 2, and 3 represent the regions below, in, and above the permeable layer, respectively, and the κ's are the respective matrix thermal conductivities as defined by (2.8.6). If the upper impermeable region is of the same mineralogy as the lower one, $\kappa_1 = \kappa_3$. Below the layer, the temperature distribution approaches the geothermal value

$$\frac{\partial T}{\partial z} = -G, \tag{7.3.3}$$

which will be supposed horizontally uniform. At the overlying land surface $z = Z$, $T = 0$, say.

The boundary condition (7.3.2) can be simplified by neglecting terms of order $(h/l)^2$ to be consistent with (7.3.1). In general, the slope of the isotherms as well as that of the layer is of order (h/l); hence, the first term in each set of parentheses of (7.3.2) is of order $(h/l)^2$ relative to the second, so that approximately

$$T_1 = T_2, \qquad \kappa_1 \frac{\partial T_1}{\partial z} = \kappa_2 \frac{\partial T_2}{\partial z} \qquad \text{at} \quad z = \zeta_B,$$
$$T_2 = T_3, \qquad \kappa_2 \frac{\partial T_2}{\partial z} = \kappa_3 \frac{\partial T_3}{\partial z} \qquad \text{at} \quad z = \zeta_T. \tag{7.3.4}$$

From (7.3.1) the temperature distribution in each region must be of the form

$$T = A(x) + zB(x), \tag{7.3.5}$$

where A and B are slowly varying functions of horizontal position. In region 1, from (7.3.3) $B_1 = -G$, so that

$$T_1 = A_1(x) - zG.$$

After substitution of expressions such as (7.3.5) into the conditions (7.3.4) and with the condition that $T_3(Z) = 0$, we have in all five algebraic relations for the five other coefficient functions A_1, A_2, A_3, B_2, B_3, and after a little algebra, the solution for the temperature distribution is found to be

$$T_1 = G\left\{(Z - z) - \left(1 - \frac{\kappa_1}{\kappa_2}\right)(\zeta_T(x) - \zeta_B(x))\right\},$$
$$T_2 = G\left\{Z - \frac{\kappa_1}{\kappa_2} z - \left(1 - \frac{\kappa_1}{\kappa_2}\right)\zeta_T(x)\right\}, \tag{7.3.6}$$
$$T_3 = G(Z - z),$$

as given by Davis et al. (1985).

Of greatest interest is the temperature distribution in the permeable layer, 2. If the height of its upper surface varies with horizontal position, the permeable region has a horizontal temperature gradient

$$\Delta_h T = -G\left(1 - \frac{\kappa_1}{\kappa_2}\right) \nabla_h \zeta_T, \tag{7.3.7}$$

which vanishes only when the top of the layer is horizontal or when the thermal conductivities κ_1 and κ_2 happen to be equal. In this exceptional case, the isotherms are everywhere horizontal, the solution (7.3.6) reducing to $T = G(Z - z)$ throughout. More generally, the isotherms slope and convection occurs. The slope of the isotherms and the sense of the circulation reverse when κ_1/κ_2 is greater than, or less than, unity (Figure 7.7).

In a two-dimensional situation in which the permeable layer is a sheet

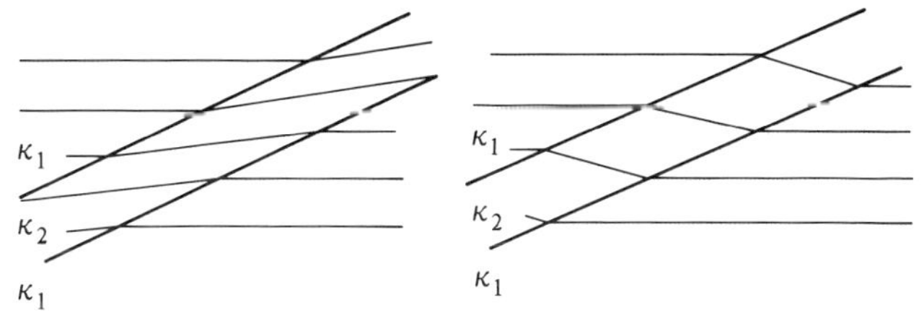

Figure 7.7. Inclined isotherms in a sloping layer with a thermal conductivity different from that above and below and a uniform vertical heat flux.

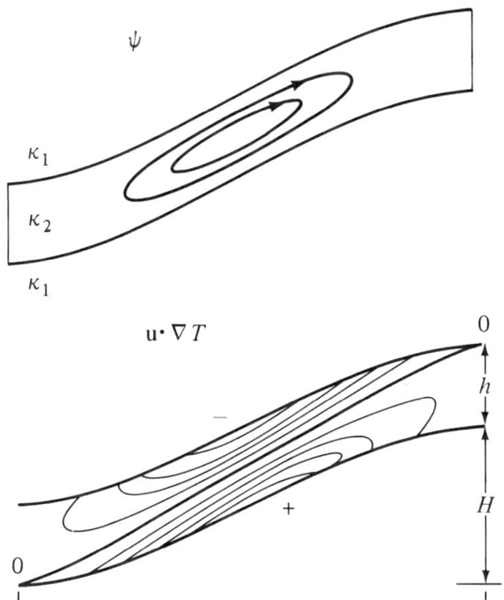

Figure 7.8. Streamlines and contours of $\mathbf{u} \cdot \nabla T$ in a permeable anticline when $\kappa_2 = 1.25\kappa_1$, calculated by Davis et al. (1985). When $\kappa_1 > \kappa_2$, the sense of the circulation is reversed, and the regions of positive and negative $\mathbf{u} \cdot \nabla T$ are interchanged.

folded to form an anticline, the stream function ψ can be found as in the preceding section. Thus,

$$\frac{\partial^2 \psi}{\partial z^2} = -\frac{g\alpha k_{\mathrm{H}}}{\nu} \nabla_{\mathrm{h}} T. \tag{7.3.8}$$

If there is no external hydraulic gradient to produce a net flow along the layer, then $\psi = 0$ at $z = \zeta_T, \zeta_B$, and if k_H is independent of position across the layer, then (7.3.8) integrates to

$$\psi = \frac{g\alpha k_H G}{2\nu}\left(1 - \frac{\kappa_1}{\kappa_2}\right)\frac{\partial \zeta_T}{\partial x}(z - \zeta_B)(\zeta_T - z). \qquad (7.3.9)$$

Streamline patterns, as calculated by Davis et al. from this expression, and distributions of $\mathbf{u}\cdot\nabla T$ for this flow are shown in Figure 7.8. Note that the most active regions of diagenesis are on the flanks, with regions of opposite sign occupying the upper and lower parts of the permeable layer. When the thermal conductivity of the layer κ_2 is greater than that outside, κ_1, fluid in the upper part of the layer moves toward the crest, so that $\mathbf{u}\cdot\nabla T < 0$, and fluid in the lower parts sinks toward the trough, and $\mathbf{u}\cdot\nabla T > 0$. Quartz, for example, tends to precipitate along the upper flanks while dissolving along the lower slopes. When $\kappa_2 < \kappa_1$, the distributions are reversed.

Note that from the scaling of Section 3.5, the validity of this low Rayleigh number solution requires that

$$\left(\frac{h}{l}\right)^2 \frac{\alpha g k_H h^2 G}{\kappa\nu} = \left(\frac{h}{l}\right)^2 \mathrm{Ra} \ll 1.$$

When h/l is small enough, the solution remains valid even though Ra may be numerically moderate.

The symmetry of the patterns of Figure 7.8 is disturbed if there is a nonhydrostatic pressure gradient along the layer to produce a net volume transport. If the transport velocity is greater than those characteristics of these circulations, that is, $(\kappa/l)\mathrm{Ra}$, where κ is the thermal diffusivity, then the flow is everywhere unidirectional though varying somewhat in speed in the vertical. At any horizontal location, $\mathbf{u}\cdot\nabla T$ will have the same sign throughout the thickness of the layer but opposite signs on rising and falling flanks, so that dissolution may occur on one flank and deposition on the other.

7.4 COMPACT PLATFORMS OR REEFS AT LOW RAYLEIGH NUMBERS

As mentioned in Section 7.1 carbonate reefs or sedimentary banks are frequently strongly layered, so that their vertical permeability is considerably less than that in the horizontal direction, whereas their horizontal extent may be only a moderate multiple of their thickness. The ratio $(h/l)^2 k_H/k_V$ is large, and the solutions given in the preceding two sections are then inappropriate. In those previous cases, the large horizontal extent allows fluid to move over the relatively short vertical distances more freely than over the large horizontal distances, despite the different permeabilities. Now the situation is reversed; fluid can move more freely over the short horizontal distances, and the convective circulation is limited by the rise of buoyant fluid through the less permeable layers.

In general, when the permeability is locally uniform, interstitial velocity gradients are proportional to the horizontal temperature gradient. It was found in Section 7.2 that at small Ra when $S = (h/l)^2 k_H/k_V \ll 1$, the convection velocities are proportional to the horizontal *gradient* of temperature. When this parameter is large, however, it appears that at any Ra the vertical velocity in an interior flow (outside any boundary layers) is proportional to the *temperature itself,* not its gradient. In the generalized vorticity balance (7.2.7) or (3.4.4), the magnitude of the second term relative to the first is $(h/l)^2 k_H/k_V$ and so is now much larger and

$$\nabla_h \left(\frac{w}{k_V} \right) = \frac{g\alpha}{\nu} \nabla_h T,$$

whence

$$w(\mathbf{x}, z) = \frac{g\alpha k_V T(\mathbf{x}, z)}{\nu} + f(z), \tag{7.4.1}$$

where $f(z)$ is an arbitrary function of z (which, physically, represents the contribution from any hydraulic forcing). If $w = 0$ when $T = 0$ at, say, the edges of a bank, then $f(z) = 0$ also. It is evident from the Darcy balance that in these interior regions, the fluid buoyancy then balances the resistance to vertical flow, a conclusion that is independent of Ra. When Ra *is* small, however, the vertical temperature gradient $\partial T/\partial z = -G$ is unaffected by the flow and

$$\frac{\partial w}{\partial z} = -\nabla_h \cdot \mathbf{u} = -\frac{\alpha g k_V G}{\nu} \tag{7.4.2}$$

when $f = 0$, so that in the interior regions of flows of this kind with $\mathrm{Ra} \ll 1$ and $(h/l)^2 k_H/k_V \gg 1$, the vertical velocity always decreases toward the surface and the horizontal flow always *diverges*.

Consider, then, a submerged permeable reef or bank on a horizontal basement; the configuration of the sediment–water interface is specified by $\zeta(\mathbf{x})$. If the ratio of width to thickness is not too small (i.e., not less than 2 or 3 or so) and the Rayleigh number is small, the internal temperature distribution driving the convection is still given to sufficient accuracy by (7.2.4), so that the vertical temperature gradient is $-G(\mathbf{x})$. Unless the reef or sedimentary bank is atop a volcanically active basement, the horizontal variation in G can be neglected.

The analysis of these situations is most revealing in terms of the re-

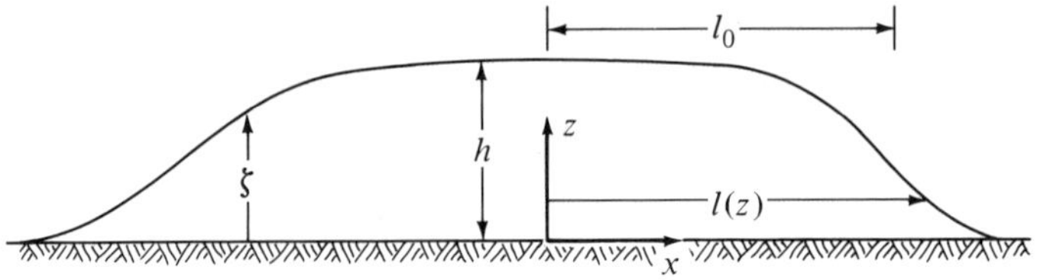

Figure 7.9. Definition sketch.

duced pressure, which vanishes at the upper surface $z = \zeta(\mathbf{x})$ of the region. With $\rho = \rho_0(1 - \alpha T)$, the horizontal and vertical force balances expressed by the Darcy equations are

$$\nabla_h p = -\frac{\mu}{k_H}\mathbf{u},$$

$$\frac{\partial p}{\partial z} = -\frac{\mu}{k_V} w + \rho_0 \alpha g T,$$

whence, by use of the incompressibility condition

$$k_H \nabla_h^2 p + k_V \frac{\partial^2 p}{\partial z^2} = \rho_0 \alpha g k_V \frac{\partial T}{\partial z} = -\rho_0 \alpha g k_V G,$$

where ∇_h^2 represents the horizontal Laplacian operator. If we perform the overall scaling as before, with l_0 and h as vertical and horizontal length scales (Figure 7.9), so that $x = Xl_0$, $y = Yl_0$, $z, = Zh$, the preceding equation becomes

$$\frac{\partial^2 p}{\partial X^2} + \frac{\partial^2 p}{\partial Y^2} + \left(\frac{l_0^2}{h^2}\frac{k_V}{k_H}\right)\frac{\partial^2 p}{\partial Z^2} = -\frac{\rho_0 \alpha g k_V G l_0^2}{k_H}, \tag{7.4.3}$$

which is subject to the boundary conditions that $p = 0$ on $z = \zeta(\mathbf{x})$ and that there is no flow across the lower boundary so that $w = 0$ on $z = 0$. From the vertical Darcy balance above, this is equivalent to

$$\frac{\partial p}{\partial Z} = \rho_0 g h \alpha T|_{z=0} \quad \text{at} \quad Z = 0, \tag{7.4.4}$$

where, from (7.2.4),

$$T|_{z=0} = T(\zeta(\mathbf{x})) + G\zeta(\mathbf{x}), \tag{7.4.5}$$

which varies with horizontal position.

Two-dimensional reefs or banks

These geometries are the simplest. When the Y coordinate is taken along the bank, the middle term on the left of (7.4.3) is zero, and when $(h/l_0)^2 k_H/k_V \gg 1$, the coefficient of the third term is very small and, overall, can be neglected. The equation reduces simply to

$$\frac{\partial^2 p}{\partial X^2} = -\frac{\rho_0 g \alpha G k_V l_0^2}{k_H}.$$

Since the right-hand side is a constant, the reduced pressure is a quadratic function of X (or x); for a bank or reef symmetrical about $x = 0$,

$$p = \frac{\rho_0 g \alpha G k_V}{2k_H}(l^2(z) - x^2), \tag{7.4.6}$$

where $l(z)$ is the half-width of the reef at a height z above the basement as in Figure 7.9 – $p = 0$ when $x = l(z)$.

This is a fine solution for the interior flow but, fairly obviously, it cannot be reconciled with the lower boundary condition (7.4.4). The mathematical reason for the difficulty is our neglect (indicated by the overall scaling) of the vertical derivative term in (7.4.3), which reduces the order of the equation from second to zeroth in Z. As described in Section 3.5, we have a singular perturbation situation that points to the existence of an important lower boundary layer across which the *pressure* (here) adjusts rapidly to connect the interior distribution (7.4.6) to the boundary pressure gradient (7.4.4). If the upper surface is flat, an upper boundary layer can also develop, as will be seen later.

The lower boundary layer is necessary not only mathematically (to find a solution) but also physically, since it is necessary to drive the flow. In the interior region, the pressure is greatest at any depth along the axis of symmetry $x = 0$; its horizontal gradient drives the fluid *outward* with the velocity

$$u = -\frac{k_H}{\mu}\frac{\partial p}{\partial x} = \frac{g \alpha G k_H}{\nu} x, \tag{7.4.7}$$

which increases linearly with distance from the symmetry axis. This is consistent with the more general conclusion (7.4.2). Fluid is supplied by convective upwelling from below, from the lower boundary layer, but in order for the system to work, there must be a *negative* reduced pressure in the boundary layer near the bottom so that fluid can be sucked in from the sides before being lifted vertically by the buoyancy. The boundary layer pressure must then change rapidly across its thickness as the force balance near the basement (where the vertical velocity is small and the pressure gradient is established by the buoyancy) changes to that in the interior (where the buoyancy is balanced by the flow resistence, the vertical pressure gradient being smaller).

The analysis of this lower boundary layer proceeds by the assignment to it of a vertical scale δ, as yet unknown, and the rescaling of equation (7.4.3) in two dimensions with $x = XL$, $z = Z\delta$, where L is the half-width of the bank at the boundary layer height above the basement. If the dimensionless pressure P is defined such that

$$p(x, z) = \frac{\rho_0 \alpha g k_V G l^2}{k_H} P(X, Z), \tag{7.4.8}$$

the scaled equation becomes

$$\frac{\partial^2 P}{\partial X^2} + S_\delta^{-1} \frac{\partial^2 P}{\partial Z^2} = -1, \tag{7.4.9}$$

where

$$S_\delta = \frac{\delta^2}{L^2} \frac{k_H}{k_V}. \tag{7.4.10}$$

Outside the boundary layer (in terms of the boundary layer variables, as $Z \to \infty$), the pressure must match the interior solution (7.4.6), so that

$$P \to \tfrac{1}{2}(1 - X^2) \quad \text{as} \quad Z \to \infty. \tag{7.4.11}$$

If we neglect for the moment the influence of any oceanic thermocline, the lower boundary condition (7.4.4) becomes

$$\frac{\partial P}{\partial Z} = S_\delta \frac{h}{\delta} \hat{\zeta}(x), \tag{7.4.12}$$

where $\hat{\zeta}(x) = \zeta(x)/h$ is the dimensionless thickness of the bank and is in the range (0, 1). Let

$$P = \tfrac{1}{2}(1 - X^2) + \varpi,$$

and substitution into (7.4.9) gives

$$\frac{\partial^2 \varpi}{\partial X^2} + S_\delta^{-1} \frac{\partial^2 \varpi}{\partial Z^2} = 0 \tag{7.4.13}$$

with

$$\varpi \to 0 \quad \text{as} \quad Z \to \infty,$$
$$\frac{\partial \varpi}{\partial Z} = S_\delta \frac{h}{\delta} \hat{\zeta}(x) \quad \text{at} \quad Z = 0. \tag{7.4.14}$$

The detailed solution can be found by representing the bank configuration $\hat{\zeta}(x)$ as a Fourier series over the interval $-2 \le X \le 2$ with $\hat{\zeta}$ continued beyond $|X| = 1$ such that $\zeta(X) = -\zeta(2 - |X|)$; we are interested in the solution only in the range $|X| \leq 1$. Thus,

$$\hat{\zeta}(X) = \sum_{1}^{\infty} b_n \cos \tfrac{1}{2}n\pi X,$$

where (7.4.15)

$$b_n = \frac{1}{2}\int_{-2}^{2} \hat{\zeta}(x) \cos \tfrac{1}{2}\, n\pi \mathrm{X}\, dX.$$

If $\hat{\zeta}(\pm 1) = 0$, it can be shown that $b_n = 0$ when n is even. If also

$$\varpi(X, Z) = \sum_{1}^{\infty} a_n(Z) \cos \tfrac{1}{2}\, n\pi X, \qquad (7.4.16)$$

the substitution of (7.4.16) into (7.4.13) reduces the partial differential equation to an ordinary one in Z that can be solved for the a_n subject to the conditions (7.4.14). Restoration of the dimensional factors in the solution gives finally for the pressure field in the boundary layer

$$p(x, z) = \frac{\rho_0 \alpha g G k_V L^2}{2k_H}\left\{\left(1 - \frac{x^2}{L^2}\right) - \frac{2}{\pi}\left[\frac{h^2}{L^2}\frac{k_H}{k_V}\right]^{1/2} \sum_{1}^{\infty} \frac{b_n}{n} \exp\left[-\frac{n\pi}{2}\left(\frac{k_H}{k_V}\right)^{1/2}\frac{z}{L}\right] \cos\left(\frac{n\pi x}{2L}\right)\right\}. \qquad (7.4.17)$$

This solution looks a little complicated, but its salient features can be discerned readily. The first term in braces is the outer solution continued into the boundary layer; its magnitude is unity at the center, $x = 0$, decreasing quadratically to zero at the ends, $x = \pm L$. The coefficient of the summation is large, since $(h/L)^2 k_H/k_V \gg 1$; in the boundary layer near $z = 0$ there is, as we anticipated, a large suction that draws in fluid from the sides. This attenuates with distance from the boundary in accordance with the exponential terms in the sum. The extent of the boundary layer disturbance is determined by the $n = 1$ term (others drop off more rapidly), which is significant only when

$$z \sim \delta = L\left(\frac{k_V}{k_H}\right)^{1/2} \qquad \text{or less,}$$

so that from (7.4.10) $S_\delta = 1$. The boundary layer pressure balance involves all three terms: the horizontal and vertical pressure forcing and the buoyancy. The boundary layer thickness, as a fraction of that of the bank,

$$\frac{\delta}{h} = \frac{L}{h}\left(\frac{k_V}{k_H}\right)^{1/2}, \qquad (7.4.18)$$

is small.

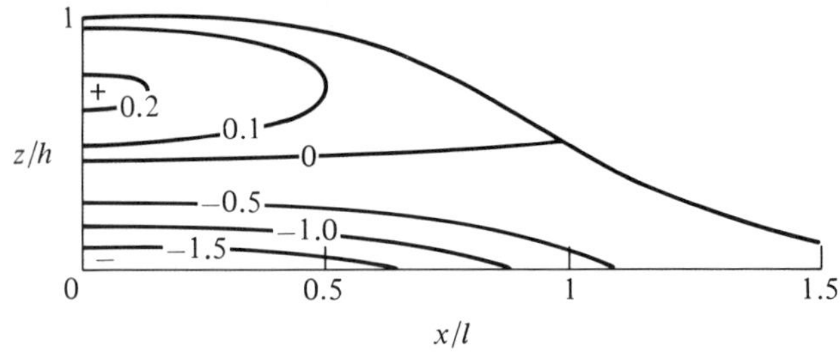

Figure 7.10. Contours of reduced pressure in units of $\rho g h^2 \alpha G$ for low Rayleigh number convection in a two-dimensional submerged bank with $S = 25$. The bank profile is specified by $\zeta = h/(1 + (x/l)^4)$. Note the strong negative pressure in the pressure boundary layer above the basement, which provides suction to draw in fluid, as shown in Figure 7.11. In the interior region, the pressure gradients are weak and the fluid rises essentially under the influence of its buoyancy.

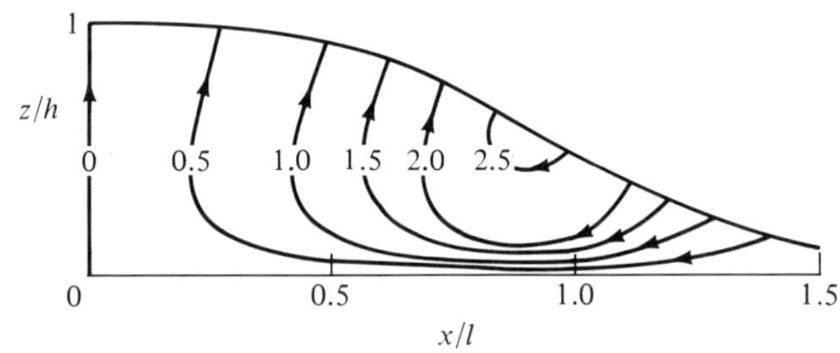

Figure 7.11. Flow streamlines corresponding to the pressure distributions of Figure 7.10. The units of the stream function are κ Ra(l/h).

The complete solutions for the pressure field and the associated streamlines are illustrated in Figures 7.10 and 7.11 for a bank of simple shape. The inward flow is relatively intense in the boundary layer, but above, it is outward and more diffuse. The total volume flux Q per unit length of the bank can be estimated most simply from the interior solution just above the lower boundary layer. Here, the temperature $T \simeq \zeta(x)G$, so that from (7.4.1) (with $f = 0$, as appropriate here)

$$Q = \int_{-l}^{l} w \, dx = \frac{\alpha g k_V G}{\nu} \int_{-l}^{l} \zeta(x) \, dx = \frac{\alpha g k_V G A}{\nu}, \qquad (7.4.19)$$

where A is the cross-sectional area of the bank. This can be expressed alternatively in terms of the Rayleigh number $\mathrm{Ra} = \alpha g k_H G h^2/\kappa\nu$ as

$$Q = \kappa \, \mathrm{Ra} \left\{ \frac{A k_V}{h^2 k_H} \right\}. \qquad (7.4.20)$$

Axisymmetrical geometries

In an axisymmetrical bank where the height contours are circular, the flow patterns are quite similar. The internal pressure distribution is given by

$$p = \frac{\rho_0 \alpha g k_V G}{4k_H}(R^2(z) - r^2),$$

where R is the radius at height z and r is measured outward from the center. A rather similar boundary layer occurs above the basement with thickness

$$\delta = R(0)\left(\frac{k_V}{k_H}\right)^{1/2}$$

(where $R(0)$ is the radius of the bank at its base), in which flow converges toward the center before rising under the influence of its buoyancy to move upward and outward. In either case, if the top of the bank is submerged and horizontal over an extended region of dimension L_T, fluid at the top can seep upward as readily as it can move outward, so that there is a second upper boundary layer there across which the internal pressure distribution is relieved, to become zero at the flat upper surface. This upper layer, however, has much less influence on the overall flow than does the lower one.

The total volume flux passing through the bank is found as before by integration over the lower surface of the interior region, where (7.4.1) holds,

$$Q_T = \int w \, \mathbf{dA} = \frac{\alpha g k_V G}{\nu} \int \zeta(\mathbf{x}) \, dA,$$

and the integral represents simply the volume V of the bank. Thus,

$$Q_T = \frac{\alpha g k_V G V}{\nu} = \kappa \, \mathrm{Ra}\left\{\frac{V}{h^2}\frac{k_V}{k_H}\right\}. \qquad (7.4.21)$$

Since $V \sim \pi R^2 h$,

$$Q_T \sim \pi \kappa h \, \mathrm{Ra}\left\{\frac{R^2}{h^2}\frac{k_V}{k_H}\right\},$$

and since both Ra and the factor in braces are small for these solutions, the total volume of fluid flowing through the bank per unit time is necessarily smaller than the combination κh.

7.5 HIGH RAYLEIGH NUMBER PLUMES

When the Rayleigh number is large, as it may be in a thick, highly permeable deposit, especially when the geothermal heat flux is locally high, the flow patterns and temperature distributions are qualitatively different from those discussed in the preceding sections. As described in Section 3.5, the convection is characterized by the presence of relatively thin thermal boundary layers and plumes, where the temperature variations are concentrated. Outside these regions, $\mathbf{u} \cdot \nabla T \simeq 0$, and the temperature is in essence constant along the streamlines; if seawater is drawn into a submerged bank at high Rayleigh number, it will remain cold until it enters either a plume or a thermal boundary layer. Plumes, in particular, are the engines that drive the convection – the buoyancy of the hot fluid forces it upward and reduces the pressure at depth, so that the surrounding fluid is sucked in (from the sides mostly, when $k_H \gg k_V$), as illustrated in Figure 7.12. The location of a plume may be determined by any one of several factors. It may rise from a local basement high, since heated interstitial fluid just above the basement will tend to converge toward the high point. A local maximum in the heat flux across the basement will have a similar effect. A vertical or nearly vertical fracture zone in the permeable bank may offer less resistance to fluid motion than the surrounding matrix. Only when real-world aspects like these are absent – in the experimenter's laboratory or on the theoretician's page – can symmetry be expected to prevail.

Nonetheless, it is instructive to consider two archetypical but extreme ($\mathrm{Ra} \gg 1$) situations that can be analyzed simply and give considerable insight: the two-dimensional plume and the axisymmetrical one. In either of these, the rates of variation of velocity and temperature in horizontal directions are large compared with those in the vertical direction; the vertical velocity w is generally larger than the horizontal components. Also, k_H is generally large compared with k_V. Consequently, (7.4.1) is pertinent here, and since both w and T vanish far from the plume, $f(z) = 0$ and

$$w = \frac{g\alpha k_V T(\mathbf{x})}{\nu}, \qquad (7.5.1)$$

where $T = T(\mathbf{x})$ is the distribution of internal temperature above that of the surrounding matrix outside the plume.

The temperature distribution itself is specified by

$$\mathbf{u} \cdot \nabla T = \kappa \, \nabla^2 T, \qquad (7.5.2)$$

and at high Rayleigh numbers, neither set of terms can be neglected in the plume. The only simplification that occurs is that $\delta^2 T/\delta z^2$ is small

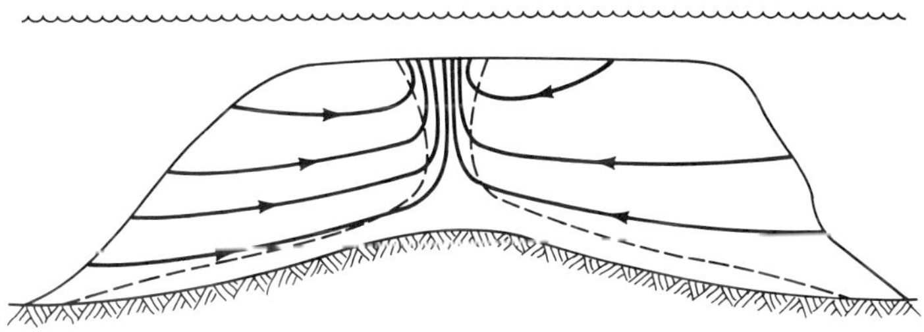

Figure 7.12. A schematic illustration of the flow in a submerged bank at high Ra, with a plume rising from a basement high. The heated region lies inside the broken lines, the temperature outside being very close to that of the ambient seawater.

compared with the corresponding second derivatives in the horizontal directions, and (7.5.2) becomes

$$\mathbf{u} \cdot \nabla T = \kappa \, \nabla_{\mathrm{h}}^2 T. \tag{7.5.3}$$

Two-dimensional plumes

When the flow is two-dimensional, the stream-function representation can conveniently be used, and (7.5.1) and (7.5.2) become

$$\frac{\partial \psi}{\partial x} = -\frac{g\alpha k_{\mathrm{V}} T}{\nu} \tag{7.5.4}$$

and

$$\frac{\partial \psi}{\partial z}\frac{\partial T}{\partial x} - \frac{\partial \psi}{\partial x}\frac{\partial T}{\partial z} = \kappa \frac{\partial^2 T}{\partial x^2}. \tag{7.5.5}$$

An important result that can be established immediately is that the convective heat flux

$$F_{\mathrm{h}} = (\rho C)_{\mathrm{f}} \int_{-\infty}^{\infty} wT \, dx \tag{7.5.6}$$

is constant with height and equal to the total rate of heat input at the bottom. This is shown by intergration of (7.5.5), which can be written as

$$\frac{\partial}{\partial x}\left(\frac{\partial \psi}{\partial z} T\right) - \frac{\partial}{\partial z}\left(\frac{\partial \psi}{\partial x} T\right) = \kappa \frac{\partial^2 T}{\partial x^2},$$

$$-\frac{\partial}{\partial z}\int_{-\infty}^{\infty} \frac{\partial \psi}{\partial x} T \, dx = \int_{-\infty}^{\infty} \frac{\partial}{\partial x}\left(\kappa \frac{\partial T}{\partial x} - \frac{\partial \psi}{\partial z} T\right) dx$$

$$= 0,$$

since $T \to 0$ as $x \to \pm\infty$ at all z. Consequently,

$$-\int_{-\infty}^{\infty} \frac{\partial\psi}{\partial x} T\,dx = \int_{-\infty}^{\infty} wT\,dx = \frac{F_{\rm h}}{(\rho C)_{\rm f}} = Q, \text{ say,} \tag{7.5.7}$$

where Q is independent of z. In view of (7.5.4) this is equivalent to

$$\int_{-\infty}^{\infty} T^2\,dx = \frac{\nu Q}{g\alpha k_{\rm V}}. \tag{7.5.8}$$

The governing equations (7.5.4) and (7.5.5), subject to the condition (7.5.7), allow similarity solutions, as demonstrated by Wooding(1963). They are of the form*

$$\psi = -\kappa\lambda^{-1/3}z^{1/3}f(\eta), \tag{7.5.9}$$

$$T = \frac{Q}{\kappa}\lambda^{1/3}z^{-1/3}f'(\eta), \tag{7.5.10}$$

where the similarity variable

$$\eta = \frac{x}{(\lambda z^2)^{1/3}} \tag{7.5.11}$$

and

$$\lambda = \frac{\nu\kappa^2}{g\alpha k_{\rm V} Q}, \tag{7.5.12}$$

which has the physical dimensions of a length. Since $T \to 0$ when $x \to \pm\infty$ (far outside the plume),

$$f'(\eta) \to 0 \quad \text{as} \quad \eta \to \pm\infty. \tag{7.5.13}$$

Clearly $f(\eta)$ is constant outside the plume (though it has different values on the two sides), so let

$$f(\eta) \to -c \quad \text{as} \quad \eta \to -\infty, \tag{7.5.14}$$

where the constant c is to be determined. Finally, the condition (7.5.8) requires that

$$\int_{-\infty}^{\infty} f'^2(\eta)\,d\eta = 1. \tag{7.5.15}$$

The form of the function $f(\eta)$ is found by substitution of (7.5.9) and (7.5.10) into (7.5.5), whence after reduction,

$$ff'' + f'^2 + 3f''' = 0. \tag{7.5.16}$$

* The procedure for finding similarity expressions is explained in books on fluid mechanics, such as that of Batchelor (1967).

This nonlinear equation occurs in the analysis of a viscous two-dimensional jet in fluid mechanics (see, e.g., Batchelor 1967, p. 345) and can be integrated exactly.

$$\frac{d}{d\eta}(ff') + 3f''' = 0,$$

so that

$$ff' + 3f'' = 0,$$

the constant of integration vanishing since f' (and so f'') vanishes when $\eta \to \pm\infty$. A further integration from $-\infty$ to η gives

$$\tfrac{1}{2}f^2 + 3f' = \tfrac{1}{2}c^2,$$

from (7.5.14), so that

$$\frac{df}{d\eta} = \tfrac{1}{6}(c^2 - f^2)$$

and

$$f(\eta) = c\tanh(\tfrac{1}{6}c\eta). \tag{7.5.17}$$

The constant c is determined by the condition (7.5.15); after a small calculation it is found that

$$c = \left(\frac{9}{2}\right)^{1/3} \simeq 1.65.$$

Thus, the solutions for the stream function and the temperature distribution are given by

$$\psi = -c\kappa\lambda^{-1/3}z^{1/3}\tanh(\tfrac{1}{6}c\eta), \tag{7.5.18}$$

$$T = \tfrac{1}{6}c^2\left(\frac{Q}{\kappa}\right)\lambda^{1/3}z^{-1/3}\,\mathrm{sech}^2(\tfrac{1}{6}c\eta), \tag{7.5.19}$$

and are as illustrated in Figure 7.13. The total volume flux per unit horizontal length of the plume is the difference between the values of ψ at $\eta = \pm\infty$, that is, $3.3\kappa(z/\lambda)^{1/3}$.

These represent the properties of a plume arising from a line source with finite heat flux $(\rho C)_f Q$ per unit length and zero volume flux, the singularity in the temperature field at $z = 0$ representing the singularity of the line source. When the base of the plume is of finite width (but small compared with the layer thickness), the virtual origin for the flow, $z = 0$, lies below the basement, its actual location being determined by matching the heat and volume fluxes into the base of the plume with those in the thermal boundary layer above the basement. When the horizontal extent of the basement heat source is comparable to or larger than the

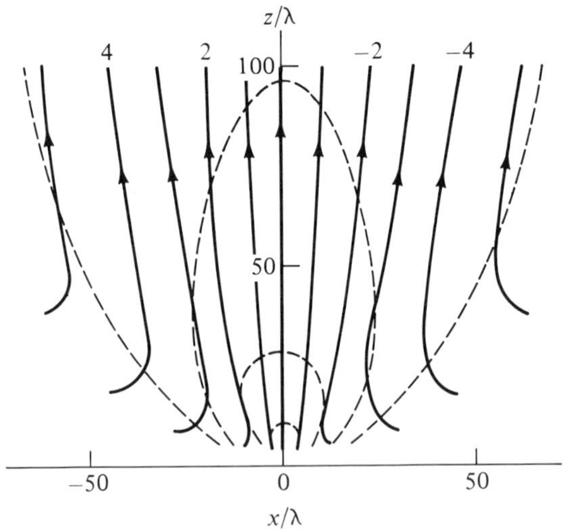

Figure 7.13. A two-dimensional plume above a line source of heat at $x = z = 0$. The streamlines are in units of κ, and the isotherms, shown as dashed lines, are at intervals $0.05\ Q/\kappa$.

layer depth, this solution is not very useful, but the flow patterns can be found by methods described in the next section.

In these solutions, the plume width δ may be characterized by the region $-1 < \frac{1}{6} < 1$, beyond which the temperature and velocity fields drop off rapidly. From (7.5.11)

$$\begin{aligned} \delta &= \frac{12}{c} (\lambda z^2)^{1/3} \\ &\approx 7.3(\lambda z^2)^{1/3}, \end{aligned} \tag{7.5.20}$$

which increases with height above the virtual origin as $z^{2/3}$. The temperature and the vertical velocity along the centerline decrease rather slowly as $z^{-1/3}$, while the total volume flux increases as $z^{1/3}$, as a result of fluid being drawn into the plume from either side. In view of (7.5.3), the general magnitude of $\mathbf{u} \cdot \nabla T$ is of order $\kappa T/\delta^2$, where T is the axial temperature and δ the plume half-width; it is positive on the flanks of the plume as cold water is drawn into it and negative on the plane of symmetry as fluid moves upward, down the temperature gradient.

When the upper surface of the bank is submerged, the plume may vent directly into the overlying water and a very strong temperature gradient develops just below the surface, as shown in the next section. If, however, it is exposed, the plume cannot pass through the water table or any freshwater lens that may have collected there, so that it is forced to spread out horizontally, gradually cooling as it does so. There does not seem to be any simple solution for this part of the flow, but the experiments described by Elder (1967) and the numerical solutions of the next section confirm this qualitative description. Along the axis of the plume, the temperature increases monotonically with depth, but on the flanks, the temperature along a vertical traverse has a maximum in the spreading region, below which it *decreases* until the lower thermal boundary layer is reached. The Florida Plateau seems to be the site of a natural circulation of this kind, as pointed out by Kohout, (1965) and Kohout, Henry and Banks, (1977).

Axisymmetrical plumes

When the plume rises from a localized maximum in the basement heat flux or a local basement high, it may be approximately axisymmetrical, and solutions can be found for this case also. The flow can now be represented in terms of the Stokes stream function (3.1.5),

$$u = \frac{1}{r}\frac{\partial \psi_s}{\partial z}, \qquad w = -\frac{1}{r}\frac{\partial \psi_s}{\partial r},$$

where r and z are the radial and vertical coordinates, respectively. From (7.5.1) the vertical velocity is again proportional to the temperature,

$$w = -\frac{1}{r}\frac{\partial \psi_s}{\partial r} = \frac{\alpha g k_V T}{\nu}, \tag{7.5.21}$$

while the thermal energy balance becomes

$$\frac{\partial \psi_s}{\partial z}\frac{\partial T}{\partial r} - \frac{\partial \psi_s}{\partial r}\frac{\partial T}{\partial z} = \kappa \frac{\partial}{\partial r}\left(r\frac{\partial T}{\partial r}\right). \tag{7.5.22}$$

The constancy of the vertical convective heat flux in the plume can be established as before by integration over a horizontal plane intersecting the plume:

$$2\pi \int_0^\infty rwT\,dr = Q = \text{const}, \tag{7.5.23}$$

where $(\rho C)_f Q$ is the total heat flux in the plume (not the heat flux per unit length as it was in the two-dimensional case previously).

Similarity solutions can again be found for this case. They are of the form

$$\psi_s = \kappa z f(\eta), \qquad T = \frac{Q}{\kappa z} g(\eta), \tag{7.5.24}$$

where f and g are functions to be determined and

$$\eta = \left(\frac{\alpha g k_V Q}{\nu\kappa^2}\right)^{1/2} \frac{r}{z} = \mathrm{Ra}^{1/2}\left(\frac{r}{z}\right), \qquad (7.5.25)$$

where, as suggested by the notation, Ra is a variant of the usual Rayleigh number, here based on the total heat flux in the plume. The precise forms of the functions f and g are found by substitution of (7.5.24) into (7.5.21) and (7.5.22), whence

$$\begin{aligned} g(\eta) &= -\frac{1}{\eta} f'(\eta), \\ f(\eta)g'(\eta) + f'(\eta)g(\eta) &= g'(\eta) + \eta g''(\eta). \end{aligned} \qquad (7.5.26)$$

Since the right-hand side can be written as the derivative of $\eta g'$, the preceding equation integrates to

$$fg = \eta g', \qquad (7.5.27)$$

the constant of integration vanishing since f, $g \to 0$ as $\eta \to \infty$. The equations (7.5.26) and (7.5.27) can now be combined and integrated further to give

$$f(\eta) = -\frac{4A\eta^2}{1 + A\eta^2}, \qquad (7.5.28)$$

$$g(\eta) = \frac{8A}{(1 + A\eta^2)^2}, \qquad (7.5.29)$$

where A is a numerical constant of integration that is determined by the condition (7.5.23). One finds that

$$A = \frac{3}{64\pi} \simeq 0.015. \qquad (7.5.30)$$

These solutions are illustrated in Figure 7.14.

The properties of the axisymmetrical solutions differ somewhat from those for the two-dimensional case. The half-width of the plume is characterized by $A\eta^2 = 1$, so that the total width

$$\delta = 2(A\,\mathrm{Ra})^{-1/2} z \simeq 16\,\mathrm{Ra}^{-1/2}\, z \qquad (7.5.31)$$

now increases linearly with height z above the virtual origin. Note that δ/z must be small, so that these solutions require the Rayleigh number Ra of equation (7.5.25) to be large. The temperature along the vertical axis $\eta = 0$, namely,

$$T(0, z) = \frac{8AQ}{\kappa z} \simeq 0.12 \frac{Q}{\kappa z},$$

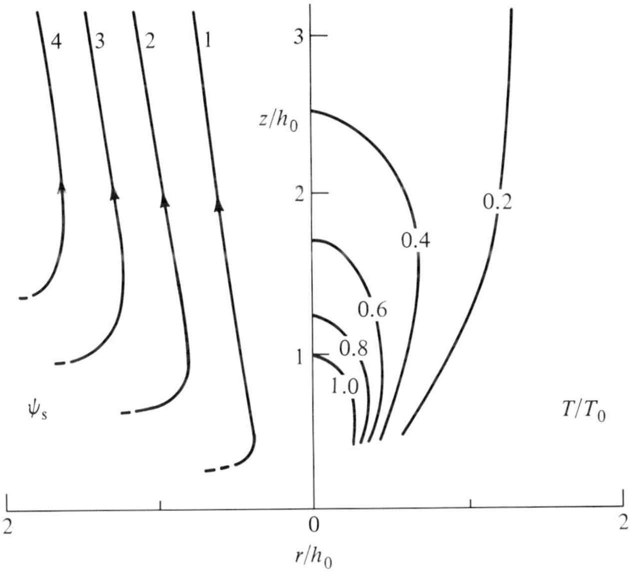

Figure 7.14. The axisymmetrical plume above a point heat source at Ra = 100. The temperature scale T_0 is the temperature on the axis at an arbitrarily chosen height h_0. The distribution of Stokes stream function ψ_s on the left has units κh_0; equal increments of ψ_s specify equal increments of total transport through the axisymmetrical volume contained.

decreases inversely with height (more rapidly than the two-dimensional case). The vertical velocity component

$$w = \frac{\alpha g k_V T}{\nu} = \frac{\kappa}{z}\,\mathrm{Ra}\; g(\eta) = \frac{\kappa\,\mathrm{Ra}}{z}\left(\frac{8A}{(1 + A\eta^2)^2}\right) \tag{7.5.32}$$

also decreases inversely with height. The radial inflow velocity, which is

$$u = -\frac{1}{r}\frac{\partial\psi_s}{\partial z} = -\frac{\kappa}{r}\left\{\frac{4A\eta^2(A\eta^2 - 1)}{(A\eta^2 + 1)^2}\right\}, \tag{7.5.33}$$

approaches $-4\kappa/r$ outside the plume and is independent of height. The volume flux in the plume therefore increases linearly with z; by integration of (7.5.32)

$$\begin{aligned} V = \text{volume flux} &= 2\pi \int_0^\infty rw\,dr \\ &= 8\pi\kappa z. \end{aligned} \tag{7.5.34}$$

This expression can be found alternatively by balancing the rate of change of volume flux by the radial inflow far from the plume: $2\pi r \cdot (4\kappa/r) = 8\pi\kappa = dV/dz$, from which (7.5.34) follows.

Although the plume solutions of this section are accurate only for Rayleigh numbers that are unrealistically large for most geological applications, their qualitative characteristics will be seen in Secton 7.6 to appear as the Rayleigh number increases and the temperature distribution is distorted by the flow.

7.6 FLOW, TEMPERATURE, AND GRADIENT REACTION PATTERNS AT INTERMEDIATE RAYLEIGH NUMBER AND SCALE RATIO

Two-dimensional numerical solutions

The analytical solutions of the previous few sections pertain to small or large values of the Rayleigh number and of the scale ratio $S = (k_H/k_V)(h/l)^2$ and represent different asymptotic limits in which the flow patterns and distributions of diagenesis are qualitatively different from one another. At intermediate values of these parameters, the corresponding patterns of flow, temperature, and so on form families whose characteristics lie between those at the limits and which are best established by numerical analysis.

For two-dimensional flow in a horizontally laminated bank, the complete generalized vorticity balance is

$$\frac{1}{k_H}\frac{\partial^2\psi}{\partial z^2} + \frac{1}{k_V}\frac{\partial^2\psi}{\partial x^2} = -\frac{g\alpha}{\nu}\frac{\partial T}{\partial x}, \qquad (7.6.1)$$

which we can express in a form suitable for calculation by defining the dimensionless horizontal and vertical coordinates X and Z, temperature θ, and stream function φ such that

$$\begin{gathered} x = Xl, \qquad z = Zh, \\ T(x, z) = T_0\theta(X, Z), \\ \psi(x, z) = \frac{\alpha g T_0 k_V l}{\nu}\varphi = \kappa\left(\frac{l}{h}\right) \mathrm{Ra}\, \varphi(X, Z), \end{gathered} \qquad (7.6.2)$$

where

$$\mathrm{Ra} = \frac{\alpha g T_0 k_V h}{\nu\kappa} \qquad (7.6.3)$$

is the Rayleigh number based on the vertical permeability and the layer depth and T_0 is the maximum basement temperature. In terms of these variables, equation (7.6.1) becomes

$$\frac{\partial^2 \varphi}{\partial x^2} + S^{-1}\frac{\partial^2 \varphi}{\partial z^2} = -\frac{\partial \theta}{\partial x}, \quad (7.6.4)$$

where

$$S = \frac{k_H}{k_V}\left(\frac{h}{l}\right)^2.$$

The temperature field is specified by

$$\frac{\partial \psi}{\partial z}\frac{\partial T}{\partial x} - \frac{\partial \psi}{\partial x}\frac{\partial T}{\partial z} = \kappa\left\{\frac{\partial^2 T}{\partial x^2} + \frac{\partial^2 T}{\partial z^2}\right\}, \quad (7.6.5)$$

and in terms of the same variables, this becomes

$$\mathrm{Ra}\left\{\frac{\partial \varphi}{\partial Z}\frac{\partial \theta}{\partial X} - \frac{\partial \varphi}{\partial X}\frac{\partial \theta}{\partial Z}\right\} = \left(\frac{h}{l}\right)^2 \frac{\partial^2 \theta}{\partial X^2} + \frac{\partial^2 \theta}{\partial Z^2}. \quad (7.6.6)$$

For the purpose of illustrating the evolution of the flow patterns as the parameters are varied, we will consider a bank of rectangular cross section on an impermeable basement at $Z = 0$, so that $\varphi = 0$ there. The plane $X = 0$ is the axis of symmetry, so that $\varphi = 0$ also when $X = 0$. If the upper surface of the bank is exposed, it might be supposed that a shallow freshwater lens inhibits flow across the surface, so that $\varphi = 0$ at $Z = 1$. If, however, convecting fluid may move freely across the upper surface as it can if the bank is submerged or if there is no freshwater lens and the convecting fluid is removed at the surface by evaporation or runoff, then because the surface pressure gradient vanishes, fluid rises vertically through the upper surface and $\partial\varphi/\partial Z = 0$ at $Z = 1$. In either case, fluid can enter or leave the bank at the side walls, so that $\partial\varphi/\partial X = 0$ there. In all of the calculations of this section, the scale l represents the half-width of the basement temperature distribution, which is taken as $T = T_0 \exp[-(x/l)^2]$, so that $\theta = \exp(-X^2)$ at $Z = 0$; the temperature at the other interfaces is the ambient temperature, taken as zero. The total width of the bank is $4l$, so that the sides are at $X = \pm 2$; since the flows are symmetrical about $X = 0$, only the right-hand half of each pattern is illustrated.

First, consider the variations in flow pattern at low Rayleigh number

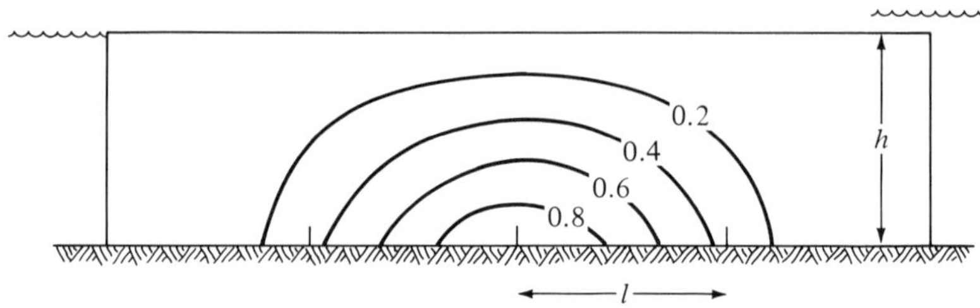

Figure 7.15. Isotherms T/T_0 in a bank with a basement temperature distribution $T_0 \exp[-(x/l)^2]$.

as the scale ratio $S = (k_H/k_V)(h/l)^2$ varies from small to large values. When Ra $\rightarrow 0$, the terms on the left of (7.6.6) are negligible and the temperature field, conductively determined, is independent of S (and the actual value of Ra, provided that it is always small), as shown in Figure 7.15. Figures 7.16a–f are concerned with the case of no flow across the upper surface; they show on the left the distributions of streamlines and on the right the rock alteration index for gradient reactions $\mathbf{u} \cdot \nabla T$ as S increases from 0.1 to 33. (Isothermal reactions, of course, follow the streamlines at rates proportional to the local speed.) If the Rayleigh number (7.6.3) is small and fixed, the sequence of patterns can be considered to result from increases in the horizontal permeability k_H keeping all other quantities fixed. When S is 0.1, the flow pattern and the distribution of the rock alteration index $\mathbf{u} \cdot \nabla T$ are almost identical to those illustrated in Figure 7.5 for $S \rightarrow 0$, with a feeble closed circulation in each half of the bank. As S increases to 0.33, the cells become elongated horizontally and the velocities increase; $\mathbf{u} \cdot \nabla T$ is positive near the basement as the fluid, flowing inward, moves up the temperature gradient and negative near the axis of symmetry and near the top as the rising and spreading fluid moves down the temperature gradient. By the time S is equal to 1, a significant amount of fluid is being drawn into the bank over the lower part of the side wall and leaving it near the top; the circulation has intensified further. The pattern of $\mathbf{u} \cdot \nabla T$ is similar, but the magnitudes increase. The beginnings of the lower boundary layer discussed in Section 7.4 are evident when $S = 3.3$ and the streamlines begin to crowd toward the basement, though the internal circulation persists. These trends continue as S (or k_H) increases further; fluid can move much more freely in the hori-

Figure 7.16. Low Rayleigh number streamlines in units of κ Ra $(l/h) \times 10^{-1}$ and distribution of $\mathbf{u} \cdot \nabla T$ in units of κT_0 Ra$/h^2 \times 10^{-1}$, where Ra involves the vertical permeability, for various values of $S = (k_H/k_v)(h/l)^2$. No flow across the upper surface.

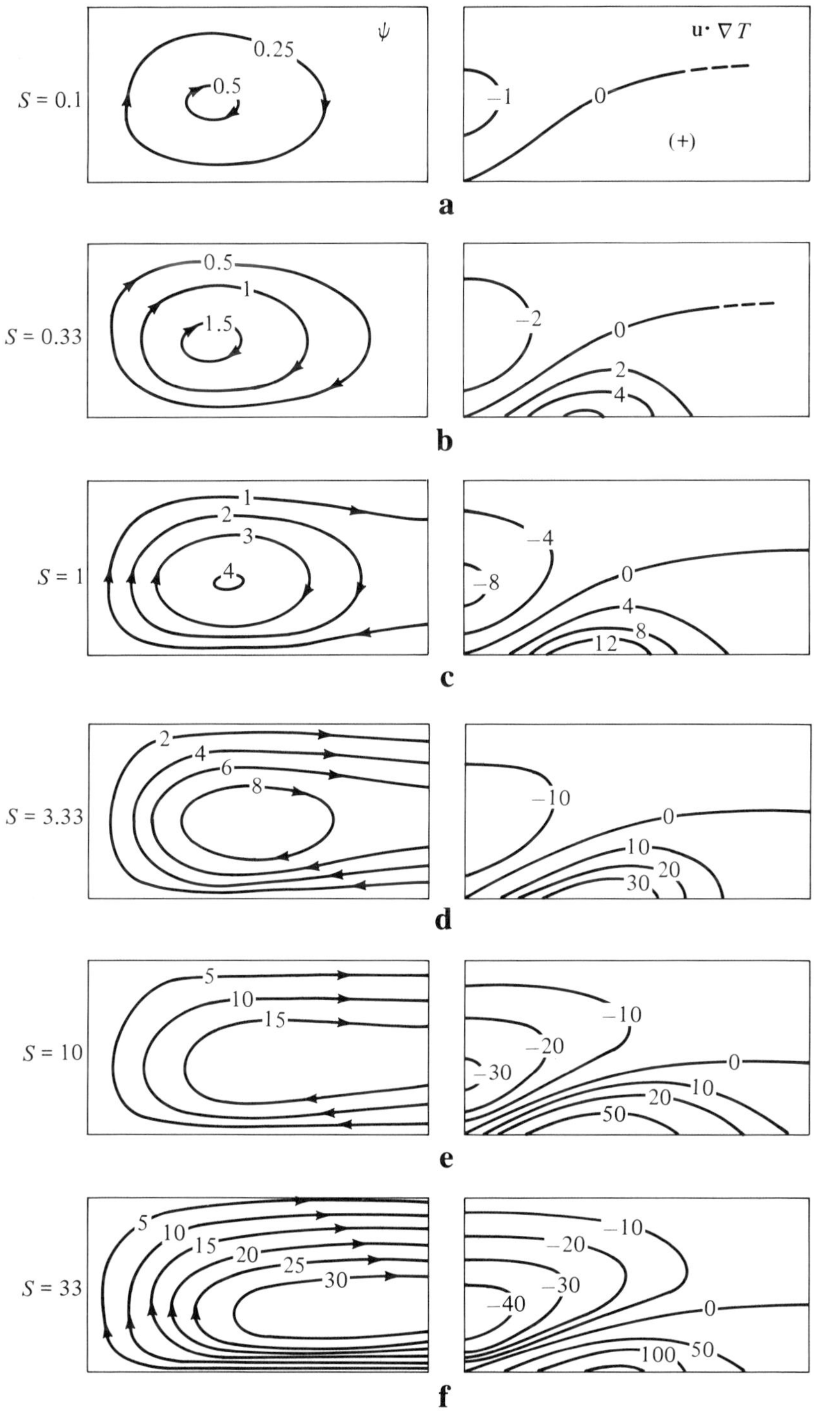
S = 0.1
ψ
0.25
0.5
u· ∇ T
−1
0
(+)
a
S = 0.33
0.5
1
1.5
−2
0
2
4
b
S = 1
1
2
3
4
−4
−8
0
4
8
12
c
S = 3.33
2
4
6
8
−10
0
10
20
30
d
S = 10
5
10
15
−10
−20
−30
0
10
20
50
e
S = 33
5
10
15
20
25
30
−10
−20
−30
−40
0
50
100
f

zontal direction than the vertical, and the circulation develops the well-developed lower boundary layer described analytically in Section 7.4 with relatively high velocities inward (indicated by the closeness of the streamlines there) and a more diffuse upward and outward motion above. For $S = 33$, the boundary layer thickness is about one-fifth of the overall thickness of the bank, generally consistent with the value of $S^{-1/2}$ indicated by equation (7.4.18). The pattern of $\mathbf{u} \cdot \nabla T$ is, however, not very different from that at lower values of S, though the positive regions are more concentrated in the lower boundary layer and negative regions above are somewhat more diffuse. The magnitudes, measured in units of κ Ra T_0/h^2, are much larger.

Somewhat similar trends are evident in the sequence of patterns for a fully submerged bank, or one in which fluid can move freely across the upper surface (Figure 7.17). At low values of $S = (k_H/k_V)(h/l)^2$ fluid enters the bank mostly at the top flanks and rises from it near the plane of symmetry, but as S increases beyond unity, fluid increasingly enters from the sides. The region of negative RAI, where fluid is moving down the temperature gradient, spreads across the top of the layer and down into it, positive regions becoming more concentrated and more intense in a zone above the basement on either side the symmetry plane. As S (or k_H) increases further, fluid ceases to be drawn in from the top but enters from the sides, becoming increasingly concentrated near the bottom as the lower boundary layer develops. When $S = 33$ (i.e., the horizontal permeability being then large), the lower part of the flow pattern is virtually indistinguishable from that of Figure 7.16f for an impermeable upper surface; in the upper part of the region about half of the fluid flow through the bank shortcuts to the upper surface and half emerges from the upper part of the side. The region of negative $\mathbf{u} \cdot \nabla T$ occupies about 70 percent of the whole cross-sectional area; it is relatively diffuse though numerically largest in the region low on the plane of symmetry. Positive regions, with magnitudes about four times as large, are concentrated above the basement in the region of the maximum horizontal temperature gradient and high flow rate.

All these calculations are for cases when the Rayleigh number is substantially less than unity and the flow field does not disturb the temperature distribution of Figure 7.15. As the Rayleigh number increases, the isotherms become increasingly distorted by the flow; the temperature field and the flow pattern become more strongly interdependent, as the next

Figure 7.17. Low Rayleigh number convection in a submerged bank in which fluid can move across the upper surface, for various values of S. The units are the same as for Figure 7.16.

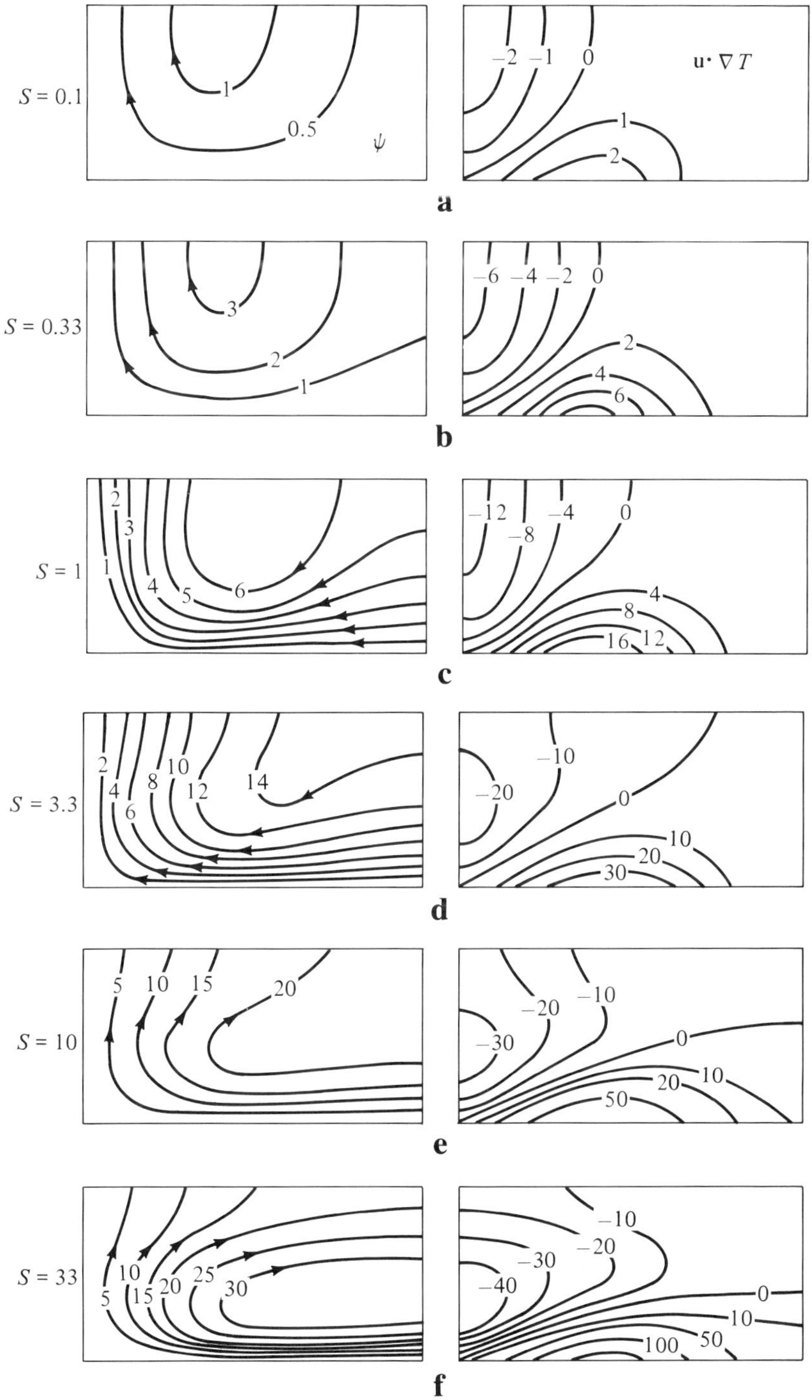
S = 0.1
1
0.5
ψ
−2 −1 0
u· ∇T
1
2
a
S = 0.33
3
2
1
−6 −4 −2 0
2
4
6
b
S = 1
2
3
1
4
5
6
−12
−8
−4
0
4
8
16 12
c
S = 3.3
2
4
6
8
10
12
14
−10
−20
0
10
20
30
d
S = 10
5
10
15
20
−20
−10
−30
0
20
10
50
e
S = 33
10
5
15
20
25
30
−10
−20
−30
−40
0
10
100
50
f

sequence of calculations illustrates. In them, the basement temperature distribution is maintained as before and so extends over the half-width l, and k_H/k_V and h/l (and so S) are held constant while the Rayleigh number increases.

In Figure 7.18, the upper surface is impermeable; it shows streamline patterns, temperature distributions, and contours of $\mathbf{u} \cdot \nabla T$ at various values of Ra, taking $k_H/k_V = 10$ and $h/l = 1$ so that $S = 10$. For Ra = 1, the flow pattern is very similar to that of Figure 7.16e for Ra $\ll 1$, but the circulation (the volume transport being measured in units of κ Ra(h/l)) is a little more intense. The temperature field is distorted somewhat from that shown in Figure 7.15, the isotherms being drawn inward near the base and upward in the upper region as a result of the flow. When Ra = 3, this thermal distortion is greater; hot fluid near the axis is carried higher, so that the vertical temperature gradient is *reduced* in the interior along the axis and increased near the surface as a surface conductive layer develops that is not unlike that discussed in Section 6.4. This is reflected in the patterns of $\mathbf{u} \cdot \nabla T$, where the more intense negative region migrates upward. As the Rayleigh number increases to 10 and 30, the upward flow and the temperature distribution become more concentrated near the plane of symmetry, even though the basement temperature distribution extends horizontally over the same distance. The basement is overlain by a thermal boundary layer, above which the temperature is close to that of the ambient water at the sides of the bank. The overall pattern becomes increasingly more mushroom-shaped, with the lower thermal boundary layer merging into the central plume sheet, then spreading outward as an overhanging warm region, surrounding on three sides the nearly isothermal interior. Except near the bottom and the top, the isotherm shapes are close to those of the streamlines, so that the temperature gradient in the direction of flow becomes small. Regions of significant negative values of the rock alteration index $\mathbf{u} \cdot \nabla T$ become confined to the lower part of the plane of symmetry where the fluid "turns the corner" and, most extensively, just below the upper surface in the upper thermal boundary layer where the vertical temperature gradient is large. At mid-depths, the patterns of isotherms and streamlines are so similar that $\mathbf{u} \cdot \nabla T$ is small. Large positive values of $\mathbf{u} \cdot \nabla T$ are confined

Figure 7.18. Convection patterns at $S = 10$ for increasing Ra with no flow across the upper surface. Streamlines in units of κ Ra$(l/h) \times 10^{-1}$ are shown in the top panels of each set (the flow becomes more vigorous as Ra increases, since the units are larger in proportion), isotherms in the center panels, and distributions of rock alteration index, $\mathbf{u} \cdot \nabla T$, in units of κT_0 Ra$/h^2 \times 10^{-1}$ in the lower ones.

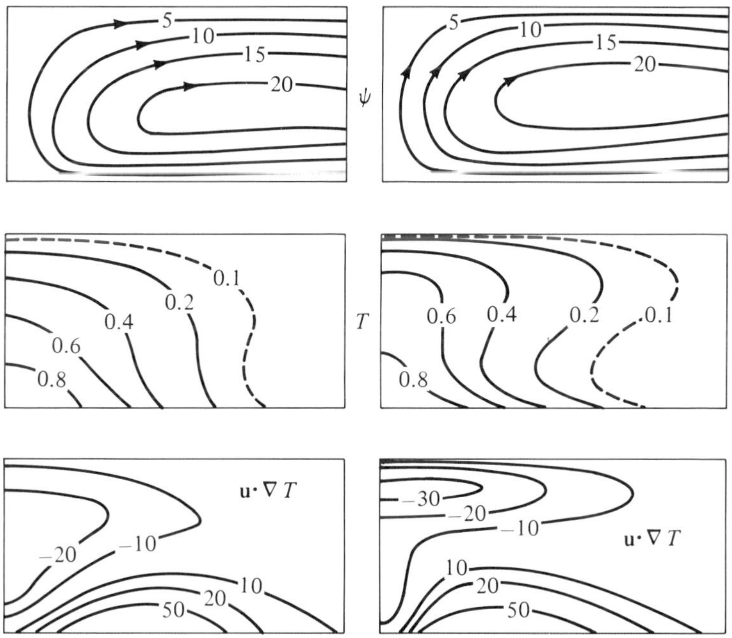

Ra = 1 **a** Ra = 3

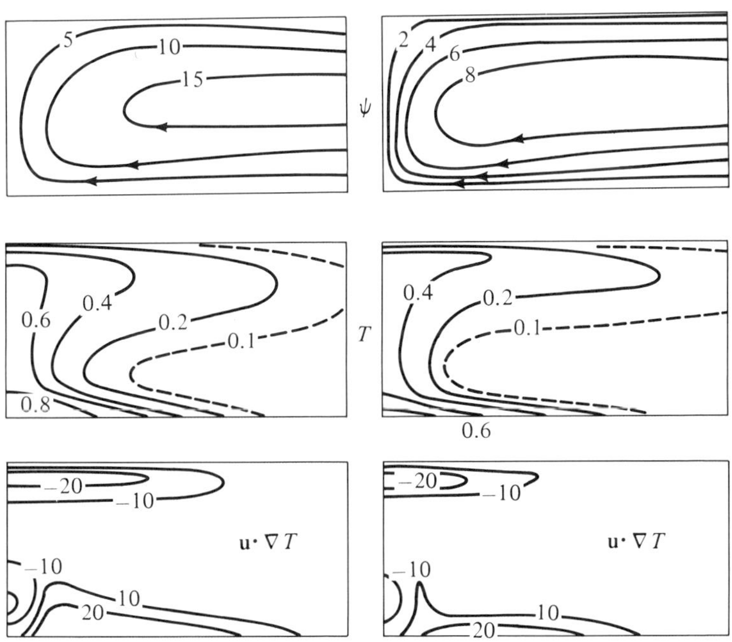

Ra = 10 **b** Ra = 30

to the thermal boundary layer above the basement as fluid in this region moves essentially horizontally to higher temperatures.

Figure 7.19 shows corresponding patterns found when the fluid is free to move across the upper surface. In these calculations, $S = 1$, so that Figure 7.19a for Ra = 1 can be compared with Figure 7.17c for Ra $\ll$ 1 and the internal temperature distribution of Figure 7.15. They are clearly very similar. When the Rayleigh number is 3, the isotherms are, as before, somewhat distorted, inward near the base and upward near the symmetry plane. When Ra = 10, the central truncated plume structure begins to emerge, and by Ra = 30 it is clearly evident; since the fluid moves vertically across the upper surface, the plume is capped now by a convective–diffusive layer of the same genre as that in a pressure-driven discharge zone where the near-surface temperature gradient is large. The rock alteration index $\mathbf{u} \cdot \nabla T$ has very large negative values close to the surface over a horizontal region that corresponds to the plume width; this is very much less than the horizontal extent of the heated basement.

These calculations are intended to be illustrative but not exhaustive. The effect, for example, of variations in S at large Rayleigh number can be inferred without further calculation: As k_H increases with all the other quantities remaining the same, S increases in proportion. The fluid moves horizontally much more readily, so that in the isothermal interior the streamlines are more nearly horizontal and the lower thermal layer somewhat thinner. The central plume structure is not altered much, since the flow is essentially vertical and the overall distribution of $\mathbf{u} \cdot \nabla T$ is similar. (In the interior region, the flow is quite different, but the temperature gradients are negligible in each case.)

Two-dimensional flow scaling analysis

A characteristic property of these solutions at moderate to large values of S and Ra is the existence of lower boundary regions where the temperature variations are large and flow is predominantly horizontal, converging toward the central plume. This observation provides the basis for a scaling theory that gives simple analytical expressions for the variations of boundary layer and plume width and total volume transport in terms of the geometry, the thermal diffusivity κ, and the parameters S and Ra.

Near the symmetry plane and the basement, the vertical velocity is small and the vertical gradient in reduced pressure is balanced largely by the fluid buoyancy $\rho \alpha g T_0$, so that if the boundary layer thickness is δ, the central suction is

Figure 7.19. Convection pattern at $S = 1$ for increasing Rayleigh number when fluid is free to move across the upper surface. The scales are the same as in Figure 7.18.

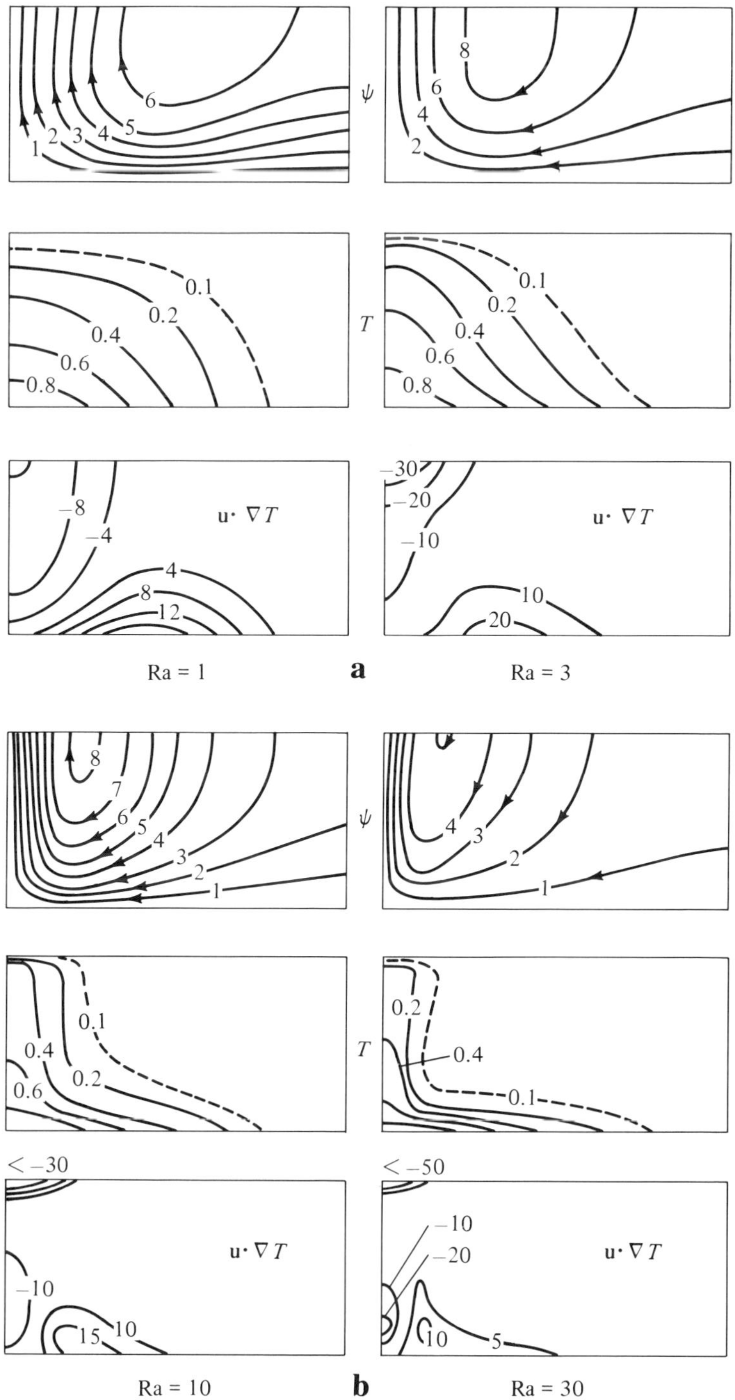
ψ
T
u· ∇T
Ra = 1
a
Ra = 3
ψ
T
< −30
< −50
u· ∇T
Ra = 10
b
Ra = 30

$$p \sim \rho\alpha g T_0 \delta,$$

which produces an inward horizontal velocity over the distance l of

$$u \sim \frac{k_H \alpha g T_0 \delta}{\nu l}. \tag{7.6.7}$$

In the plume itself, however, the buoyancy is balanced by the resistence to vertical flow, so that the vertical velocity in the plume

$$w \sim \frac{k_V \alpha g T_0}{\nu} = \frac{\kappa}{h}\,\mathrm{Ra}. \tag{7.6.8}$$

The volume flux of fluid in the boundary layer, proportional to uδ, must be the same as that in (half of) the rising plume, that is, wd, where d is the plume half-width, so that

$$u\delta = wd. \tag{7.6.9}$$

Finally, the basement heat flux is conducted across the lower boundary at the rate (in kinematic units) $\kappa\, \partial T/\partial z \sim \kappa T_0/\delta$, and the total heat flux over the half-width l is $\kappa T_0 l/\delta$. Since at large Ra, longitudinal heat conduction is negligible, this is equal to the heat convected by flow in the plume, that is, $wT_0 d$, so that

$$Q_h \sim wT_0 d \sim \kappa T_0 l/\delta. \tag{7.6.10}$$

These physical balances express the essence of the dynamics involved in these flows. Our interest here is in parametric variations that are not evident from the numerical calculations, rather than in accurate numerical values that are, so that the numerical coefficients in these balances (which are generally of order unity) are ignored. Algebraic manipulation of equations (7.6.7) to (7.6.10) provides the scaling relations sought.

From (7.6.10) $wd \sim \kappa l/\delta$ and

$$\delta d \sim \frac{\kappa l}{w}. \tag{7.6.11}$$

But from (7.6.9), on the one hand, and (7.6.7) and (7.6.8), on the other,

$$\frac{\delta}{d} \sim \frac{w}{u} \sim \frac{k_V}{k_H}\frac{l}{\delta},$$

so that

$$\frac{\delta^2}{dl} \sim \frac{k_V}{k_H}, \tag{7.6.12}$$

which, incidentally, is small. Now d can be eliminated between (7.6.11) and (7.6.12), giving

$$\frac{\delta^3}{l} \sim \frac{\kappa l}{w}\frac{k_V}{k_H} \sim lh\left(\frac{k_V}{k_H}\right) \mathrm{Ra}^{-1},$$

from (7.6.8), or

$$\frac{\delta^3}{h^3} \sim \frac{l^2}{h^2}\frac{k_V}{k_H} \mathrm{Ra}^{-1} = S^{-1}\,\mathrm{Ra}^{-1},$$

where $\mathrm{Ra} = \alpha g T_0 k_V h/\nu\kappa$, as defined previously. Thus, the lower boundary layer thickness as a fraction of the total height h is

$$\frac{\delta}{h} \sim S^{-1/3}\,\mathrm{Ra}^{-1/3} \qquad (7.6.13)$$

at large Rayleigh numbers. Similarly, the plume half-thickness d is, from (7.6.12) and (7.6.13),

$$\frac{d}{l} \sim S^{-1/3}\,\mathrm{Ra}^{-2/3}. \qquad (7.6.14)$$

The total volume flux per unit length of the two-dimensional plume

$$F \sim wd \sim \frac{\kappa}{h}\,\mathrm{Ra}\,S^{1/3}\,\mathrm{Ra}^{-2/3}\,l$$
$$= \kappa\left(\frac{l}{h}\right)\mathrm{Ra}^{1/3}\,S^{1/3}, \qquad (7.6.15)$$

for moderate to large values of Ra and S. The numerical coefficient of proportionality can be found by fitting (7.6.15) to the results of numerical calculation and it is found that

$$F \simeq 0.3\kappa\left(\frac{l}{h}\right)\mathrm{Ra}^{1/3}\,S^{1/3} = 0.3\kappa(\mathrm{Ra})_{\mathrm{H}}^{1/3}, \qquad (7.6.16)$$

for $(\mathrm{Ra})_{\mathrm{H}} \gtrsim 100$, where

$$(\mathrm{Ra})_{\mathrm{H}} = \frac{\alpha g T_0 k_H l}{\nu\kappa}$$

is a Rayleigh number based on the *horizontal* extent of the domain and the *horizontal* permeability.

Axisymmetrical flows

A similar scaling analysis for axisymmetrical flows has been given by Wilson et al. (1990). If R represents the radius of the platform, the radial inflow to the base of the plume is given by (7.6.7) with R replacing l, and the continuity relation (7.6.9) is replaced by

$$u\delta \cdot \pi\Delta \sim w \cdot \tfrac{1}{4}\pi\Delta^2 \tag{7.6.17}$$

when Δ is the plume diameter at its base. The heat balance (7.6.10) becomes

$$Q_h \sim wT_0 \cdot \tfrac{1}{4}\pi\Delta^2 \sim \frac{\kappa T_0}{\delta} \cdot \pi R^2, \tag{7.6.18}$$

so that

$$w \sim \frac{4\kappa R^2}{\delta\Delta^2}. \tag{7.6.19}$$

Moreover, if the radius $\frac{1}{2}\Delta$ of the rising plume is such that the local (internal) value of the parameter

$$S_\Delta = \frac{k_H}{k_V}\left(\frac{h}{\frac{1}{2}\Delta}\right)^2$$

is large, we have

$$w \sim \frac{\kappa}{h}\,\mathrm{Ra}, \tag{7.6.20}$$

where $\mathrm{Ra} = \alpha g T_0 k_V h/\kappa\nu$. Finally, analogously to (7.6.7)

$$u \sim \frac{k_H \alpha g T_0 \delta}{\nu R} = \kappa \frac{k_H}{k_V}\frac{\delta}{\mathrm{hR}}\mathrm{Ra}. \tag{7.6.21}$$

From (7.6.17) (7.6.20), and (7.6.21),

$$\frac{k_V}{k_H} \sim \frac{4\delta^2}{\Delta R}, \tag{7.6.22}$$

and from (7.6.19) and (7.6.20),

$$\mathrm{Ra} \sim 4\frac{h}{\delta}\left(\frac{R}{\Delta}\right)^2, \tag{7.6.23}$$

so that if the plume and boundary layer thicknesses can be inferred, so can the Rayleigh number and the permeability anisotropy. The total volume transport in the central plume is found to be

$$V \simeq \tfrac{1}{4}\pi\Delta^2 w \simeq \pi\kappa R\left(\frac{R}{\delta}\right) \simeq \pi\kappa R\left(\frac{k_H}{k_V}\right)^{1/5}(\mathrm{Ra})_H^{1/5} \tag{7.6.24}$$

from (7.6.19).

7.7 PALEOCONVECTION AND DOLOMITIZATION IN THE LATEMAR MASSIF

A particularly striking and well-documented example of the consequences of paleoconvection is provided by the intensity and distribution of dolomitization in the Middle Triassic Latemar Buildup of the Dolomites in northern Italy. The Latemar Massif itself is a small, isolated carbonate platform, about 5 km in diameter and some 700 m thick, that preserves, in its vertically stacked "layer cake" deposits, and exhibits, in its remarkable exposure resulting from preferential erosion, a stratigraphic record of shallow marine carbonate deposition from the middle Triassic (c. 200 million years ago). Studied extensively by Bosellini and Rossi (1974) Gaetani, Fois, Jadoul, and Hicora (1981), Bosellini (1984), Goldhammer (1987), Wilson et al. (1990), and others, the Latemar Buildup is unusual in its proximity to the Predazzo volcanic–intrusive complex that was active late in its formation, at which time lava flows filled basins adjacent to the Latemar and a ring–dike complex intruded into it.

The stratigraphy of the formation is evident (to the trained eye) in the remarkable cliff exposures. The lower portion, consisting of approximately 300 m of horizontally bedded grainstones (averaging about 0.5 m in thickness), exhibits abundant early calcite cementation and represents periods of shallow-water subtidal deposition punctuated by intervals of submarine cementation. The lower Middle Triassic portion is called the "Lower Edifice" by Gaetani et al. (1981). Above it, the upper platform consists of some 400 m of horizontal cyclically bedded carbonates containing evidence of repeated subaerial exposure (Hardie, Bosellini, and Goldhammer 1986; Goldhammer, Dunn and Hardie 1987) alternating with shallow submergences. At the top of each cycle there is a thin (5–15 cm) layer containing dolomite with evidence of incipient soil formation. Cross-cutting these layers are small dispersed veinlets of dolomite cement, but the total volume of dolomite in these microstructures is insignificant. There is almost no lateral variation in lithography in the interior of the Latemar Platform until, at its edges, it is bounded by steeply sloping foreslope breccias and grainstones deposited as the platform aggregated.

The original limestone of the formation has been massively but incompletely dolomitized in a zone restricted to the core of the massif and with well-defined boundaries with the undolomitized limestone, and it is possible to map the geometry of the dolomite body that cross-cuts the primary bedding and fans out toward the top. A total of 2 to 3 km^3 of massive porous dolomite has been generated over a time interval estimated to be a few million years (certainly not more than 10 or 20) – this represents roughly 2×10^{13} mol dolomite. The most credible source of the magnesium required for the conversion of limestone to dolomite is the seawater that surrounded the platform. Assuming an initial magnesium concentration in the seawater of 5.3×10^{-2} molal (Garrels and Thompson, 1962), Wilson et al. (1990) calculate that about 500 km^3 of seawater would be required to furnish the necessary amount of magnesium if all of it were available for dolomitization – several hundred times the total pore volume of the Latemar. A more refined calculation based on the amount of magnesium present in excess of the Mg^{2+}/Ca^{2+} ratio at calcite–dolomite equilibrium leads to a rather larger value of seawater needed – some 1,000 km^3 at an internal temperature in the massif of 100°C. Quite evidently, a continuing circulation of seawater from the surroundings, through the platform and out, over a considerable time period, is required to account for the extent of this replacement.

The existence of the volcanic intrusions that occurred after the formation of the platform indicates the presence at the time of elevated temperatures near the basement that would drive a convective circulation in the permeable bank of one kind or another, as described earlier in this chapter. The questions of interest are whether the pattern of circulation, as evidenced by the pattern of the massively dolomitized region, can unambiguously indicate the type of flow-controlled reaction involved, whether the pattern points to values of Ra and S during convection that are consistent with the overall geometrical and geological constraints, and whether, with these values, the convection can deliver sufficient Mg^{2+} ions to provide for the extent of the dolomitization in a geologically reasonable time interval, a few million years.

Figure 7.20 is a simplified map given by Wilson (1989) of the present horizontal extent of the dolomitized region in relation to the depositional facies of the Latemar Buildup (omitting dikes and volcanic breccias) indicating its generally centered configuration in the larger limestone platform. Figure 7.21 gives the distribution of intense dolomitization projected onto a cross section showing the overall mushroom-shaped distribution. At the base, massive dolomite is in vertical contact with the limestone and a core of brecciated dolomite. This grades upward into a zone of large, closely spaced pods of massive dolomites, which become more widely spaced in the mushroom cap. On a smaller scale, it is seen in Figure 7.22 that while the dolomite pods clearly cross-cut the depositional bedding, fingers of dolomite reach out along more permeable bedding planes, the transition surfaces from limestone to dolomite remaining sharp. These characteristics indicate clearly that the reactions occurred in isothermal reaction fronts that went to completion and moved through the fabric at speeds proportional to the local flow speeds and, consequently, proportional to the local permeability. In some places, these left behind pockets of less permeable, unaltered limestone. The pattern is quite inconsistent with the result of gradient reactions, in which the reactions are more diffuse and ubiquitous, though more intense in high-flow regions than in those where the flow is slow. (The more generalized and dispersed

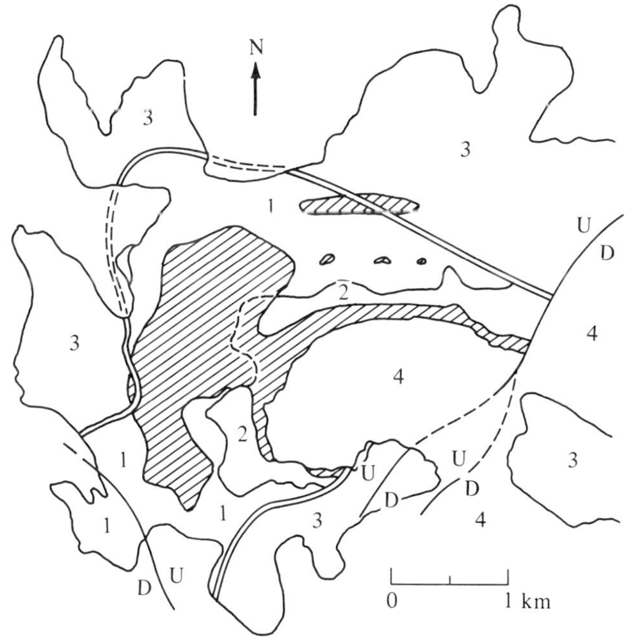

Figure 7.20. A schematic geological map of the Latemar Buildup near Predazzo in the Italian Dolomite region after Wilson (1989). The region of dolomitization is cross-hatched. Key: 1, the Ladinian Platform of Latemar limestone, whose margin in indicated by the double line; 2, the Lower Edifice; 3, foreslopes (Marmolada limestone); 4, volcanics. Faults are indicated as the single lines with the directions of relative displacement (up, down).

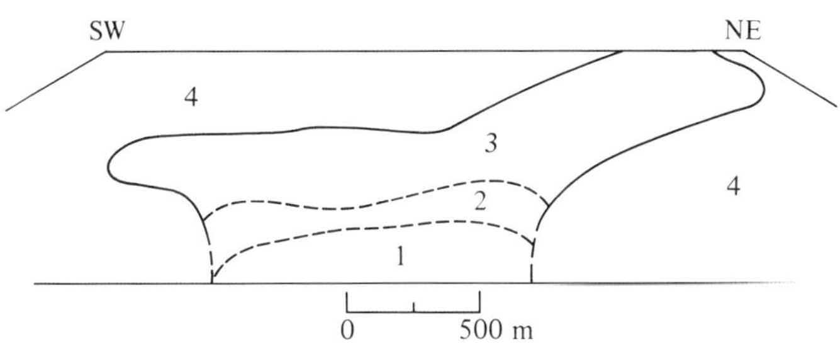

Figure 7.21. The intensity of dolomitization in the Latemar Buildup, projected onto a cross section, after Wilson (1989). In region 1, massive dolomite is in vertical contact with limestone with a core of brecciated dolomite. Region 2 contains large, closely spaced pods of massive dolomite, grading into region 3, where the pods are more widely spaced. Region 4 contains limestone, only sparsely dolomitized.

Figure 7.22. A field photograph by Wilson (1989) showing dolomite lenses in the upper, sparsely dolomitized zone (3 in Figure 7.21). Dolomite is identified in the line drawing by stippling.

centimeter-scale veinlets noted earlier are probably the result of gradient reactions.)

Isothermal reaction fronts follow the streamlines, so that the pattern of massive dolomitization indicates the pattern of the ancient flow – a central rising core of fluid fanning out toward the top. The general mushroom shape of the configuration is evidence that the circulating fluid did not generally move freely across the upper surface of the platform and is consistent with the geological indications of repeated subaerial exposure. An examination of the patterns of Figures 7.16 and 7.18 points to a moderate value of S with fluid circulating in and out of the perimeter of the structure and the concentration of the central core to a moderate value of Ra.

More quantitative estimates can be made by the use of the algebraic relations of Section 7.6 together with the measured geometry of the massively dolomitized region. In these flows, the thickness of the inaccessible lower boundary layer δ is close to that of the cap, which in the Latemar is about 300 m. The plume diameter Δ shown in Figure 7.21 is about 1,300 m and the radius R of the platform is some 2,500 m. With these values, the permeability ratio, given by equation (7.6.22), would be

$$\frac{k_V}{k_H} \sim \frac{4\delta^2}{\Delta R} \approx 0.12, \tag{7.7.1}$$

a reasonable value in view of the bedding present. The Rayleigh number of the convection is given by (7.6.23),

$$\mathrm{Ra} \sim 4\frac{h}{\delta}\left(\frac{R}{\Delta}\right)^2 \approx 35, \tag{7.7.2}$$

with the figures given above, which is sufficient to produce a significant advective distortion of the isotherms. With a thermal diffusivity $\kappa = 5.7 \times 10^{-7}$ m^2 s^{-1}, the total volume flux

$$V = \pi\kappa R(R/\delta) \sim 10^{-3}\ \mathrm{km\ yr^{-1}}, \tag{7.7.3}$$

so that within 10^6 years, 1,000 km^3 of seawater would pass through the system, as required by the geochemistry.

Although the Rayleigh number for the convection, the permeability *ratio,* and this total transport rate can be found fairly accurately from the geometry of the massive dolomitization, the temperatures cannot, since T_0 occurs in Ra in combination with k_V, whose explicit value is subject to much more uncertainty than is the temperature. However, since

$$\mathrm{Ra} = \frac{\alpha g h T_0 k_V}{\nu\kappa} \sim 35,$$

then

$$T_0 k_V \sim 35 \frac{\kappa \nu}{\alpha g h} \sim 1.2 \times 10^{-13}\ \mathrm{m^2\ K}.$$

with $\nu \approx 0.12 \times 10^{-6}\ \mathrm{m^2\ s^{-1}}$ and $\alpha \approx 3 \times 10^{-3}\ \mathrm{K^{-1}}$. Fluid inclusion data described by Wilson et al. (1990) indicate homogenization temperatures (a lower limit for that at the time of formation) of 75°C near the top of the formation and 220°C in the stem. If we take $T_0 \sim 300$°C, then $k_V \sim 4 \times 10^{-16}\ \mathrm{m^2} = 4 \times 10^{-12}\ \mathrm{cm^2}$ and, in view of (7.7.1), $k_H \sim 3 \times 10^{-11}\ \mathrm{cm^2}$, very reasonable values for a limestone structure such as this. It is important to recognize that in this example, the flow characteristics and the numerical value of the total transport rate were found *without* having to guess the appropriate values of the permeabilities, which may be uncertain to one or two orders of magnitude. It is therefore gratifying to find that the flow rate found accounts so easily for the transport required in the formation of this massive dolomite body.

The exploration of these questions is just beginning. The very few examples that could be considered in this book merely suggest the insights that can be gained by combining the principles of geochemistry and fluid flow in contexts where the mineralogy and structures are adequately defined. Carefully observed and documented field measurements are central to guide theory that must be correctly argued and numerical modeling that must be pertinent. Each should illuminate the others. The repeated interpretation of field observations in terms of soundly based and relevant theory and the repeated confirmation of prior predictions by field observation are necessary in the years to come if the conjunction is to be fruitful. Our concepts will certainly continue to evolve. One would hope that gradually our ability to anticipate or predict the occurrences and distributions of fabric alterations and mineral deposits will grow. We will understand the workings of our planet a little better. Though modern prospecting will probably always be largely an art, it is an art that needs all the science at its disposal.

References

Aagaard, P., and Helgeson, H. C. 1982. Thermodynamic and kinetic constraints on reaction rates among minerals and aqueous solutions. I. Theoretical considerations. *Amer. J. Sci., 282,* 273–85.

Adams, J. E., and Rhodes, M. L. 1960. Dolomitization by seepage refluxion. *Bull. Amer. Assoc. Petrol. Geol., 44,* 1912–20.

Barnes, H. L. (ed.). 1979. *Geochemistry of hydrothermal ore deposits.* New York: Holt, Rinehart & Winston.

Batchelor, G. K. 1967. *An introduction to fluid mechanics.* Cambridge University Press.

Bathurst, R. G. C. 1975. *Carbonate sediments and their diagenesis.* New York: Elsevier.

Baumgartner, L. P., and Ferry, J. M. In press. A physical model for infiltration-driven mixed volatile reactions and its application to regional metamorphism. *Contrib. Mineral. & Petrol.*

Bear, J. 1979. *Dynamics of fluids in porous media.* New York: Elsevier.

Bethke, C. M. 1985. A numerical model of compaction-driven groundwater flow and heat transfer and its application to the paleohydrology of intracratonic sedimentary basins. *J. Geophys. Res., 90,* 6817–28.

Bird, R. B., Stewart, W. E., and Lightfoot, E. N. 1960. *Transport phenomena.* New York: Wiley.

Bjørlykke, K., Mo, A., and Palm, E. 1988. Modelling of thermal convection in sedimentary basins and its relevance to diagenetic reactions. *Mar. Petrol. Geol., 5,* 338–51.

Boles, J. R., and Ramseyer, K. 1987. Diagenetic carbonate in miocene sandstone reservoir, San Joaquin Basin, California. *Amer. Assoc. Petrol. Geol. Bull., 71,* 1475–87.

Bonham, L. C. 1980. Migration of hydrocarbons in compacting basins, pp. 69–88 in *Problems of Petroleum Migration.* Amer. Assoc. Petrol. Geol. Studies in Geol. no. 10, Houston.

Bories, S. A., and Combarnous, M. A. 1973. Natural convection in a sloping porous layer. *J. Fluid Mech., 57,* 63–79.

Borkowska-Pawlak, B., and Kordylewski, W. 1985. Cell-pattern sensitivity to box configuration in a saturated porous medium. *J. Fluid Mech., 150,* 169–81.

Bosellini, A. 1984. Progradation geometries of carbonate platforms: Examples from the Triassic of the Dolomites, northern Italy. *Sedimentology, 31,* 1–24.

Bosellini, A., and Rossi, D. 1974. Triassic buildup of the Dolomites, northern

Italy, pp. 209–33, in *Reefs in time and space,* ed. L. Laporte. Soc. Econ. Paleontol. Mineralog. Special Publ. no. 18, Tulsa, Okla.

Brace, W. F. 1980. Permeability of crystalline and argillaceous rocks. *Int. J. Rock Mech. Min. Sci., 17,* 241–51.

Burley, S. D., Mullis, J., and Matter, A. 1989. Timing diagenesis in the Tartan Reservoir (UK North Sea): Constraints from combined cathodoluminescence microscopy and fluid inclusion studies. *Mar. Petrol. Geol., 6,* 98.

Caltagirone, J. P., and Bories, S. A. 1985. Solutions and stability criteria of natural convective flow in an inclined porous layer. *J. Fluid Mech., 155,* 267–87.

Carslaw, H. S., and Jaeger, J. C. 1959. *Conduction of heat in solids.* Oxford: Clarendon Press.

Cathels, L. M. 1977. An analysis of the cooling of intrusives by ground-water convection which includes boiling. *Econ. Geol., 72,* 804–26.

Cathels, L. M., and Smith, A. T. 1983. Thermal constraints on the formation of Mississippi Valley-type lead–zinc deposits and their implications for episodic basin dewatering and deposit genesis. *Econ. Geol., 78,* 983–1002.

Chadham, J., Hoff, D., Merino, E., Ortoleva, P., and Sen, A. 1986. Reactive infiltration instabilities. *IMA J. Appl. Math., 36,* 207–21.

Chandler, R., Koplik, J., Lerman, K., and Willemsen, J. F. 1982. Capillary displacement and percolation in porous media. *J. Fluid Mech., 119,* 249–67.

Chouke, R. L., van Meurs, P., and van der Poel, C. 1959. The instability of slow, immiscible viscous liquid–liquid displacements in permeable media. *Trans. AIME, 216,* 188–94.

Combarnous, M. A., and Bories, S. A. 1975. Hydrothermal convection in saturated porous media. *Adv. Hydrosci., 10,* 231–307.

Combarnous, M. A., and LeFur, B. 1969. Transfert de chaleur par convection naturelle dans une couche poreuse horizontale. *C. R. Acad. Sci. Paris B, 269,* 1009–12.

Dagan, G. 1984. Solute transport in heterogeneous porous formations. *J. Fluid Mech., 145,* 151–77.

Daly, R. A. 1951. Elastic properties of materials of the earth's crust, pp. 50–86 in *Internal constitution of the earth,* ed. B. Gutenberg. New York: Dover.

Davis, S. H., Rosenblat, S., Wood, J. R., and Hewett, T. A. 1985. Convective fluid flow and diagenetic patterns in domed sheets. *Amer. J. Sci., 285,* 207–23.

Drever, J. I. 1982. *The geochemistry of natural waters.* Englewood Cliffs, NJ: Prentice-Hall.

Elder, J. W. 1965. Physical processes in geothermal areas, pp. 211–39 in *Terrestrial heat flow,* ed. W. K. H. Lee. Amer. Geophys. Union, monogr. ser. 8, Washington, D.C.

1967. Steady free convection in a porous medium heated from below. *J. Fluid Mech., 27,* 29–48.

Evans, D. G., and Nunn, J. A. 1989. Free thermohaline convection in sediments surrounding a water column. *J. Geophys. Res., 94,* 12,413–22.

Ferry, J. M. 1987. Metamorphic hydrology at 13 km depth and 400–550°C. *Amer. Min., 72,* 39–58.

Ferry, J. M., and Dibble, G. M. In press. Fluid flow, mineral reactions, and metasomatism. *Geology.*

Freeze, R. A. 1975. A stochastic-conceptual analysis of one-dimensional groundwater flow in non-uniform homogeneous media. *Water Resources Res., 11,* 725–41.

Gaetani, M. E., Fois, E., Jadoul, F., and Hicora, A. 1981. Nature and evolution of middle Triassic carbonate buildups in the Dolomites, Italy. *Marine Geol., 74,* 25–57.

Garrels, R. M., and Christ, C. L. 1965. *Solutions, minerals and equilibria.* New York: Harper & Row.

Garrels, R. M., and Thompson, M. E. 1962. A chemical model for seawater at 25°C and one atmosphere total pressure. *Amer. J. Sci., 260,* 57–66.

Garven, G., and Freeze, R. A. 1984a. Theoretical analysis of the role of groundwater flow in the genesis of stratabound ore deposits. 1. Mathematical and numerical model. *Amer. J. Sci., 284,* 1085–1124.

1984b. Theoretical analysis of the role of groundwater flow in the genesis of stratabound ore deposits. 2. Quantitative results. *Amer. J. Sci., 284,* 1125–74.

Ge, S., and Garven, G. 1989. Tectonically induced transient groundwater flow in foreland basins, pp. 145–57 in *Origin and evolution of sedimentary basins and their energy and mineral resources,* ed. R. A. Price. Geophys. Monograph 48, Amer. Geophys. Un/IUGG, Washington, D.C.

Gill, A. E. 1969. A proof that convection in a vertical slab is stable. *J. Fluid Mech., 35,* 545–48.

Goldhammer, R. K. 1987. Platform carbonate cycles, middle Triassic of northern Italy. Ph.D. dissertation, Johns Hopkins University.

Goldhammer, R. K., Dunn, P. A., and Hardie, L. A. 1987. High frequency glacio-eustatic sea level oscillations with Milankovitch characteristics recorded in Middle Triassic platform carbonates in northern Italy. *Amer. J. Sci., 287,* 853–92.

Griffiths, R. W. 1981. Layered double-diffusive convection in porous media. *J. Fluid Mech., 102,* 221–48.

Habermann, B. 1960. The efficiency of miscible displacement as a function of mobility ratio. *Trans. AIME, 219,* 264–72.

Hanor, J. S. 1984. Variation in the composition of oil-field brines with depth in northern Louisiana and southern Arkansas: Implications for mechanisms and rates of mass transport and diagenetic reaction. *Gulf Coast Assoc. Geol. Soc. Trans., 34,* 55–61.

1987. Kilometre-scale thermohaline overturn of pore waters in the Louisiana Gulf Coast. *Nature, 327,* June 11, p. 501.

Hardie, L. A., Bosellini, A., and Goldhammer, R. K. 1986. Repeated subaerial exposure of subtidal carbonate platforms, Triassic, northern Italy: Evidence for high frequency sea level oscillations on a 10^4 year scale. *Paleooceanography, 1,* 447–57.

Harned, H. S., and Owen, B. B. 1958. The physical chemistry of electrolytic solutions. New York: Reinhold.

Haszeldine, R. S., Samson, I. M., and Cornford, C. 1984. Quartz diagenesis and convective fluid movement: Beatrice Oilfield, UK North Sea. *Clay Minerals, 19,* 391–402.

Helgeson, H. C., Murphy, W. M., and Aagaard, P. 1984. Thermodynamic and kinetic constraints on reaction rates among minerals and aqueous solutions. II. Rate constants, effective surface area and the hydrolysis of feldspar. *Geochim. Cosmochim. Acta, 48,* 2405–32.

Hewett, T. A. 1986. Porosity and mineral alteration by fluid flow through a temperature field, pp. 83–94 in *Reservoir characterization,* ed. L. W. Lake and H. B. Carroll, Jr. Orlando, FL: Academic Press.

Hill, S. 1952. Channelling in packed columns. *Chem. Eng. Sci., 1,* 247–53.

Hinch, E. J., and Bhatt, B. S. 1990. Stability of an acid front moving through porous rock. *J. Fluid Mech., 212,* 279–88.

Hogg, N. G., Katz, E. J., and Sanford, T. B. 1978. Eddies, islands and mixing. *J. Geophys. Res., 83,* 2921–38.

Homsy, G. M. 1987. Viscous fingering in porous media. *Ann. Rev. Fluid Mech., 19,* 271–311.

Horton, C. W., and Rogers, F. T., Jr. 1945. Convection currents in a porous medium. *J. Appl. Phys., 16,* 367–70.

Hubbert, M. K. 1940. The theory of ground-water motion. *J. Geol., 48,* 785–944.

1967. Application of hydrodynamics to oil exploration. *Proc. 78th World Petrol. Congress, 1B,* 59–75.

Imhoff, P. T., and Green, T. 1988. Experimental investigation of double-diffusive groundwater fingers. *J. Fluid Mech., 188,* 363–82.

Joseph, D. D. 1976. *Stability of fluid motions.* New York: Springer-Verlag.

Katto, Y., and Masuoka, T. 1967. Criterion for onset of convection flow in a fluid in a porous medium. *Int. J. Heat Mass Transfer, 10,* 297–309.

Kestin, J., Khalifa, H. E., and Correia, R. J., 1981. Tables of the dynamic and kinematic viscosity of aqueous NaCl solutions in the temperature range of 20°–150°C and the pressure range 0.1–35 MPa. *J. Phys. Chem. Ref. Data, 10,* 71–87.

Kohout, F. A. 1965. A hypothesis concerning cyclic flow of salt water related to geothermal heating in the Floridian aquifer. *Trans. N.Y. Acad. Sci., 28,* 249–71.

Kohout, F. A., Henry, H. R., and Banks, J. E. 1977. Hydrology related to geothermal conditions of the Florida Plateau, pp. 1–41. in *The geothermal nature of the Florida Plateau,* ed. D. L. Smith and G. M. Griffin. Fla. Bur. Geol., Spec. Publ. 21, Tallahassee, Fla.

Kvernvold, O., and Tyvand, P. A. 1980. Dispersion effects on thermal convection in porous media. *J. Fluid Mech., 99,* 673–86.

Lamb, H. 1932. *Hydrodynamics.* Cambridge University Press.

Lapwood, E. R. 1948. Convection of a fluid in a porous medium. *Proc. Camb. Phil. Soc., 44,* 508.

Lerman, A. 1979. *Geochemical processes: Water and sediment environments.* New York: Wiley.

Lichtner, P. C. 1985. Continuum model for simultaneous chemical reactions and mass transport in hydrothermal systems. *Geochim. Cosmochim. Acta, 49,* 779–800.

1988. The quasi-stationary state approximation to coupled mass transport and fluid–rock interaction in a porous medium. *Geochim. Cosmochim. Acta, 51,* 143–65.

McBride, E. F. 1989. Quartz cement in sandstones: A review. *Earth-Sci Rev., 26,* 69–112.

McManus, K. M., and Hanor, J. S. 1988. Calcite and iron sulfide cementation of Miocene sediments flanking the west Hackberry salt dome, southwest Louisiana, USA. *Chem. Geol., 74,* 99–112.

Merino, E. 1984. Survey of geochemical self-patterning phenomena, pp. 305–28 in *Chemical instabilities,* ed. G. Nicolis and F. Baras. Dordrecht: Reidel.

Merino, E., Ortoleva, P., and Strickholm, P. 1983. Generation of evenly spaced pressure-solution seams during (late) diagenesis: A kinetic theory. *Contrib. Mineral. Petrol., 82,* 360–70.

Murray, B. T., and Chen, C. F. 1989. Double-diffusive convection in a porous medium. *J. Fluid Mech., 201,* 147–66.

Nield, D. A. 1968. Onset of thermohaline convection in a porous medium. *Water Resources Res., 4,* 553–60.

1984. Non-Darcy effects in convection in a saturated porous medium, pp. 129–40 in *Convective flows in porous media,* ed. F. A. Wooding and I. White. DSIR Sci. Info. Publ. Centre, Wellington, N.Z.

Nield, D. A., and Joseph, D. D. 1985. Effects of quadratic drag on convection in a saturated porous medium. *Phys. Fluids, 28,* 995–7.

Ortoleva, P., Chadham, J., Merino, E., and Sen, A. 1987. Geochemical self-organization. II. Reactive-infiltration instability. *Amer. J. Sci., 287,* 1008–40.

Ortoleva, P., Merino, E., Moore, C., and Chadham, J. 1987. Geochemical self-organization. I. Reaction-transport feedbacks and modeling approach. *Amer. J. Sci., 287,* 979–1007.

Palciauskas, V. V., and Domenico, P. A. 1976. Solution chemistry, mass transfer and the approach to chemical equilibrium in porous carbonate rocks and sediments. *Geol. Soc. Amer. Bull., 87,* 207–14.

Palm, E., Weber, J. E., and Kvernvold, O. 1972. On steady convection in a porous medium. *J. Fluid Mech., 54,* 153–62.

Park, C-W., and Homsy, G. M. 1984. Two-phase displacement in Hele–Shaw cells: Theory. *J. Fluid Mech., 139,* 291–308.

1985. The instability of long fingers in Hele–Shaw flows. *Phys. Fluids, 28,* 1583–5.

Phillips, O. M. 1970. On flows induced by diffusion in a stably stratified fluid. *Deep-Sea Res., 17,* 435–43.

1977. *The dynamics of the upper ocean.* Cambridge University Press.

1990. Flow controlled reactions in rock fabrics. *J. Fluid Mech., 212,* 263–78.

Plummer, L. N., and Wigley, T. M. L. 1976. The dissolution of calcite in CO_2-saturated solutions at 25°C and 1 atmosphere total pressure. *Geochim. Cosmochim. Acta, 40,* 191–202.

Pryor, W. A. 1973. Permeability–porosity patterns and variations in some Holocene sand bodies. *Amer. Assoc. Petrol. Geol. Bull., 57,* 162–89.

Rayleigh, Lord. 1916. On convection currents in a horizontal layer of fluid when the higher temperature is on the under side. *Phil. Mag.,* ser. 6, *32,* 529–46.

Rudraiah, N., and Srimani, P. K. 1980. Finite amplitude cellular convection in a fluid-saturated porous layer. *Proc. Roy. Soc. Lond. A, 373,* 199–222.

Saffman, P. G., and Taylor, G. I. 1958. The penetration of a fluid into a porous medium or Hele–Shaw cell containing a more viscous liquid. *Proc. Roy. Soc. A, 245,* 312–29.

Shercliff, J. A. 1975. Seepage flow in unconfined aquifers. *J. Fluid Mech., 71,* 181–92.

Sherwood, J. D. 1986. Island size distribution in stochastic simulation of the Saffman–Taylor instability. *J. Phys. A, 19,* 125–200.

Sibson, R. H. 1987. Earthquake rupturing as a mineralizing agent in hydrothermal systems. *Geology, 15,* 701–4.

1989. Earthquake faulting as a structural process. *J. Struct. Geol., 11,* 1–14.

Sibson, R. H., Robert, F., and Poulsen, K. H. 1988. High-angle reverse faults, fluid-pressure cycling and mesothermal gold-quartz deposits. *Geology, 16,* 551–5.

Slobold, R. L., and Thomas, R. A. 1963. Effects of transverse diffusion on fingering in miscible-phase displacement. *Soc. Petrol. Eng. J., 3,* 9–13.

Stoessell, R. K. 1987. Mass transport in sandstones around dissolving plagioclase grains. *Geology, 15,* 295–8.

Sverjensky, D. A. 1986. Genesis of Mississippi Valley-type lead–zinc deposits. *Ann. Rev. Earth Planet. Sci., 14,* 177–99.

Toth, D. J., and Lerman A. 1977. Organic matter reactivity and sedimentation rates in the ocean. *Amer. J. Sci., 277,* 465–85.

Turner, J. S. 1973. *Buoyancy effects in fluids.* Cambridge University Press.

Walther, J. V., and Helgeson, H. C. 1979. Calculation of the thermodynamic properties of aqueous silica and the solubility of quartz and its polymorphs at high pressures and temperatures. *Amer. J. Sci., 277,* 1315–51.

Washburn, E. W. 1929. *International critical tables.* New York: McGraw-Hill.

Wilson, E. N. 1989. Dolomitization of the Triassic Latemar buildup, Dolomites, northern Italy. Ph.D. dissertation. Johns Hopkins University.

Wilson, E. N., Hardie, L. A., and Phillips, O. M. 1990. Dolomitization front geometry, fluid flow patterns and the origin of massive dolomite: the triassic Lateman buildup, Northern Italy. *Amer. J. Sci., 290,* 741–96.

Wilson, J. C., and McBride, E. F. 1988. Compaction and porosity evolution of Pliocene sandstones. Ventura basin, California. *Amer. Assoc. Petrol. Geol. Bull., 72,* 664–81.

Wood, J. R. 1987a. A model for dolomitization by pore fluid flow, pp. 17–35 in *Physics and chemistry of porous media.,* ed. J. R. Banavar, J. Koplik, and K. W. Winker. New York: Amer. Inst. Phys.

1987b. Calculation of mass transfer coefficients for dolomitization models. *Appl. Geochem., 2,* 629–38.

Wood, J. R., and Hewett, T. A. 1982. Fluid convection and mass transfer in porous limestones: A theoretical model. *Geochim. Cosmochim. Acta, 46,* 1707–13.

Wooding, R. A. 1963. Convection in a saturated porous medium at large Rayleigh number or Peclet number. *J. Fluid Mech., 15,* 527–44.

Wooding, R. A., and Chapman, T. G. 1966. Groundwater flow over a sloping impermeable layer. 1. Application of the Dupuit–Forchheimer assumption. *J. Geophys. Res., 71,* 2895–902.

Wooding, R. A., and Morel-Seytoux, H. J. 1976. Multiphase fluid flow through porous media. *Ann. Rev. Fluid Mech., 8,* 233–74.

Index